Series in BioEngineering

The Series in Bioengineering serves as an information source for a professional audience in science and technology as well as for advanced students. It covers all applications of the physical sciences and technology to medicine and the life sciences. Its scope ranges from bioengineering, biomedical and clinical engineering to biophysics, biomechanics, biomaterials, and bioinformatics.

More information about this series at http://www.springer.com/series/10358

Leif Sörnmo
Editor

Atrial Fibrillation from an Engineering Perspective

Springer

Editor
Leif Sörnmo
Department of Biomedical Engineering
Lund University
Lund
Sweden

ISSN 2196-8861 ISSN 2196-887X (electronic)
Series in BioEngineering
ISBN 978-3-030-09838-4 ISBN 978-3-319-68515-1 (eBook)
https://doi.org/10.1007/978-3-319-68515-1

Softcover re-print of the Hardcover 1st edition 2018

Printed on acid-free paper

This Springer imprint is published by the registered company Springer International Publishing AG part of Springer Nature
The registered company address is: Gewerbestrasse 11, 6330 Cham, Switzerland

Dedicated to Steve Coleman for his groundbreaking research on functional arrhythmias

Preface

Atrial fibrillation (AF) is a complex, age-related arrhythmia which has reached global and epidemic proportions with considerable socioeconomic impact. During the last decade, substantial progress has been made in research on AF, including a better understanding of the basic mechanisms that initiate and maintain AF, low-cost technology for early detection and prevention of AF, and treatment with catheter ablation, which is clinical routine in many hospitals. However, the efficacies of antiarrhythmic drugs and catheter ablation still need to be improved and therefore the focus of much ongoing research.

Although clinical data, genetic factors, and imaging of different modalities play an important role when classifying patients with AF, electrophysiological information continues to be essential in the diagnosis and management of AF. New methodological challenges have emerged in ECG analysis as a result of new clinical findings as well as advances in technology. For example, different clinical studies suggest that frequent atrial ectopy is a precursor of AF and that brief AF episodes may be associated with increased risk of stroke—results which call for new types of signal processing algorithms to detect these events. The pervasive use of smartphone-based ECG applications is becoming an increasingly important tool in the quest for finding asymptomatic AF, implying that robust algorithms need to be developed to ensure that AF detection and characterization of atrial activity can be performed in signals of lower quality. Spatiotemporal analysis of body surface potential maps represents yet another challenge, where the limits of extracting clinically relevant information remain to be established.

The aim of this book is twofold, namely to offer a comprehensive, state-of-the-art review of methods developed for noninvasive analysis of AF, serving as a springboard for those developing new methods, and to provide a text which can be used at different levels in education. The lack of review articles on methods for detection of AF, extraction of f waves, and characterization of f waves is addressed by three chapters which consider these topics at length. This book is confined to describing aspects related to signals recorded noninvasively, whereas aspects related to invasive signals could easily form the contents of another book and therefore left out. Although this is an edited book, where each chapter is written by a different team of authors, sequential reading is still recommended since the chapters, to some degree, build on each other and contain numerous cross-references.

The title should not be interpreted as if there is a divide between clinical and engineering research. Rather, the title reflects the fact that research in AF includes an engineering signature, where mathematics is one of the cornerstones. The first chapter is entitled “A Clinical Perspective on Atrial Fibrillation” to lay the foundation for engineering-oriented research and to emphasize the importance of interdisciplinarity.

This book is intended for master students and doctoral students in biomedical engineering, electrical engineering, and computer science, as well as for researchers and practicing engineers with an interest in the analysis of cardiac arrhythmias. The unified style and standardized notations make this book suitable as a supplement to textbooks on biomedical signal processing. The chapters on detection and extraction have already been used in capstone projects at Lund University as part of a course in biomedical signal processing. In fact, the projects have turned out to be popular as the methods described in this book have varying levels of complexity and therefore let the student choose the desired level. In addition to the mandatory fundamental courses on digital signal processing and probability theory, familiarity with matrices and linear algebra and basic concepts in statistical signal processing is recommended.

With much appreciation, I would like to thank the authors for their expert contributions and for generously sharing their time with this project.

Special thanks to Pablo Laguna (Zaragoza), Vaidotas Marozas (Kaunas), and Andrius Petrėnas (Kaunas) who spent an enormous amount of time reviewing different versions of the entire book. Their engagement has significantly contributed to improve the contents.

Thanks to Pietro Bonizzi (Maastricht), Ki Chon (Storrs), Rebeca Goya (Madrid), Mikael Henriksson (Lund), Philip Langley (Hull), Jinseok Lee (Iksan), Julien Oster (Nancy), Olle Pahlm (Lund), Birutė Paliakaitė (Kaunas), Roberto Sassi (Milano), Monika Šimaitytė (Kaunas), Martin Stridh (Lund), Steven Swiryn (Chicago), Olof Sörnmo (Lund), Jean-Marc Vesin (Lausanne) for stimulating discussions and helpful explanations.

Thanks also to Marianna Meo (Bordeaux), Michela Masè (Trento), Massimo Rivolta (Milano), Roger Larsson (Lund), María de la Salud Guillem (Valencia) for generously sharing figures reproduced in this book.

Lund, Sweden
March 2018

Leif Sörnmo

Contents

Contributors

Raúl Alcaraz Innovation in Bioengineering Research Group, University of Castilla–La Mancha, Campus Universitario, Cuenca, Spain

Valentina D. A. Corino Department of Electronics, Information and Bioengineering, Politecnico di Milano, Milan, Italy

Pablo Laguna Biomedical Signal Interpretation and Computational Simulation (BSICoS), Aragón Institute of Engineering Research (I3A), Zaragoza University, Zaragoza, Spain; Centro de Investigacíon Biomédica en Red de Bioingeniería, Biomateriales y Nanomedicina (CIBER-BBN), Zaragoza, Spain

Luca T. Mainardi Department of Electronics, Information and Bioengineering, Politecnico di Milano, Milan, Italy

Vaidotas Marozas Biomedical Engineering Institute, Kaunas University of Technology, Kaunas, Lithuania

Andrius Petrėnas Biomedical Engineering Institute, Kaunas University of Technology, Kaunas, Lithuania

Pyotr G. Platonov Department of Cardiology, Clinical Sciences and Center for Integrative Electrocardiology, Lund University, Lund, Sweden

José Joaquín Rieta Biomedical Synergy, Electronic Engineering Department, Universidad Politécnica de Valencia, Gandía, Spain

Frida Sandberg Department of Biomedical Engineering and Center for Integrative Electrocardiology, Lund University, Lund, Sweden

Leif Sörnmo Department of Biomedical Engineering and Center for Integrative Electrocardiology, Lund University, Lund, Sweden

Acronyms

ABS	Average beat subtraction
ACC	American College of Cardiology
AF	Atrial fibrillation
AFDB	MIT–BIH Atrial Fibrillation Database
AFTDB	Atrial Fibrillation Termination Database
AFR	Atrial fibrillatory rate
AHA	American Heart Association
AI	Atrial impulses
ANN	Artificial neural network
APB	Atrial premature beat
ApEn	Approximate entropy
AR	Autoregressive
ARMA	Autoregressive moving average
ASIC	Application-specific integrated circuit
A:V	Atrial to ventricular
AUC	Area under the curve
AV	Atrioventricular
AZ	Azimuth
BSPM	Body surface potential mapping
CC	Correlation coefficient
CPU	Central processing unit
CSampEn	Coefficient of sample entropy
CV	Coefficient of variation
DACL	Dominant atrial cycle length
DAF	Dominant atrial frequency
DC	Direct current
ECG	Electrocardiogram
EL	Elevation
EKF	Extended Kalman filter
EMEA	European Medicines Agency

ESC	European Society of Cardiology
ESN	Echo state network
FDA	Food and Drug Administration
FN	False negative
FP	False positive
HD	Harmonic decay
HMM	Hidden Markov model
HRV	Heart rate variability
ICA	Independent component analysis
KL	Kullberg–Leibler
LC	Left clavicle
LDA	Linear discriminant analysis
LED	Light-emitting diode
LMS	Least mean square
LS	Least squares
LTAFDB	Long-Term AF Database
MAE	Mean absolute error
MESOR	Midline estimating statistic of rhythm
ML	Maximum likelihood
MITDB	MIT–BIH Arrhythmia Database
MSCPE	Mean square cross prediction error
MRI	Magnetic resonance imaging
MSE	Mean square error
NMASD	Normalized mean of absolute successive differences
NMSE	Normalized mean square error
NSR	Normal sinus rhythm
NSRDB	MIT–BIH Normal Sinus Rhythm Database
NZPP	Count of nonzero bins in the Poincare plot
PC	Principal component
PCA	Principal component analysis
PDF	Probability density function
PL	Planarity
PTBDB	Physikalisch–Technische Bundesanstalt Database
RATAF	RATe control in Atrial Fibrillation
RMS	Root mean square
RMSSD	Root mean square of successive differences
RNN	Recurrent neural network
ROC	Receiver operating characteristic
RT	Recovery time
SampEn	Sample entropy
SC	Spectral concentration
SCC	Signed correlation coefficient
SE	Spectral entropy
ShEn	Shannon entropy
SNR	Signal-to-noise ratio

SO	Spectral organization
SOBI	Second-order blind identification
SPR	Spectral power ratio
SQI	Signal quality index
SR	Sinus rhythm
SSA	Singular spectral analysis
SSAFDB	Short Single-lead AF Database
SSampEn	Simplified sample entropy
STC	Spatiotemporal cancellation
STFT	Short-term Fourier transform
SVD	Singular value decomposition
SVM	Support vector machine
TN	True negative
TP	Turning point *or* true positive
TVCF	Time-varying coherence function
VCG	Vectorcardiogram
VPB	Ventricular premature beat
VFL	Ventricular flutter
VEB	Ventricular ectopic beat
VP	Ventricular pacing
VR	Ventricular residue
WRMS	Weighted root mean square
WVD	Wigner–Ville distribution
XWVD	Cross Wigner–Ville distribution

Chapter 1
A Clinical Perspective on Atrial Fibrillation

Pyotr G. Platonov and Valentina D. A. Corino

1.1 Introduction

Atrial fibrillation (AF) is the most common arrhythmia encountered in clinical practice which requires therapeutic interventions. Its prevalence is growing and the number of patients with AF is increasing along with the aging population in the industrialized countries. Atrial fibrillation is not only affecting the quality of life due to the irregular heartbeats, palpitation attacks, or inappropriate acceleration of the heart rate, but it is also one of the most common risk factors of ischemic stroke, which may lead to irreversible handicap and death. Contrary to many other arrhythmias encountered in clinical practice, AF may require therapeutic interventions even in patients who do not have any subjective discomfort from their arrhythmia. Accordingly, this defines an unmet challenge of correct and timely arrhythmia detection.

By affecting millions of patients worldwide, AF is characterized by a palette of clinical manifestations which to some extent is defined by preexisting comorbidities, from completely asymptomatic variants to significant limitations of everyday life due to arrhythmia-related palpitations, fatigue, chest pain, or aggravation of heart failure. A more severe background clinical profile in patients with heart failure, diabetes, and hypertension is usually associated with more severe symptoms during AF. However, in many cases it remains unclear why patients with similar AF phenotype in terms of frequency of arrhythmic attacks and heart rate during AF can have completely different clinical manifestations so that some patients would be in need for hospital

P. G. Platonov (✉)
Department of Cardiology, Clinical Sciences and Center for Integrative Electrocardiology, Lund University, Lund, Sweden
e-mail: pyotr.platonov@med.lu.se

V. D. A. Corino
Department of Electronics, Information and Bioengineering, Politecnico di Milano, Milan, Italy
e-mail: valentina.corino@polimi.it

L. Sörnmo (ed.), *Atrial Fibrillation from an Engineering Perspective*, Series in BioEngineering, https://doi.org/10.1007/978-3-319-68515-1_1

admission and cardioversion, while others would be completely unaware of the heart rhythm disturbance.

An introductory chapter on clinical matters to a book whose primary focus is engineering comes with the inherent challenge of balancing between unnecessary clinical aspects and a too simplistic picture of complex biological phenomena which to a significant part are not completely understood. However, it is difficult to imagine the development of contemporary clinical medicine, and cardiac electrophysiology in particular, without the progress in many engineering disciplines. Reaching common understanding of basic electrical phenomena in the heart, their importance for health, and risks associated with complications of heart rhythm abnormalities is therefore vital to guide the development of technology aimed at facilitating patient care, prolonging life, and improving its quality.

This chapter provides an overview of basic concepts related to the mechanisms underlying AF, its impact on the human health, and basic principles used by clinicians to prevent AF and minimize its ominous impact on the human organism. The main areas of uncertainties in clinical decision-making will be highlighted, where the existing knowledge gaps make further translational research efforts highly warranted.

1.2 Atrial Fibrillation: Definition

Atrial fibrillation is a supraventricular tachyarrhythmia, characterized by uncoordinated atrial electrical activation and, consequently, ineffective atrial contractions. In the vast majority of cases, AF diagnosis is based on an ECG demonstrating

1. irregular RR intervals,
2. absence of distinct repeating P waves, and
3. presence of undulating atrial activity, also known as fibrillatory waves or f waves,

see Fig. 1.1a. While seemingly straightforward, there are situations in clinical practice when this definition is not always simple to apply. On the other hand, a healthy person would present with an ECG in sinus rhythm, characterized by the presence of P waves originating from the sinus node and reflecting atrial depolarization and regular RR intervals, see Fig. 1.1b.

Irregularity of RR intervals, being a key feature of AF, will only be present in patients with preserved atrioventricular (AV) conduction. Even though this is the case for the vast majority of patients with AF, patients with complete AV block (either induced by cardio-active medications or developed as a result of disease or cardiac surgery) or escape rhythm, originating from the conduction system segments located below the site of block, will have regular RR intervals which may mislead ECG assessment. RR interval irregularity in AF lacks any organization detectable by the human eye. This type of RR behavior differs from other situations in which irregularity of RR intervals is observed, e.g., in patients with atrial or ventricular premature contractions, variable conduction through the AV node during regular atrial tachycardias, or a second-degree AV block, characterized by nonconducted

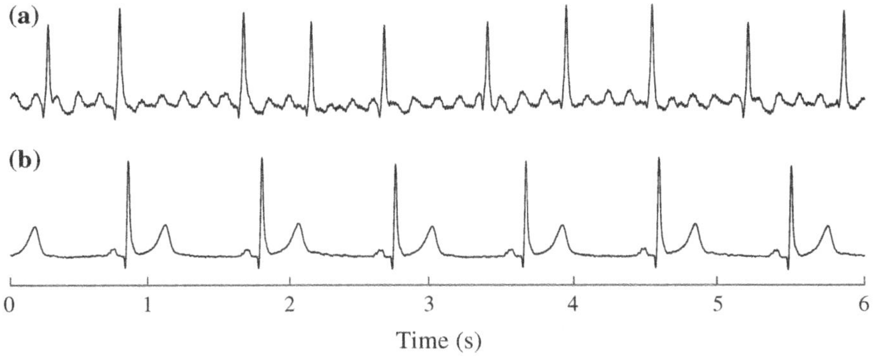

Fig. 1.1 ECG recorded during **a** AF and **b** normal sinus rhythm. In atrial fibrillation, the RR intervals are irregular and the P waves replaced by an undulating atrial activity, known as f waves

atrial beats resulting in pauses occurring on the background of fairly regular rhythm driven by the sinus node or other atrial sources.

The absence of P waves is rarely controversial in patients with low-amplitude f waves, see Fig. 1.2a. However, large-amplitude f waves mimicking P waves, especially in the right precordial leads V_1 and V_2, may represent a challenge, see Fig. 1.2b. In such situations, one should check whether the atrial waves occur at the same time in several ECG leads and have a distinct and repetitive morphology, which then would contradict an AF diagnosis. On the contrary, indistinct atrial wave morphology, variable and short intervals between successive atrial waves (measurable in the leads with distinct atrial waves, often the right precordial leads), and the lack of isoelectric line between them would support AF diagnosis.

1.3 Classification of Atrial Fibrillation

Several classifications of AF exist and are used in clinical practice. The most common way to describe AF is based on the duration and the recurrent nature of arrhythmic episodes, which make patients contact healthcare providers, see Table 1.1.

It is, however, important to appreciate that AF is not a static condition: a patient with paroxysmal AF may develop persistent AF episodes that do not cease spontaneously. Moreover, allocation of a given arrhythmic episode to either a paroxysmal or persistent condition largely depends on the subjective judgement used to administer or withhold cardioversion attempt early on in the course of the AF attack. When sinus rhythm is restored by cardioversion, the ultimate duration of the AF episode is unknown. Thus, patients with highly symptomatic AF seeking care soon after arrhythmia breakthrough may have their arrhythmia assessed differently compared to patients with less symptomatic AF, who may have to wait longer with greater likelihood of spontaneous conversion within several hours or days.

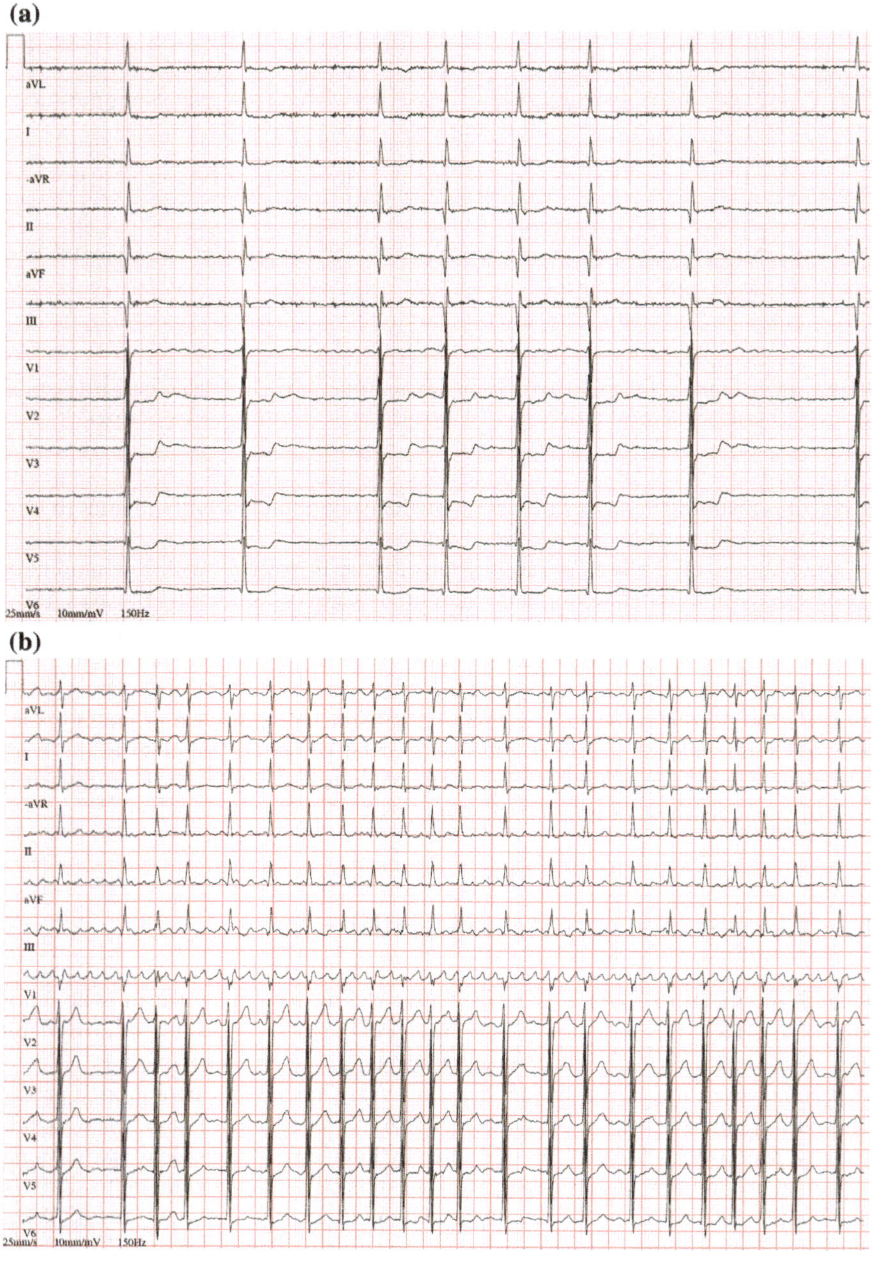

Fig. 1.2 Different ECG manifestations of AF. **a** Low-amplitude ("fine") f waves which are nearly isoelectric between the QRS complexes in all leads, and **b** large-amplitude ("coarse") f waves, particularly in the right precordial leads

The clinical relevance of the classification presented in Table 1.1 is based on the relationship between the persistence of AF episodes and the efficiency of therapeutic options aimed at terminating AF and preventing its recurrence. As a rule, the more persistent AF is, the more difficult it is to achieve arrhythmia freedom by therapeutic interventions.

The alternative way to describe AF is based on the presence of comorbidities, which may have etiological links to the arrhythmia itself. In this regard, description of AF as valvular or nonvalvular is one of the most commonly used in clinical practice. The reason for this description is also related to a rather specific and therapy-resistant course of the disease in patients with valvular disease, associated with volume and pressure overload of the left atrium, leading to severe left atrium dilatation and extensive fibrotic replacement of the atrial myocardium.

Lone AF is a historical descriptor that has been variably applied to predominantly younger persons without clinical or echocardiographic evidence of cardiopulmonary disease, hypertension, or diabetes mellitus. Because of the high variability in the use of this descriptor, lone AF is today considered as potentially confusing and rarely used to guide arrhythmia management.

1.4 Epidemiology of Atrial Fibrillation

Atrial fibrillation increases in prevalence with advancing age. Given the intermittent nature of the arrhythmia and indistinct symptoms present in a considerable proportion of patients with AF, the exact assessment of AF prevalence depends largely on the methodology used for screening of this condition. The use of routine resting ECG, or clinically motivated physical exams applied to epidemiological cohorts, estimates AF prevalence as 0.5% in individuals <50 years of age, 1.5–2% in 50–60 years of age, and 3% and higher in patients above 70 years of age (Fig. 1.3) [1]. However, dedicated population screening for AF using thumb-ECG, performed in Sweden,

Table 1.1 Clinical classification of atrial fibrillation

Type	Definition
New onset AF	is defined by the occurrence of the first episode, irrespective of its duration and severity of AF-related symptoms
Paroxysmal AF	is recurrent (≥2 episodes) and self-terminates in less than seven days, usually within 24 h
Persistent AF	fails to self-terminate within seven days. Episodes require termination by cardioversion
Long-standing persistent AF	has lasted for one year or more when it is decided to adopt a rhythm-control strategy
Permanent AF	exists when the presence of arrhythmia is accepted by the patient and the physician

Fig. 1.3 Prevalence of AF according to age and gender [1]

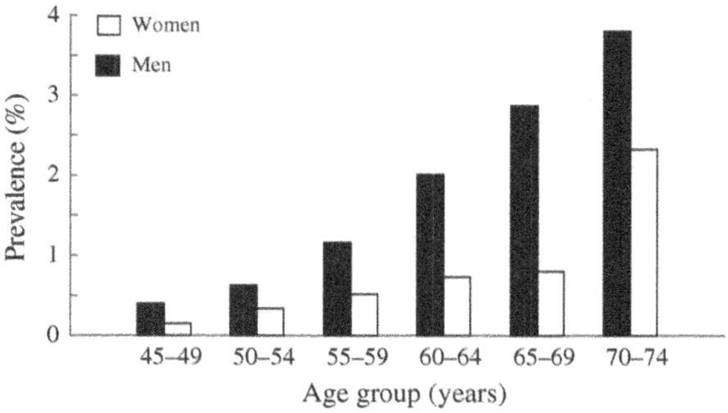

demonstrated that 7% of the population older than 65 years have AF, of whom individuals >75 years of age have particularly high prevalence, reaching 12% [2]. Atrial fibrillation is significantly more common among men, particularly at young age, however, gender-related differences diminish with increasing age.

If we consider the age distribution among patients with AF, then approximately 1% of them are <60 years of age, whereas up to 12% are 75–84 years of age [3]. In the United States, the percentage of Medicare fee-for-service beneficiaries with AF was in 2010 reported to be 2% for those <65 years of age and 9% for those ≥65 years of age [4]. The lifetime risk of developing AF after 40 years of age was shown to be about 25%, being slightly higher among men than women [5].

1.5 Mechanisms of Atrial Fibrillation

Advances in clinical and fundamental research promoted over the last decades have led to a well-established understanding of AF as an epiphenomenon which despite similar manifestations may have different underlying mechanisms, thus requiring individualized treatment [6].

The mechanisms of AF are complex and require a combination of triggers, commonly represented by ectopic atrial firing and a vulnerable atrial substrate which promotes perpetuation of AF. The relative importance of trigger mechanisms and atrial substrate characteristics for the development of AF may vary and, to a large extent, affect clinical manifestations of the arrhythmia. Less advanced atrial substrate in the presence of rapidly firing atrial foci may be found in patients with paroxysmal AF with high likelihood of spontaneous conversion. On the other hand, age- or disease-related changes in atrial myocardium may lead to increased vulnerability of atrial myocardium and longer duration of AF episodes, which may become long-standing or permanent.

Focal ectopic firing originating from the myocyte sleeves within the pulmonary veins was first proposed in the late ’90s as the triggering mechanism of AF [7], which led to the development of catheter-based ablation therapy, resulting in reduction of AF burden. Contemporary understanding of the pulmonary veins role in the genesis of AF is based on the presence of myocytes exhibiting unique electrical properties, such as pacemaker cells, transitional cells, and Purkinje cells [8], and a complex fiber architecture which together promote reentry and ectopic activity initiating AF [9]. However, the presence of a trigger in atrial myocardium per se is not sufficient for initiation of AF, but requires prerequisites for stabilization of reentry in atrial myocardium in order to maintain AF. With rare exceptions of AF caused by mutations in genes coding ion channels in patients with structurally normal atria, fibrotic replacement of atrial myocardium remains the cornerstone of atrial pathology in patients with AF. However, the exact mechanisms underlying the structural abnormalities in the atrial walls observed in patients with the arrhythmia and its relationship to the arrhythmia mechanisms still remain poorly understood.

The common perception of AF as a result of the interplay between the structural changes in the atrial myocardium, induced by the well-described cardiovascular risk factors, and structural remodeling, induced by the arrhythmia itself, has recently been challenged by observations of progressive structural abnormalities in the atrial walls that occur independently of the cardiovascular comorbidities and persistence of AF [10]. It is also well-known that lone AF is not an uncommon clinical entity that may manifest early in life without any apparent risk factors, which would explain development of atrial fibrosis in patients with structurally normal hearts [11]. To what extent fibrotic atrial cardiomyopathy represents a "common cause" of AF or a mechanism responsible for arrhythmia development in a subgroup of patients with AF phenotype remains, however, uncertain.

1.6 Atrial Myocardium Characteristics in Atrial Fibrillation

An indirect indication of the link between cardiovascular comorbidities and AF comes from epidemiological studies in which potentially fibrosis-causing conditions such as hypertension, ischemic heart disease, and diabetes were highly predictive of incident AF [12]. Age-related increase in the prevalence of AF has also been well-documented [1], and explained by growing cardiovascular disease burden in the elderly as well as age-related increase in the extent of atrial fibrosis [13]. However, attempts to provide a quantitative assessment of atrial structural abnormalities associated with AF have shown a more complex picture. Even though catheter-based techniques of endocardial voltage mapping and emerging noninvasive magnetic resonance imaging (MRI) have shown their value in visualization of atrial structural abnormalities, histological evaluation of atrial tissue samples remains the gold standard for tissue characterization. This approach, however, is often limited to a small volume of tissue samples collected in patients undergoing atrial biopsy, or confined to right or left atrial appendages in patients undergoing open-chest heart surgery, thus imposing a significant bias on patient selection and leaving large portions of the atrial walls, in which AF perpetuates, outside reach.

One of the first observations of the structural substrate of AF in patients without apparent structural heart disease came from studies where biopsies were collected from atrial septum as well as from ventricles in patients with lone AF [11, 14], reporting on a consistent finding of myocardial inflammation and fibrosis confined to the atrial myocardium, but not present in the ventricular walls. These studies were the first to suggest the presence of occult myocardial disease that may have direct causal relationship with development of AF.

The concept of atrial cardiomyopathy has been further expanded by studying histology specimens from multiple sampling locations in the right and left atrium collected post mortem from deceased patients with common cardiovascular comorbidities with previous paroxysmal, permanent AF, and those without AF history enrolled in three equal groups according to prespecified inclusion criteria [15]. The extent of fibrosis and fatty tissue in the atrial myocardium showed strong and significant correlation with the presence of AF at all tissue sampling locations in the left and right atria. Notably, patients with and without AF did not differ in regard to cardiovascular comorbidities, and no age-related increase in the extent of atrial fibrosis was observed. Similar observations were made in patients with persistent or long-standing AF referred for surgical ablation [16], thus suggesting that development of structural abnormalities in the atria is not a result of concomitant diseases, but rather a phenomenon associated with AF. Indirect assessment of atrial fibrosis using MRI in a large cohort has further supported this theory by not finding any significant differences in the estimated fibrosis extent between AF patients with and without comorbidities [17]. So far, however, there is no histology data that would specifically address the question of causal relationships between the burden of concomitant cardiovascular diseases and atrial fibrosis in patients with AF.

Contrary to the findings in lone AF, a similar extent of fibrotic replacement and inflammatory infiltration in the free walls of the right and left ventricles was observed in patients with common cardiovascular comorbidities [14]. In a controlled study, ventricular fibrosis demonstrated strong correlation with AF history and extent of fibrosis in the major atrial conduction routes such as Bachmann's bundle and terminal crest [18]. These findings may be interpreted as indicating an underlying occult cardiomyopathy with significant inflammatory component in patients with AF.

Whether or not structural abnormalities observed in the atria are the cause or consequence of AF remains an open question. The presence of a relationship between the extent of fibrosis and AF burden can be explained both ways: expansive fibrotic process in the atria may promote persistent AF, or be a consequence of the long-standing fibrillatory process. The lack of this relationship, however, would favor the concept of the primary fibrotic atrial cardiomyopathy underlying AF development. Available data suggest that the extent of fibrosis tends to be larger in patients with permanent AF than in patients with paroxysmal AF [15], see Fig. 1.4, but the relationship between extent of structural abnormalities and duration of AF seems to disappear in patients with persistent AF [16]. In another study that quantified the expression of extracellular matrix proteins in atrial tissue samples collected during

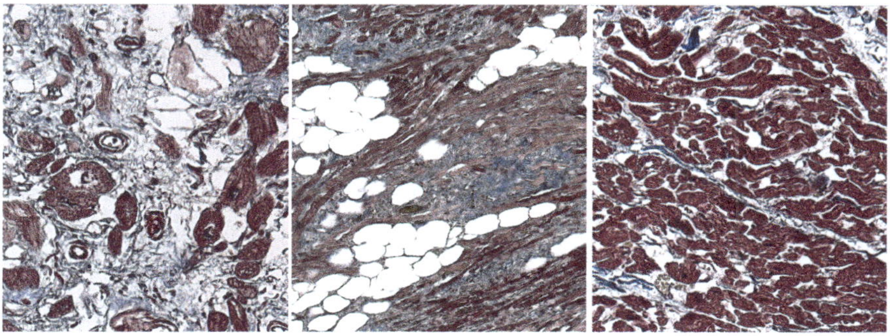

(a) Permanent AF (b) Paroxysmal AF (c) No history of AF

Fig. 1.4 Light microscopy of crista terminalis specimens in patients with or without history of AF. **a** Fibrosis extent 51%, fat 15%, capillary density 2%, mean cardiomyocyte diameter 12 m, **b** fibrosis extent 14%, fat 24%, capillary density 0.4%, mean cardiomyocyte diameter 11 m, and **c** fibrosis extent 5%, fat 1%, capillary density 1%, mean cardiomyocyte diameter 15 m. (Masson's trichrome stain; original magnification 200). (Reprinted from [15] with permission)

heart surgery, no systematic difference between patients with paroxysmal and permanent AF was documented [19]. Even though this does not address the unresolved causality issue, one can speculate that fibrosis extent in the atrial walls may be linked to AF burden and clinical manifestations of the arrhythmia at the early stages of the disease. However, upon reaching a certain level, fibrosis would no longer affect AF phenotypes in patients who develop persistent AF.

1.7 Atrial Fibrillation and Stroke

Ischemic stroke is a devastating complication of AF. One in five of all strokes is attributed to AF [20], and AF in stroke patients confers an increased risk of morbidity and mortality when compared with non-AF-related stroke [21]. The main mechanism of AF-related stroke is considered to be thrombus formation in the left atrium in condition of irregular contractility. When a blood clot is formed it can be pumped out of the heart to the brain, leading to cerebral artery occlusion.

Clinical risk factors for development of ischemic stroke in patients with AF are well-known. The CHA_2DS_2–VASc score is a clinical tool developed to assess the risk of ischemic stroke in patients with AF and to guide administration of oral anticoagulation therapy, with proven effect on reduction of the risk of ischemic stroke [22, 23]. The letters of the score stand for individual risk factors known to predispose to ischemic stroke, involving the following risk factors:

- **C**ongestive heart failure (1 point),
- **H**ypertension (1 point),
- **A**ge $\geq$75 years (2 points),

- **D**iabetes mellitus (1 point),
- **S**troke/Transient ischemic attack/Thromboembolism (2 points),
- **V**ascular disease (1 point),
- **A**ge 65–74 years (1 point),
- **S**ex **c**ategory (female) (1 point).

Advanced age of ≥75 years and history of ischemic stroke are the most powerful predictors of ischemic stroke in patients with AF, receiving double points in the risk score calculation. In general, patients with a high CHA_2DS_2–VASc score are recommended life-long oral anticoagulation therapy.

An important aspect of the CHA_2DS_2–VASc score applicability is its dependence on AF diagnosis, which means that patients fulfilling one or more criteria listed in the risk score would not be offered stroke prevention therapy unless AF is documented. However, AF is often asymptomatic, and sometimes ischemic stroke may be the first clinical presentation of underlying AF.

It has been reported that at least one third of patients with AF had asymptomatic AF [24, 25]. In patients with implantable devices, subclinical AF was quite common and associated with increased risk of stroke [26]. In this context, sensitivity and specificity of AF screening techniques as well as the reasonable balance between the associated costs, the need for surgical interventions (as in the case of implantable subcutaneous monitors), and the risk of false positive AF detection become the factors defining the clinical utility of AF screening.

As the history of ischemic stroke automatically places a patient with AF in the high-risk group regarding ischemic stroke recurrence, screening for AF becomes particularly important in ischemic stroke survivors. Using the standard ECG at admission with ischemic stroke, AF is documented in 20–25% of those who survived ischemic stroke [21, 27]. Additional, repeated conventional snapshot ECG recordings after stroke onset appeared to increase AF detection rate by 1.4–6.7% [28–30]. Diagnostic yield of 24–48-h continuous, ambulatory ECG monitoring in patients with ischemic stroke and sinus rhythm at admission has been reported to be 1–6.4% [28, 30, 31], increasing to 12.5% when monitoring was continued for a week [31]. In stroke patients who underwent 30-day ambulatory, automatically triggered AF detection, AF was documented in 6–11% of cases [2, 32]. Outpatient cardiac telemetry during 3–4 weeks in patients with cryptogenic stroke revealed 17–20% of new AF cases [33, 34]. However, the highest detection rate of AF in patients with cryptogenic stroke was reported in patients with insertable cardiac monitors and appeared to be 30% [35]. Though the superiority of the latter strategy for AF detection is obvious, its cost-effectiveness is largely affected by proper selection of patients who would benefit from continuous screening for AF.

While AF most certainly is a risk factor for ischemic stroke, it is not necessarily the direct cause of it. The causality of association between AF and ischemic stroke was questioned by the reported lack of temporal relationship between stroke events and symptomatic AF paroxysms or atrial high-rate episodes detected by an implantable loop recorder [36–38] or an implantable device [39–42]. In different studies, only 2% of patients had subclinical AF episodes lasting more than 6 min at the time of stroke or systemic embolism [43]. Among the plausible explanations for this change of paradigm is the recently proposed concept of fibrotic atrial cardiomyopathy [44], according to which AF may result from an underlying progressive disease affecting atrial myocardium and resulting in replacement of myocardium with fibrosis and fat, increasing atrial thrombogenic properties, and the risk of stroke, which in this situation does not have to express a temporal relationship with the arrhythmia episodes.

Finally, availability of diagnostic information recorded directly from the atria in patients with implanted dual-chamber pacemakers or cardioverter–defibrillators poses a new challenge of interpretation. Our knowledge regarding clinical importance of AF and its relationship to stroke has been built on clinical AF episodes, i.e., AF detectable by conventional means of ECG diagnostics, while implantable device-detected arrhythmias, also called atrial high-rate episodes, are often subclinical, asymptomatic, and short in duration, which in some cases may only last several seconds. Even though these episodes may have electrogram characteristics indistinguishable from AF, it is yet unknown whether such brief AF episodes are associated with a risk of stroke similar to that of conventionally defined AF.

1.8 Principles of Atrial Fibrillation Management

Management of patients with AF is aimed at reaching two fundamental goals: prolonging life and improving its quality by reducing arrhythmia-related symptoms. Three major treatment strategies have been developed and implemented:

1. Ischemic stroke prevention (oral anticoagulation)
2. Heart rate control during AF (rate-control strategy)
3. Prevention of AF (rhythm-control strategy).

Only ischemic stroke prevention was shown to reduce mortality in patients with AF, while rate- and rhythm-control strategies remain the key elements of AF patients care with the primary objective to improve quality of life.

1.8.1 Ischemic Stroke Prevention

Prevention of thromboembolic complications of AF is achieved by administration of medications attenuating blood-clotting capacity (i.e., anticoagulants) to individuals at high risk of ischemic stroke. The challenge to be met when using anticoagulation therapy is to maintain the fine balance between the benefit of reducing propensity to clotting and the potential harm of the drugs related to their inherent property of prolonging bleeding time and the risk of bleeding complications, which in some cases may be fatal, such as intracranial bleedings. Clinical decision-making tools have been proposed to facilitate administration of stroke-prevention measures to individuals in need. The most commonly used risk quantification tool recommended by the management guidelines is the above-mentioned CHA_2DS_2–VASc score, which estimates the probability of ischemic stroke. However, balancing the risk of bleeding is not an easy task since a number of factors included in the CHA_2DS_2–VASc score increase both the risk of ischemic stroke and the risk of major bleeding complications of anticoagulant therapy: hypertension, advanced age, and the history of stroke.

Recent studies that questioned the causal relationships between the arrhythmia and embolic events, reviewed earlier in this chapter, have led to a paradigm shift suggesting that AF may be a marker of the increased risk of ischemic stroke rather than its direct cause. Interestingly, a number of studies have shown that the CHA_2DS_2–VASc score not only predicts the risk of ischemic stroke, but it is also a reasonably accurate tool to predict the development of AF in patients without known AF history. This strategy is based on the findings documented in cohorts of ischemic stroke survivors [45, 46] and in selected cohorts of patients evaluated for palpitations [47, 48]. Therefore, it is plausible to suggest that patients with a high CHA_2DS_2–VASc score may benefit from oral anticoagulation therapy without need for documentation of AF. The clinical utility of this approach remains to be proven in ongoing clinical trials. It is likely, however, that its risk–benefit ratio will largely depend on the CHA_2DS_2–VASc score cut-off selected for making the decision to initiate anticoagulation and the underlying risk of bleeding.

1.8.2 Rate-Control Strategy

Whether a patient would benefit from implementation of rate-control measures is the question that needs to be asked for every patient with AF, regardless of the severity of clinical manifestations of the arrhythmia. Some patients with high ventricular rate during AF may not have any distinct symptoms associated with fast and irregular heartbeats. However, if left untreated, high ventricular rate may lead to deterioration of ventricular contractile function, reduction of cardiac pumping capacity, and dilatation of the ventricular chambers (known as tachycardiomyopathy) with development of heart failure as the ultimate consequence. Adequate rate-control improves quality of life, reduces morbidity, and decreases the potential for developing tachycardia-induced cardiomyopathy.

The degree of rate control and the thresholds defining adequately controlled ventricular response during AF remain, however, an area of uncertainty. According to a "strict rate-control approach", the goal of rate-control therapy is to bring the heart rate down to $\leq$80 beats per minute (bpm) at rest, or the average heart rate $\leq$100 bpm in ambulatory monitoring. These thresholds, however, may be difficult to achieve in clinical practice in a considerable part of patients with AF. In a single study, an alternative "lenient rate-control approach", using instead 110 bpm at rest, was tested and shown to be noninferior to the strict rate-control strategy [49]. Additional independent confirmatory studies are needed to fully understand the impact of the lenient rate-control approach on mortality, heart failure symptoms, hospitalizations, and quality of life.

Rate control may be achieved by different means, including the use of pharmaceuticals or catheter-based therapies. Some antiarrhythmic drugs, including beta-blockers, calcium antagonists, or cardiac glycosides, have proven efficient in slowing down ventricular response during AF through their blocking effect on the AV node. The choice of the drug remains empirical and is driven mainly by the presence of contraindications of or intolerance to certain compounds. As a last resort, the powerful antiarrhythmic drug amiodarone can be used to control the heart rate. Due to the risk of serious side effects that may appear during long-term administration, the use of amiodarone is restricted to rare occasions when other drugs fail to achieve the therapeutic goals.

In rare cases when rate control cannot be achieved by medication due to either drug intolerance or inefficiency, catheter-based therapy may be applied. Ablation of the atrioventricular junction, achieved by local application of either radiofrequency current or deep freezing (cryoablation), leads to destruction of the functional connection between the atria, which may continue to fibrillate, and the ventricles. Obviously, this approach is only feasible in patients with implanted pacemaker, either received for other indication or implanted specifically to enable AV junctional ablation. Patients undergoing AV junctional ablation become pacemaker-dependent, which is an important limiting factor, and, therefore, this approach is the last one in the armamentarium of AF therapies. On the bright side, however, remains the fact that ventricular contractions become regular and steered exclusively by a programmable pacemaker, eliminating the concern of high rate without the need for rate-control pharmacological therapies.

1.8.3 Rhythm-Control Strategy

Rhythm-control strategy includes all therapeutic interventions aimed at prevention of AF recurrences and restoration of sinus rhythm using a combination of approaches, including cardioversion (electrical or pharmacological), antiarrhythmic drugs, and catheter ablation in the setting of appropriate anticoagulation and rate control.

It may seem surprising that rhythm-control strategy aimed at restoration or maintenance of sinus rhythm is the last item on the list of AF therapies. However, this approach, still reserved for the most symptomatic patients, has not been associated with mortality reduction and may be associated with increased number of hospital admissions. While stroke prevention and rate control are considered obligatory components of care for patients with AF, rhythm-control measures are generally reserved for patients who remain symptomatic despite adequate rate control.

In some situations, however, rhythm-control strategy may be prioritized over rate-control measures. This may be important in situations when it is difficult to achieve adequate rate control in younger patients, in tachycardia-mediated cardiomyopathy, during the first episode of AF, in AF precipitated by an acute illness, or patient preference. Another argument favoring rhythm-control strategy over limited rate-control intervention is that AF progresses from paroxysmal to persistent in many patients and subsequently results in electrical and structural remodeling which eventually becomes irreversible. For this reason, acceptance of AF as permanent in a patient may render future rhythm-control therapies, if needed, less effective. This observation may be more relevant for a younger patient who wishes to remain a candidate for future development in rhythm-control therapies. Early intervention with a rhythm-control strategy to prevent progression of AF may therefore be beneficial.

Restoration of sinus rhythm may be achieved by means of either antiarrhythmic drugs (pharmacological cardioversion) or delivery of electrical shock through the electrodes applied on the chest of the patient (electrical cardioversion). Whether to choose pharmacological or electrical approach to cardiovert a patient with AF depends on a number of factors.

Electrical cardioversion is generally considered the most efficient way to restore sinus rhythm. Compared to pharmacological cardioversion, which may have a success rate varying from 30 to 60% depending on the choice of drug, degree of atrial remodeling, and duration of the AF episode, electrical cardioversion, electrically resetting cardiomyocytes using appropriate shock settings, terminates AF immediately in more than 90% of all patients. On the other hand, restoration of sinus rhythm using electrical cardioversion requires sedation, which itself is not a risk-free intervention and does not prevent immediate recurrence of AF.

Pharmacological cardioversion, on the other hand, does not require sedation, but is more time-consuming, in part due to the time required for an antiarrhythmic drug to reach therapeutic concentration in the body, and in part due to the need for rhythm observation after drug administration in order to monitor potential proarrhythmic effects of the potent antiarrhythmic drugs used for pharmacological cardioversion. With few exceptions, most of the drugs suitable for pharmacological cardioversion are also efficient as rhythm-control agents which may be administered over long periods of time to reduce the frequency and duration of AF and to improve quality of life. Since rhythm-control strategy is the approach aimed at improving quality of life, rather than reducing mortality, the choice of drug is, to a greater extent, guided by safety concerns related to proarrhythmic side effects, expressed by nearly all potent antiarrhythmic drugs, than by drug efficacy. Patients with coronary artery disease, heart failure, and significant left ventricular hypertrophy have more restricted options than those with minimal or no structural heart disease.

Catheter ablation has evolved from an experimental technique, having emerged at the end of '90s [50], to become an efficient treatment modality, which in selected patient populations has demonstrated efficacy superior to antiarrhythmic drugs [51]. The approach is based primarily on the creation of electrically impenetrable boundaries surrounding the ostia of pulmonary veins, using either radiofrequency current or deep freezing delivered with a catheter placed in the left atrium. Depending on the degree of atrial remodeling and arrhythmia persistence, pulmonary vein isolation may be combined with additional lines in the left or right atrium aimed at further hampering propagation of fibrillatory waves in the atrial myocardium, and thus to reduce the likelihood of AF maintenance. The evidence supporting the efficacy of catheter ablation is strongest for paroxysmal AF in young patients with little or no structural heart disease, and in procedures performed in highly experienced centers. As of today, the effect of catheter ablation on reduction of mortality, stroke, or heart failure is insufficient. The ongoing randomized clinical trials "Catheter Ablation Versus Antiarrhythmic Drug Therapy for Atrial Fibrillation" (CABANA) and "Early Therapy of Atrial Fibrillation for Stroke Prevention Trial" (EAST) are expected to provide new information to assess whether catheter ablation is superior to standard therapy with either rate- or rhythm-control drugs for reducing total mortality, and whether early application of rhythm-control strategy can impact the risk of ischemic stroke, cardiovascular death, or the development of heart failure. This research will help us to understand whether catheter ablation provides benefit beyond improvements in quality of life in patients with AF.

In clinical practice, the ability to predict the likelihood of AF conversion (either spontaneous or induced by antiarrhythmic drugs) and the risk of AF recurrence after cardioversion or catheter ablation would be useful for planning rhythm-control therapies and avoiding unnecessary interventions. In patients prone to regain sinus rhythm spontaneously within a reasonable time frame, cardioversion attempt, for example, can be withheld or postponed. Patients who are unlikely to respond to an antiarrhythmic drug can be scheduled for electrical cardioversion instead, while those who are unlikely to maintain sinus rhythm over a considerable time span may not be good candidates for rhythm-control strategy at all. Therapeutic efforts could instead be focused on achieving appropriate rate control. The proper stratification tool, however, is still lacking.

The degree of structural and electrical remodeling of the atria in patients with AF is considered the factor which, to a large extent, defines the probability of success in applying rhythm-control strategy. In general, patients with long-lasting persistent AF, enlarged atria, and extensive fibrotic replacement of myocardium in the atrial walls are less likely to benefit from cardioversion. Clinical assessment of these factors may involve using different diagnostic modalities such as echocardiography, MRI, and endocardial voltage mapping, which may not be practical for all patients, is associated with significant costs and catheterization-related risks. On the other hand, characteristics of the atrial fibrillatory process, retrievable from the ECG in terms of frequency content and degree of organization, may contain important prognostic information [52, 53].

1.9 Electrocardiography in Atrial Fibrillation Diagnosis

Novel wearable devices can greatly improve the diagnosis of AF in ambulatory outpatients. For screening purposes, low-cost and easy-to-use handheld or wearable devices can be used to record the ECG, either with a dedicated device or a smartphone. In both cases, a single-lead ECG is recorded between the thumbs, fingers, or palms. Using handheld devices, systematic screening has been performed in a population at risk under certain circumstances, e.g., in primary care during seasonal influenza vaccination in the Dutch population aged 65 years and older [54], or in stroke and transient ischemic attack patients [55]. For example, the smartphone has been used for screening purposes in community pharmacies [56] and in subjects identified via general practitioner records [57]. Section 2.3 provides an overview of different technologies available for recording the ECG.

A large number of studies suggest that handheld or wearable devices are well-suited for AF screening, particularly in high-risk patients. This would be of great importance from a clinical point of view as more patients with AF would be detected at an early stage. However, early identification/detection of AF is compounded by the silent nature of AF in about one third of all patients [24]. Since the risk of stroke is the same for silent AF and symptomatic AF [58], it is important to detect the arrhythmia at an early stage so that therapies can be introduced which protect the patient from progression of AF as well as from the consequences of the arrhythmia.

In ECG-based detection of AF, the analysis of RR intervals has received the most attention since such information is readily available in most applications. However, such analysis is problematic when an AF episode is preceded by some other type of arrhythmia which is also manifested by an irregular ventricular rhythm resembling AF. Therefore, the analysis of P and f waves, although being more complex, is receiving more attention since morphologic information is essential when distinguishing AF from other arrhythmias. Morphologic information is also essential to the detection of brief AF episodes since a handful of RR intervals does not provide accurate quantification of rhythm irregularity. Chapter 4 provides a comprehensive review of AF detectors based on rhythm information as well as on rhythm and morphologic information.

Despite its disadvantages, rhythm-based detection offers the possibility to detect AF in single-lead pacemakers or defibrillators. From a clinical point of view, new onset or new recognition of AF in patients with reduced left ventricular systolic function is common, and therapeutic decisions are made easier by accurate estimation of AF burden. Another group of patients which would benefit from early detection of AF are those with heart failure and a biventricular pacemaker implanted, in whom long episodes of undetected AF may substantially reduce biventricular pacing up to causing decompensation [59].

1.10 Standardization of Atrial Fibrillation ECG Characteristics Assessment

There is considerable variation in the methodology and definitions of parameters used to characterize AF between different clinical and engineering research groups. While some of the ECG-based parameters for characterizing f waves, e.g., the atrial fibrillatory rate (AFR) and the closely related dominant AF frequency (DAF), have been tested in clinical contexts for more than a decade, the clinical significance of novel descriptors characterizing the degree of AF organization with, e.g., signal entropy or harmonic decay in the power spectrum, is limited to small-cohort studies which are either cross-sectional or retrospective by design, and thus stay even farther away from clinical routine than do the parameters which characterize the spectral properties of the f waves. The interdependence of different parameters of AF and their relation to methodological issues also remains to be further investigated.

It is not known to what extent differences in the methodology used for AF assessment may affect the results. Differences in signal processing algorithms, not always apparent from the descriptions of published methods which may involve proprietary information, can influence the result of studies comparing different methodologies. There is a paucity of data that would compare performance of individual methods of AF complexity assessment on the same patient cohort performed by different groups, which further reinforces the need for exchange of data and methodology. There is an unmet need for reproducibility studies of AF parameters in the clinical context. However, methodological differences may hamper interpretation and comparability of study findings, thus suggesting that reproducibility studies should ideally be preceded by applying an alternative methodology to the same patient cohort as used in the original study before testing performance of biomarkers on a different patient cohort.

One can draw a parallel with studies assessing the value of biochemical or genetic markers for diagnosis or prognosis in the clinical context. It is unlikely that the results of such studies would be comparable and generalizable in clinical practice if different non-standardized laboratory procedures would be used for estimation of biomarker values or identification of genetic variants. Until similar logic is applied to ECG-derived AF parameters, it would be unrealistic to expect that results of the studies involving ECG signal processing will be widely implemented in clinical practice.

From the clinical point of view, AF is such a multicausal rhythm disorder so that its clinical impact and interpretation of ECG characteristics should not be assumed to have similar meaning in different clinical contexts [6]. More importantly, despite years of clinical research and refinement of methodologies aimed at characterizing AF complexity, we are still far from understanding the natural course of different markers which can be derived from the ECG, their intra-individual reproducibility during recurrent AF episodes, propensity to showing circadian behavior, relation to the time from AF onset, and the degree of AF persistence and evolution during long-time observation in patients with permanent AF.

Taking AFR as an example of a spectral parameter that was assessed using the same methodology in the context of sinus rhythm restoration and maintenance after cardioversion, it was found to lack predictive value in patients with long-standing AF [60], to have significant association between a lower AFR and a higher probability of sinus rhythm during follow-up in patients with shorter AF duration [61], and to be highly predictive of spontaneous conversion in patients with AF duration less than 48 h [62]. At the same time, AFR is known to express circadian fluctuations, at least in patients with permanent AF [63, 64], and to demonstrate pronounced and rapid acceleration over the course of several minutes [65] to 3–4 h from the onset of an AF episode [66].

Even though a number of issues remain to be clarified, it is quite clear that the interpretation of AFR is highly dependent of the clinical context, and more studies are needed to define its limits of applicability as a biomarker suitable for risk stratification and prediction of intervention effect. Few other quantifiable characteristics of AF can be compared with AFR with respect to the level of understanding of biomarker "behavior" in different clinical contexts.

1.10.1 Electrocardiographic Characteristics as Biomarkers

As long as AF parameters are considered in the context of characterization of disease process and prediction of intervention effect, they fall under the definition posed by regulatory authorities governing drug development and approval. The European Medicines Agency (EMEA) defines biomarkers as "tests that can be used to follow body processes and diseases in humans and animals. They can be used to predict how patients will respond to a medicine or whether they have, or are likely to develop, a certain disease" [67].

Both EMEA and Food and Drug Administration (FDA) have developed biomarker qualification procedures that need to be fulfilled in order to accept the use of a biomarker in the context of a clinical trial, e.g., as a patient selection or stratification criterion, a measure of intervention effect or an endpoint. Biomarkers being considered for qualification should be "conceptually independent of the specific test performing the measurement [but] cannot become qualified without a reliable means to measure it" [68]. Signal characteristics of AF should therefore be seen as potential biomarkers used for guidance of therapy and should have a reliable means of their assessment. Therefore, already at an early stage of their development and validation, care should be taken to make sure that studies in which AF signal characteristics are assessed provide important information that defines the context of biomarker use, i.e., the specific populations within the AF continuum, the clinical type of AF, the potential impact of concomitant therapies, and interventions.

1.10.2 Roadmap for Standardization of AF Parameters

With regard to the above-mentioned observations of inter-individual variability and context dependence of AF signal characteristics, as well as the strict requirements posed by regulatory authorities, several issues need to be dealt with on the way to implementation of ECG-derived AF parameters in clinical decision-making.

Cross-Validation of Signal Processing Techniques

Reproducibility studies are needed to find out whether differences in the predictive value of AF parameters in different studies are caused by differences in clinical and demographical characteristics of studied populations or differences in signal processing methodologies. There is a continued need to establish open databases that would promote collaboration between initiative participants who in turn would be encouraged to share their data and developed algorithms. PhysioNet (www.physionet.org) [69] and Telemetric and Holter ECG Warehouse (THEW, www.thew-project.org) [70] are important repositories where large AF-ECG data sets are available, usually accompanied by clinical information. The former repository is supported by National Institute of General Medical Sciences (NIGMS) and National Institute of Biomedical Imaging and Bioengineering (NIBIB), and the latter by FDA and National Heart, Lung, and Blood Institute (NHLBI).

On the other hand, any data set containing different types of AF ECG recorded at different ventricular rates, f wave amplitude, signal quality, and so on, are suitable for technical assessment of agreement between different signal processing methods and their robustness.

Biomarker Behavior in Different Clinical Contexts

Since the use of ECG-derived AF parameters is highly dependent on the clinical context, there is a need for studies which specifically target different patient populations in order to obtain reference values valid in specific clinical situations, in which different degrees of atrial electrical or structural remodeling are expected. For example, the performance of AF parameters should be assessed separately in patients with short versus long duration of AF, with recurrent versus permanent AF, treated versus not treated with oral antiarrhythmic drugs, etc. Since any intervention on the atria may affect both spectral content and organization, studies should specifically address AF behavior in patients with a history of cardiac surgery and/or catheter ablation.

Comprehensive Data Processing

Researchers should be encouraged to assess and present the full range of AF characteristics, thus not limiting the results to presentation of significant findings only. Publication bias is a known phenomenon which hampers further development of biomarkers, and may lead to potentially unnecessary and time- and resource-consuming analysis performed by independent groups unaware of negative findings obtained by others in similar clinical contexts.

Prospective Studies

While retrospective studies and studies on selected patient populations may be suitable for initial assessment of novel AF markers, only prospective studies may answer questions concerning the suitability of these markers for prediction of intervention effects or long-term prognosis. A similar approach is exercised by regulatory authorities that include review of clinical studies demonstrating the value of a novel parameter in the biomarker qualification procedure [67, 68].

References

1. J.G. Smith, P.G. Platonov, B. Hedblad, G. Engström, O. Melander, Atrial fibrillation in the Malmö Diet and Cancer study: a study of occurrence, risk factors and diagnostic validity. Eur. J. Epidemiol. **25**, 95–102 (2010)
2. P. Doliwa Sobocinski, E. Anggårdh Rooth, V. Frykman Kull, M. von Arbin, H. Wallén, M. Rosenqvist, Improved screening for silent atrial fibrillation after ischaemic stroke. Europace **14**, 1112–1116 (2012)
3. P.A. Wolf, E.J. Benjamin, A.J. Belanger, W.B. Kannel, D. Levy, R.B. D'Agostino, Secular trends in the prevalence of atrial fibrillation: the Framingham study. Am. Heart J. **131**, 790–795 (1996)
4. D.D. McManus, M. Rienstra, E.J. Benjamin, An update on the prognosis of patients with atrial fibrillation. Circulation **126**, e143–146 (2012)
5. D.M. Lloyd-Jones, T.J. Wang, E.P. Leip, M.G. Larson, D. Levy, R.S. Vasan, R.B. D'Agostino, J.M. Massaro, A. Beiser, P.A. Wolf, E.J. Benjamin, Lifetime risk for development of atrial fibrillation: the Framingham Heart study. Circulation **110**, 1042–1046 (2004)
6. P. Kirchhof, G. Breithardt, E. Aliot, S. Al Khatib, S. Apostolakis, A. Auricchio, C. Bailleul, J. Bax, G. Benninger, C. Blomström-Lundqvist, L. Boersma, G. Boriani, A. Brandes, H. Brown, M. Brueckmann, H. Calkins, B. Casadei, A. Clemens, H. Crijns, R. Derwand, D. Dobrev, M. Ezekowitz, T. Fetsch, A. Gerth, A. Gillis, M. Gulizia, G. Hack, L. Haegeli, S. Hatem, K.G. Häusler, H. Heidbüchel, J. Hernandez-Brichis, P. Jais, L. Kappenberger, J. Kautzner, S. Kim, K.-H. Kuck, D. Lane, A. Leute, T. Lewalter, R. Meyer, L. Mont, G. Moses, M. Mueller, F. Münzel, M. Näbauer, J.C. Nielsen, M. Oeff, A. Oto, B. Pieske, R. Pisters, T. Potpara, L. Rasmussen, U. Ravens, J. Reiffel, I. Richard-Lordereau, H. Schäfer, U. Schotten, W. Stegink, K. Stein, G. Steinbeck, L. Szumowski, L. Tavazzi, S. Themistoclakis, K. Thomitzek, I.C. Van Gelder, B. von Stritzky, A. Vincent, D. Werring, S. Willems, G.Y.H. Lip, A.J. Camm, Personalized management of atrial fibrillation: proceedings from the fourth Atrial Fibrillation competence NETwork/European Heart Rhythm Association consensus conference. Europace **15**, 1540–1556 (2013)
7. M. Haïssaguerre, P. Jaïs, D.C. Shah, L. Gencel, V. Pradeau, S. Garrigues, S. Chouairi, M. Hocini, P. Le Métayer, R. Roudaut, J.C.J., Right and left atrial radiofrequency catheter therapy of paroxysmal atrial fibrillation. J. Cardiovasc. Electrophysiol. **7**, 1132–1144 (1996)
8. M. Hocini, S.Y. Ho, T. Kawara, A.C. Linnenbank, M. Potse, D. Shah, P. Jaïs, M.J. Janse, M. Haïssaguerre, J.M.T. De Bakker, Electrical conduction in canine pulmonary veins: electrophysiological and anatomic correlation. Circulation **105**, 2442–2448 (2002)
9. L. Staerk, J.A. Sherer, D. Ko, E.J. Benjamin, R.H. Helm, Atrial fibrillation: epidemiology, pathophysiology, and clinical outcomes. Circ. Res. **120**, 1501–1517 (2017)
10. H. Kottkamp, Fibrotic atrial cardiomyopathy: a specific disease/syndrome supplying substrates for atrial fibrillation, atrial tachycardia, sinus node disease, AV node disease, and thromboembolic complications. J. Cardiovasc. Electrophysiol. **23**, 797–799 (2012)

11. A. Frustaci, M. Caldarulo, A. Buffon, F. Bellocci, R. Fenici, D. Melina, Cardiac biopsy in patients with 'primary' atrial fibrillation. Histologic evidence of occult myocardial diseases. Chest **100**, 303–306 (1991)
12. J.G. Smith, C. Newton-Cheh, P. Almgren, J. Struck, N.G. Morgenthaler, A. Bergmann, P.G. Platonov, B. Hedblad, G. Engström, T.J. Wang, O. Melander, Assessment of conventional cardiovascular risk factors and multiple biomarkers for the prediction of incident heart failure and atrial fibrillation. J. Amer. Coll. Card. **56**, 1712–1719 (2010)
13. F. Gramley, J. Lorenzen, C. Knackstedt, O.R. Rana, E. Saygili, D. Frechen, S. Stanzel, F. Pezzella, E. Koellensperger, C. Weiss, T. Münzel, P. Schauerte, Age-related atrial fibrosis. Age (Dordrecht, Netherlands) **31**, 27–38 (2009)
14. A. Frustaci, C. Chimenti, F. Bellocci, E. Morgante, M.A. Russo, A. Maseri, Histological substrate of atrial biopsies in patients with lone atrial fibrillation. Circulation **96**, 1180–1184 (1997)
15. P.G. Platonov, L.B. Mitrofanova, V. Orshanskaya, S.Y. Ho, Structural abnormalities in atrial walls are associated with presence and persistency of atrial fibrillation but not with age. J. Am. Coll. Cardiol. **58**, 2225–2232 (2011)
16. D. Corradi, S. Callegari, L. Manotti, D. Ferrara, M. Goldoni, R. Alinovi, S. Pinelli, P. Mozzoni, R. Andreoli, A. Asimaki, A. Pozzoli, G. Becchi, A. Mutti, S. Benussi, J.E. Saffitz, O. Alfieri, Persistent lone atrial fibrillation: clinicopathologic study of 19 cases. Heart Rhythm **11**, 1250–1258 (2014)
17. C. Mahnkopf, T.J. Badger, N.S. Burgon, M. Daccarett, T.S. Haslam, C.T. Badger, C.J. McGann, N. Akoum, E. Kholmovski, R.S. Macleod, N.F. Marrouche, Evaluation of the left atrial substrate in patients with lone atrial fibrillation using delayed-enhanced MRI: implications for disease progression and response to catheter ablation. Heart Rhythm **7**, 1475–1481 (2010)
18. L.B. Mitrofanova, V. Orshanskaya, S.Y. Ho, P.G. Platonov, Histological evidence of inflammatory reaction associated with fibrosis in the atrial and ventricular walls in a case-control study of patients with history of atrial fibrillation. Europace **18**, iv156–iv162 (2016)
19. A. Boldt, U. Wetzel, J. Lauschke, J. Weigl, J. Gummert, G. Hindricks, H. Kottkamp, S. Dhein, Fibrosis in left atrial tissue of patients with atrial fibrillation with and without underlying mitral valve disease. Heart (British Cardiac Society) **90**, 400–405 (2004)
20. P. Kirchhof, S. Benussi, D. Kotecha, A. Ahlsson, D. Atar, B. Casadei, M. Castella, H.C. Diener, H. Heidbuchel, J. Hendriks, G. Hindricks, A.S. Manolis, J. Oldgren, B.A. Popescu, U. Schotten, B. Van Putte, P. Vardas, S. Agewall, J. Camm, G. Baron Esquivias, W. Budts, S. Carerj, F. Casselman, A. Coca, R. De Caterina, S. Deftereos, D. Dobrev, J.M. Ferro, G. Filippatos, D. Fitzsimons, B. Gorenek, M. Guenoun, S.H. Hohnloser, P. Kolh, G.Y. Lip, A. Manolis, J. McMurray, P. Ponikowski, R. Rosenhek, F. Ruschitzka, I. Savelieva, S. Sharma, P. Suwalski, J.L. Tamargo, C.J. Taylor, I.C. Van Gelder, A.A. Voors, S. Windecker, J.L. Zamorano, K. Zeppenfeld, 2016 ESC guidelines for the management of atrial fibrillation developed in collaboration with EACTS. Eur. Heart J. **37**, 2893–2962 (2016)
21. C. Marini, F. De Santis, S. Sacco, T. Russo, L. Olivieri, R. Totaro, A. Carolei, Contribution of atrial fibrillation to incidence and outcome of ischemic stroke: results from a population-based study. Stroke **36**, 1115–1119 (2005)
22. G.Y. Lip, R. Nieuwlaat, R. Pisters, D.A. Lane, H.J. Crijns, Refining clinical risk stratification for predicting stroke and thromboembolism in atrial fibrillation using a novel risk factor-based approach. The Euro Heart Survey on atrial fibrillation. Chest **137**, 263–272 (2010)
23. D.A. Lane, G.Y.H. Lip, Use of the CHA_2DS_2-VASc and HAS-BLED scores to aid decision making for thromboprophylaxis in nonvalvular atrial fibrillation. Circulation **126**, 860–865 (2012)
24. I. Savelieva, A.J. Camm, Clinical relevance of silent atrial fibrillation: prevalence, prognosis, quality of life, and management. J. Interv. Card. Electrophysiol. **4**, 369–382 (2000)
25. G.Y. Lip, C. Laroche, P.M. Ioachim, L.H. Rasmussen, L. Vitali-Serdoz, L. Petrescu, D. Darabantiu, H.J. Crijns, P. Kirchhof, P. Vardas, L. Tavazzi, A.P. Maggioni, G. Boriani, Prognosis and treatment of atrial fibrillation patients by European cardiologists: one year follow-up of the EURObservational Research Programme-Atrial Fibrillation General Registry Pilot Phase (EORP-AF Pilot registry). Eur. Heart J. **35**, 3365–3376 (2014)

26. J.S. Healey, S.J. Connolly, M.R. Gold, C.W. Israel, I.C. Van Gelder, A. Capucci, C.P. Lau, E. Fain, S. Yang, C. Bailleul, C.A. Morillo, M. Carlson, E. Themeles, E.S. Kaufman, S.H. Hohnloser, ASSERT Investigators, Subclinical atrial fibrillation and the risk of stroke. N. Engl. J. Med. **366**, 120–129 (2012)
27. D. Bengtsson, L. Brudin, P. Wanby, M. Carlsson, Previously unknown thyroid dysfunction in patients with acute ischemic stroke. Acta Neurol. Scand. **126**, 98–102 (2012)
28. L. Suissa, S. Lachaud, M.H. Mahagne, Optimal timing and duration of continuous electrocardiographic monitoring for detecting atrial fibrillation in stroke patients. J. Stroke Cerebrovasc. Dis. **22**, 991–995 (2013)
29. H. Kamel, K.R. Lees, P.D. Lyden, P.A. Teal, A. Shuaib, M. Ali, S.C. Johnston, Virtual International Stroke Trials Archive Investigators, Delayed detection of atrial fibrillation after ischemic stroke. J. Stroke Cerebrovasc. Dis. **18**, 453–457 (2009)
30. D. Jabaudon, J. Sztajzel, K. Sievert, T. Landis, R. Sztajzel, Usefulness of ambulatory 7-day ECG monitoring for the detection of atrial fibrillation and flutter after acute stroke and transient ischemic attack. Stroke **35**, 1647–1651 (2004)
31. R. Stahrenberg, M. Weber-Krüger, J. Seegers, F. Edelmann, R. Lahno, B. Haase, M. Mende, J. Wohlfahrt, P. Kermer, D. Vollmann, G. Hasenfuss, K. Gröschel, R. Wachter, Enhanced detection of paroxysmal atrial fibrillation by early and prolonged continuous holter monitoring in patients with cerebral ischemia presenting in sinus rhythm. Stroke **41**, 2884–2888 (2010)
32. A.C. Flint, N.M. Banki, X. Ren, V.A. Rao, A.S. Go, Detection of paroxysmal atrial fibrillation by 30-day event monitoring in cryptogenic ischemic stroke: the stroke and monitoring for PAF in real time (SMART) registry. Stroke **43**, 2788–2790 (2012)
33. D.J. Miller, M.A. Khan, R. Schultz, J.R. Simpson, A.M. Katramados, A.N. Russman, Outpatient cardiac telemetry detects a high rate of atrial fibrillation in cryptogenic stroke. J. Neurol. Sci. **324**, 57–61 (2013)
34. L. Elijovich, S.A. Josephson, G.L. Fung, W.S. Smith, Intermittent atrial fibrillation may account for a large proportion of otherwise cryptogenic stroke: a study of 30-day cardiac event monitors. J. Stroke Cerebrovasc. Dis. **18**, 185–189 (2009)
35. J. Brachmann, C.A. Morillo, T. Sanna, V. Di Lazzaro, H.C. Diener, R.A. Bernstein, M. Rymer, P.D. Ziegler, S. Liu, R.S. Passman, Uncovering atrial fibrillation beyond short-term monitoring in cryptogenic stroke patients: three-year results from the cryptogenic stroke and underlying atrial fibrillation trial. Circ. Arrhythm. Electrophysiol. **9**, e003333 (2016)
36. J. Reiffel, A. Verma, J.L. Halperin, B. Gersh, S. Tombul, J. Carrithers, L. Sherfesee, P. Kowey, Rationale and design of REVEAL AF: a prospective study of previously undiagnosed atrial fibrillation as documented by an insertable cardiac monitor in high-risk patients. Am. Heart J. **167**, 22–27 (2014)
37. D.J. Gladstone, M. Spring, P. Dorian, V. Panzov, K.E. Thorpe, J. Hall, H. Vaid, M. O'Donnell, A. Laupacis, R. Côté, M. Sharma, J.A. Blakely, A. Shuaib, V. Hachinski, S.B. Coutts, D.J. Sahlas, P. Teal, S. Yip, J.D. Spence, B. Buck, S. Verreault, L.K. Casaubon, A. Penn, D. Selchen, A. Jin, D. Howse, M. Mehdiratta, K. Boyle, R. Aviv, M.K. Kapral, M. Mamdani, Atrial fibrillation in patients with cryptogenic stroke. N. Engl. J. Med. **370**, 2467–2477 (2014)
38. T. Sanna, H.-C. Diener, R.S. Passman, V. Di Lazzaro, R.A. Bernstein, C.A. Morillo, M.M. Rymer, V. Thijs, T. Rogers, F. Beckers, K. Lindborg, J. Brachmann, Cryptogenic stroke and underlying atrial fibrillation. N. Engl. J. Med. **370**, 2478–2486 (2014)
39. J. Benezet-Mazuecos, J.M. Rubio, J. Farré, Atrial high rate episodes in patients with dual-chamber cardiac implantable electronic devices: unmasking silent atrial fibrillation. Pacing Clin. Electrophysiol. **37**, 1080–1086 (2014)
40. T.V. Glotzer, A.S. Hellkamp, J. Zimmerman, M.O. Sweeney, R. Yee, R. Marinchak, J. Cook, A. Paraschos, J. Love, G. Radoslovich, K.L. Lee, G.A. Lamas, M.O.S.T. Investigators, Atrial high rate episodes detected by pacemaker diagnostics predict death and stroke: report of the Atrial Diagnostics Ancillary Study of the MOde Selection Trial (MOST). Circulation **107**, 1614–1619 (2003)
41. T.V. Glotzer, E.G. Daoud, D.G. Wyse, D.E. Singer, M.D. Ezekowitz, C. Hilker, C. Miller, D. Qi, P.D. Ziegler, The relationship between daily atrial tachyarrhythmia burden from implantable

device diagnostics and stroke risk: the TRENDS study. Circulation. Arrhythm. Electrophysiol. **2**, 474–480 (2009)
42. S.H. Hohnloser, A. Capucci, E. Fain, M.R. Gold, I.C. van Gelder, J. Healey, C.W. Israel, C.P. Lau, C. Morillo, S.J. Connolly, A.S.S.E.R.T. Investigators, Committees, ASymptomatic atrial fibrillation and Stroke Evaluation in pacemaker patients and the atrial fibrillation Reduction atrial pacing Trial (ASSERT). Am. Heart J. **152**, 442–447 (2006)
43. D.T. Martin, M.M. Bersohn, A.L. Waldo, M.S. Wathen, W.K. Choucair, G.Y.H. Lip, J. Ip, R. Holcomb, J.G. Akar, J.L. Halperin, IMPACT Investigators, Randomized trial of atrial arrhythmia monitoring to guide anticoagulation in patients with implanted defibrillator and cardiac resynchronization devices. Eur. Heart J. **36**, 1660–1668 (2015)
44. B.J. Hirsh, R.S. Copeland-Halperin, J.L. Halperin, Fibrotic atrial cardiomyopathy, atrial fibrillation, and thromboembolism: mechanistic links and clinical inferences. J. Am. Coll. Cardiol. **65**, 2239–2251 (2015)
45. K.M. Henriksson, B. Farahmand, S. Asberg, A. Terént, N. Edvardsson, First-ever atrial fibrillation documented after hemorrhagic or ischemic stroke: the role of the $CHADS_2$ score at the time of stroke. Clin. Cardiol. **34**, 309–316 (2011)
46. M.A. Baturova, A. Lindgren, J. Carlson, Y.V. Shubik, S.B. Olsson, P.G. Platonov, Predictors of new onset atrial fibrillation during 10-year follow-up after first-ever ischemic stroke. Int. J. Cardiol. **199**, 248–252 (2015)
47. M.-L. Zuo, S. Liu, K.-H. Chan, K.-K. Lau, B.-H. Chong, K.-F. Lam, Y.-H. Chan, Y.-F. Lau, G.Y.H. Lip, C.-P. Lau, H.-F. Tse, C.-W. Siu, The $CHADS_2$ and CHA_2DS_2-VASc scores predict new occurrence of atrial fibrillation and ischemic stroke. J. Interv. Card. Electrophysiol. **37**, 47–54 (2013)
48. S. Suzuki, K. Sagara, T. Otsuka, H. Kano, S. Matsuno, H. Takai, T. Uejima, Y. Oikawa, A. Koike, K. Nagashima, H. Kirigaya, J. Yajima, H. Tanabe, H. Sawada, T. Aizawa, T. Yamashita, Usefulness of frequent supraventricular extrasystoles and a high $CHADS_2$ score to predict first-time appearance of atrial fibrillation. Am. J. Cardiol. **111**, 1602–1607 (2013)
49. I.C. Van Gelder, H.F. Groenveld, H.J.G.M. Crijns, Y.S. Tuininga, J.G.P. Tijssen, A.M. Alings, H.L. Hillege, J.A. Bergsma-Kadijk, J.H. Cornel, O. Kamp, R. Tukkie, H.A. Bosker, D.J. Van Veldhuisen, M.P. Van den Berg, R.A.C.E.I.I. Investigators, Lenient versus strict rate control in patients with atrial fibrillation. N. Engl. J. Med. **362**, 1363–1373 (2010)
50. M. Haïssaguerre, P. Jaïs, D.C. Shah, A. Takahashi, M. Hocini, G. Quiniou, S. Garrigue, A. Le Mouroux, P. Le Métayer, J. Clémenty, Spontaneous initiation of atrial fibrillation by ectopic beats originating in the pulmonary veins. N. Engl. J. Med. **339**, 659–666 (1998)
51. J.C. Nielsen, A. Johannessen, P. Raatikainen, G. Hindricks, H. Walfridsson, S.M. Pehrson, A. Englund, J. Hartikainen, L.S. Mortensen, P.S. Hansen, MANTRA-PAF Investigators, Long-term efficacy of catheter ablation as first-line therapy for paroxysmal atrial fibrillation: 5-year outcome in a randomised clinical trial. Heart **103**, 368–376 (2017)
52. P.G. Platonov, V.D.A. Corino, M. Seifert, F. Holmqvist, L. Sörnmo, Atrial fibrillatory rate in the clinical context: natural course and prediction of intervention outcome. Europace **16**, iv110–iv119 (2014)
53. T. Lankveld, C.B. de Vos, I. Limantoro, S. Zeemering, E. Dudink, H.J. Crijns, U. Schotten, Systematic analysis of ECG predictors of sinus rhythm maintenance after electrical cardioversion for persistent atrial fibrillation. Heart Rhythm **13**, 1020–1027 (2016)
54. D.D. McManus, J. Lee, O. Maitas, N. Esa, R. Pidikiti, A. Carlucci, J. Harrington, E. Mick, K.H. Chon, A novel application for the detection of an irregular pulse using an iPhone 4S in patients with atrial fibrillation. Heart Rhythm **10**, 315–319 (2013)
55. A.-S. Olsson, J. Engdahl, Detection of atrial fibrillation with intermittent handheld electrocardiogram in patients with ischemic stroke and transient ischemic attack and transient ischemic attack. J. Stroke Cerebrovasc. Dis. **25**, 2648–2652 (2016)
56. J.K. Lau, N. Lowres, L. Neubeck, D.B. Brieger, R.W. Sy, C.D. Galloway, D.E. Albert, S.B. Freedman, iPhone ECG application for community screening to detect silent atrial fibrillation: a novel technology to prevent stroke. Int. J. Cardiol. **165**, 193–194 (2013)

57. J.P.J. Halcox, K. Wareham, A. Cardew, M. Gilmore, J.P. Barry, C. Phillips, M.B. Gravenor, Assessment of remote heart rhythm sampling using the Alivecor heart monitor to screen for atrial fibrillation: the REHEARSE-AF study. Circulation **136**, 1784–1794 (2017)
58. G.C. Flaker, K. Belew, K. Beckman, H. Vidaillet, J. Kron, R. Safford, M. Mickel, P. Barrell, AFFIRM Investigators, Asymptomatic atrial fibrillation: demographic features and prognostic information from the Atrial Fibrillation Follow-up Investigation of Rhythm Management (AFFIRM) study. Am. Heart J. **149**, 657–663 (2005)
59. European Society of Cardiology, (ESC), European Heart Rhythm Association (EHRA), M. Brignole, A. Auricchio, G. Baron-Esquivias, P. Bordachar, G. Boriani, O.-A. Breithardt, J. Cleland, J.-C. Deharo, V. Delgado, P.M. Elliott, B. Gorenek, C.W. Israel, C. Leclercq, C. Linde, L. Mont, L. Padeletti, R. Sutton, P.E. Vardas, 2013 ESC guidelines on cardiac pacing and cardiac resynchronization therapy: the task force on cardiac pacing and resynchronization therapy of the European Society of Cardiology (ESC). Developed in collaboration with the European Heart Rhythm Association. Europace **15**, 1070–1118 (2013)
60. A. Bollmann, A. Tveit, D. Husser, M. Stridh, L. Sörnmo, P. Smith, S.B. Olsson, Fibrillatory rate response to candesartan in persistent atrial fibrillation. Europace **10**, 1138–1144 (2008)
61. F. Holmqvist, M. Stridh, J.E. Waktare, L. Sörnmo, S.B. Olsson, C.J. Meurling, Atrial fibrillatory rate and sinus rhythm maintenance in patients undergoing cardioversion of persistent atrial fibrillation. Eur. Heart J. **27**, 2201–2207 (2006)
62. M.B. Choudhary, F. Holmqvist, J. Carlson, H.-J. Nilsson, A. Roijer, P.G. Platonov, Low atrial fibrillatory rate is associated with spontaneous conversion of recent-onset atrial fibrillation. Europace **15**, 1445–1452 (2013)
63. C.J. Meurling, J.E. Waktare, F. Holmqvist, A. Hedman, A.J. Camm, S.B. Olsson, M. Malik, Diurnal variations of the dominant cycle length of chronic atrial fibrillation. Am. J. Physiol. **280**, H401–H406 (2001)
64. F. Sandberg, A. Bollmann, D. Husser, M. Stridh, L. Sörnmo, Circadian variation in dominant atrial fibrillation frequency in persistent atrial fibrillation. Physiol. Meas. **31**, 531–542 (2010)
65. S. Petrutiu, A.V. Sahakian, S. Swiryn, Short-term dynamics in fibrillatory wave characteristics at the onset of paroxysmal atrial fibrillation in humans. J. Electrocardiol. **40**, 155–160 (2007)
66. P.G. Platonov, M. Stridh, M. de Melis, L. Urban, J. Carlson, G. Corbucci, F. Holmqvist, Analysis of atrial fibrillatory rate during spontaneous episodes of atrial fibrillation in humans using implantable loop recorder electrocardiogram. J. Electrocardiol. **45**, 723–726 (2012)
67. European Medicines Agency, Biomarkers (2017), http://www.ema.europa.eu/ema/. Accessed 17 Oct 2017
68. Food and Drug Administration (FDA), Biomarker qualification program (2017), http://www.fda.gov/. Accessed 17 Oct 2017
69. A.L. Goldberger, L.A. Amaral, L. Glass, J.M. Hausdorff, P.C. Ivanov, R.G. Mark, J.E. Mietus, G.B. Moody, C.K. Peng, H.E. Stanley, PhysioBank, PhysioToolkit, and PhysioNet: components of a new research resource for complex physiologic signals. Circulation **101**, E215–220 (2000)
70. J.-P. Couderc, The Telemetric and Holter ECG Warehouse Initiative (THEW): a data repository for the design, implementation and validation of ECG-related technologies, in Proceedings of IEEE Conference on Engineering in Medicine and Biology (EMBS) **32**, 6252–6255 (2010)

Chapter 2
Lead Systems and Recording Devices

Andrius Petrėnas, Vaidotas Marozas and Leif Sörnmo

2.1 Introduction

The surface ECG is the most widely used clinical tool for detecting atrial fibrillation (AF), with the standard 12-lead ECG, recorded at rest, as the preferred lead system [1]. However, fewer leads are sufficient to confirm AF, being particularly advantageous when the patient is monitored for an extended period of time, for example, to detect silent, previously undocumented AF. Since extended ECG monitoring may last for several days, even a three-lead system, requiring five electrodes, may become inconvenient for the patient [2]. For this reason, various types of single-lead devices have been developed to ensure patient comfort in exchange for diagnostic reliability provided by more leads. While reduced lead systems are well-suited for detecting AF, they only provide limited information on the underlying mechanisms. On the other hand, comprehensive characterization of AF, such as identification of wavefront propagation patterns in the atria [3], benefits from using a large number of electrodes.

Commercial devices, especially those employing a small number of leads, tend to produce false positives due to electromyographic noise, motion artifacts, ectopic beats, or pronounced sinus arrhythmia [4], calling for manual review of computer-detected arrhythmic episodes. This shortcoming is particularly pronounced when

A. Petrėnas · V. Marozas
Biomedical Engineering Institute, Kaunas University of Technology, Kaunas, Lithuania
e-mail: andrius.petrenas@ktu.lt

V. Marozas
e-mail: vaidotas.marozas@ktu.lt

L. Sörnmo (✉)
Department of Biomedical Engineering and Center for Integrative Electrocardiology, Lund University, Lund, Sweden
e-mail: leif.sornmo@bme.lth.se

L. Sörnmo (ed.), *Atrial Fibrillation from an Engineering Perspective*, Series in BioEngineering, https://doi.org/10.1007/978-3-319-68515-1_2

it is of interest to analyze brief AF episodes. Since manual review of long-term continuous ECG recordings is time-consuming, and at times unreliable [5], it is essential to improve AF detection performance.

2.2 Lead Systems

Historically, the electrode placement of most lead systems has been focused on ventricular activity, and, consequently, they are suboptimal for atrial activity—an observation which applies especially to atrial activity during AF. Since the amplitudes of atrial waves, i.e., P and f waves, are much smaller than those of ventricular waves, better discrimination of atrial tachyarrhythmias, e.g., atrial tachycardia, atrial flutter, and AF, can be achieved in leads with larger atrial amplitude [6, 7]. Moreover, large-amplitude leads facilitate the characterization of f waves, especially when the f waves have been separated from ventricular activity (Chap. 5). For example, the atrial fibrillatory rate (AFR), a parameter useful for selection of treatment strategy [8], is more reliably estimated.

When the standard 12-lead ECG system is employed, limb lead II normally produces the largest P wave amplitude. On the other hand, since the atrial activity is disorganized during AF, the precordial leads V_1 and V_2 usually have the largest f wave amplitude due to their proximity to the atria. In general, the f waves in V_1 to V_6 have decreasing amplitude as the distance to the atria increases. Moreover, the precordial electrodes are placed close to each other, thus causing the f waves to be correlated.

For sake of clarity, it should be noted that the term "lead" is used for defining the voltage difference between two electrodes. For example, lead I is the voltage between the left and right arm electrodes, whereas lead V_1 is the voltage between the electrode placed at C_1, i.e., chest position #1, and Wilson's central terminal which is the average of the three limb leads.

2.2.1 Body Surface Potential Mapping

While the standard 12-lead ECG serves as the reference system for AF diagnosis, it only provides limited information on the wavefront propagation patterns of atrial activity. A larger number of electrodes, distributed over the body surface, better reflect spatial differences. Indeed, reconstruction of body surface potentials from the f waves of the 12-lead ECG is associated with a 53% reconstruction error, indicating that additional electrodes are needed to provide a more accurate representation of the f waves [9]. For this reason, body surface potential mapping (BSPM) is an important tool for reconstructing f waves [10]. The number of electrodes used in BSPM systems has ranged from 56 [3] to 252 [11], involving anterior, posterior, and sometimes lateral sites. Electrodes can either be distributed nonuniformly [9] or arranged as a uniform grid around V_1 [3, 12].

In one of the very first BSPM studies on AF, wavefront propagation patterns were identified, representing either a single wavefront, single wavefront with breakages, or multiple wavefronts [3]. Signals were recorded using a custom-made vest with 56 (40 anterior and 16 posterior) electrodes arranged around V_1 in a uniform grid with a distance of 2.2 cm between the electrodes. Despite the low quality of the f waves, the constructed maps of wavefront propagation were consistent with those observed using invasive or optical mapping. When comparing f waves in V_1 to those recorded using BSPM, f waves in V_1 were representative only when a single wavefront propagated across the whole atria; in such cases, f wave amplitude and AFR were similar, regardless of the site where the signal was recorded. However, for a single wavefront breaking or multiple simultaneous wavefronts coexisting, the f wave pattern differed and depended on the site.

Obviously, more electrodes result in higher spatial resolution. However, many electrodes are impractical in clinical routine, and, therefore, the following questions should be answered:

- What is the optimal number of electrodes?
- Which electrode placement provides the best atrial information?

To answer these questions, BSPM with 64 nonuniformly distributed electrodes (48 anterior and 16 posterior) was employed [9]. The criterion for selecting the optimal leads was given by the total root mean square (RMS) error of the reconstructed signals of the remaining leads not included in the lead system under consideration. Since the ventricular activity of the whole body surface can be reconstructed from the standard 12-lead ECG with an error of 25% [9], the optimal number of electrodes needed to record the atrial activity was defined by the same error percentage. The results showed that 23 electrodes were needed to achieve this reconstruction error, using the placement shown in Fig. 2.1. To reduce the error to 10%, at least 45 electrodes were required for reconstructing the atrial activity, while only 22 electrodes for the ventricular activity.

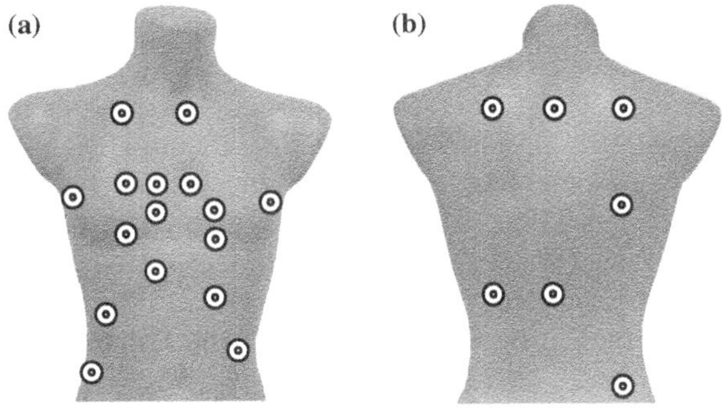

Fig. 2.1 Placement of **a** anterior and **b** posterior electrodes in AF-optimized body surface potential mapping

Assuming that the type of wavefront propagation can be successfully identified using BSPM, this technique has the potential to characterize the degree of atrial organization. The application of BSPM is limited by the time-consuming preparation procedure; however, this problem may be mitigated by integrating the electrodes into smart textiles, which would reduce the time required for electrode placement considerably.

2.2.2 Modifications of the Standard 12-Lead ECG

A lead system tailored to the analysis of AF is more likely to be clinically accepted if it can be viewed as a modified standard 12-lead ECG. Accordingly, several such modifications have been proposed [9, 13–15]. The design of a modified lead system is restricted to the 10 electrodes defining the standard 12-lead ECG. To retain Wilson's central terminal, used as the reference potential of the precordial leads, the positions of the extremity electrodes V_R, V_L, and V_F should be retained. Moreover, it is desirable to place the electrodes in relation to conventional sites to simplify electrode placement, especially since incorrect placement is a well-known problem when acquiring the standard 12-lead ECG [16]. The electrode sites are determined heuristically, either by placing the electrodes close to the atria or using some optimization criterion.

A heuristic approach to determining ECG leads with increased f wave amplitude is to place V_3 to V_6 in the vicinity of V_1 and V_2. For example, the precordial electrodes can be rearranged to form a 2×3 grid on the right side of the chest, with V_1 and V_2 unmodified and V_3 to V_6 replaced by the new electrodes V_{LS}, V_S, V_{RS}, and V_R [13]. The electrode V_{LS} (left superior, LS) is placed one intercostal space above V_2. The electrode V_S (superior, S) is placed one intercostal space above V_1. The electrode V_{RS} (right superior, RS) is moved to the right of V_S, whereas the electrode V_R is moved to the right of V_1 and aligned vertically with V_{RS}, see Fig. 2.2a. Preliminary results based on simulated f wave signals, using a biophysical model of the human atria and thorax, showed that the resulting lead system, coined as the *electroatriogram*, provides more information on atrial activity than the standard 12-lead ECG [13, 14].

Two other, heuristically derived modifications are intended for either anterior or posterior monitoring [15]. Similarly to the lead system in [13], V_1 and V_2 remain unmodified, whereas the other four leads are rearranged counterclockwise around V_1 and V_2, see Fig. 2.2b and c. The only difference compared to the placement in [13] is that the lead on the right upper side of the chest, denoted V_{RS}, is placed one intercostal space below V_1, denoted V_B. The posterior electrodes V_{1P}, V_{1PS}, V_{2P}, and V_{2PS} (posterior superior, PS) are rearranged in a similar fashion. Two electrodes are placed opposite to V_1 and V_2, whereas the remaining two are placed one intercostal space below V_1 and V_2. Following cancellation of the ventricular activity, the usefulness of the proposed electrode placement was investigated in terms of interlead dispersion of the AFR. The proximity of V_1 to V_2 gave rise to nearly the same AFR, whereas the dispersion was larger among the anterior leads. Although the posterior electrode placement was associated with 35% lower frequency dispersion than that

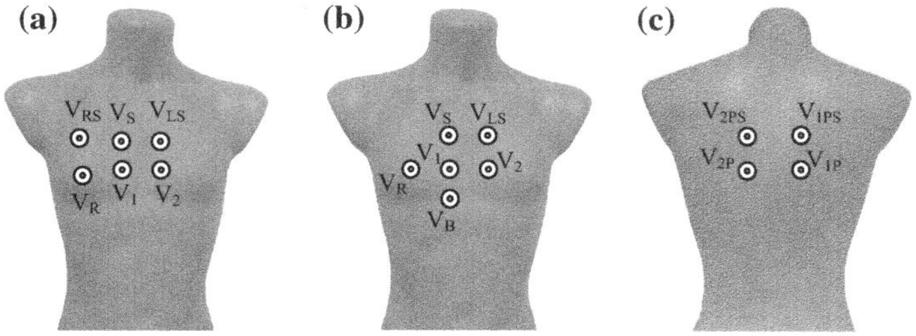

Fig. 2.2 Heuristically derived modifications of the precordial electrode placement: two anterior and one posterior electroatriogram proposed in **a** [13], **b** [15], and **c** [15], respectively

of the anterior, the combination of both approaches can be used to extract additional information. It was shown that anterior leads mostly reflect the AFR of the right atrium [17], whereas posterior leads mostly reflect the AFR of the left atrium [18]. Therefore, inclusion of both anterior and posterior leads should be applied to identify frequency gradients, and the driving atrium.

The above-mentioned heuristic approaches are useful for finding electrode placement which offers larger f wave amplitude, but not for finding electrode placement which increases the atrial information. A quantitative approach to optimal placement of precordial electrodes is to maximize the ratio of the smallest to the largest singular value of the f wave signal obtained at different sites on the body surface [14]. Similar to the heuristic approaches, four of the precordial electrodes are rearranged, while two remain unmodified. The placement of V_1 is unmodified due to its proximity to the atria, as is the placement of V_4 because its f waves are the ones which are the least correlated to those in V_1. In [14], using a biophysical model to simulate f waves, the search for optimal placement of the four precordial electrodes resulted in four distinct areas on the thorax where the electrodes should be placed to ensure additional atrial information, see Fig. 2.3a. The electrode V_S is placed one intercostal space above V_1. The electrode V_{RS} is positioned to the right of V_S at the same intercostal space. The electrode V_{LC} is placed slightly below the left clavicle (LC), whereas the electrode V_P is placed on the back behind the atria at the same level as V_1. Interestingly, two of the four new electrode sites, namely V_S and V_{RS}, were the same as those derived heuristically in [13]. In addition, the placement of V_S was the same as that used for lead S in the EASI lead system, defined by the E, A, and I electrode positions of the Frank lead system, plus an electrode S positioned over the upper end of the sternum [19].

Results obtained from simulated f wave signals showed that more information can be extracted using the optimized as well as the heuristically-derived lead systems than with the systems using conventional electrode placement. Nevertheless, the difference between the proposed lead systems in terms of gained atrial information was not large. Considering that the electrodes of the heuristically-derived placement are closer to the atria, thus producing larger f wave amplitude, the use of optimized placement, involving electrodes with smaller f waves (V_4 and V_{LC}), is questionable.

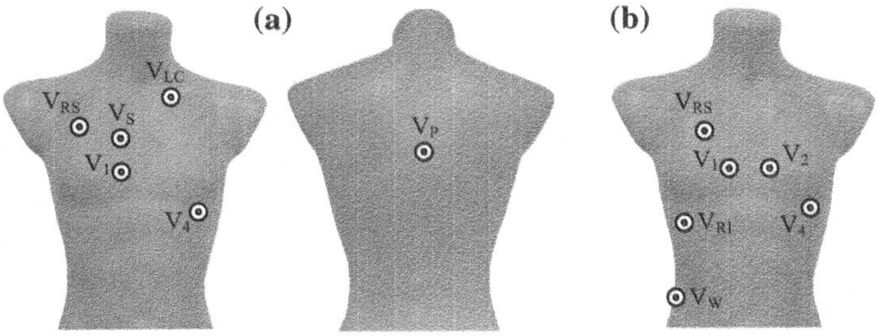

Fig. 2.3 Modifications of precordial electrode placement derived by using different optimization criterion, namely **a** the ratio of the smallest to the largest singular value of the f wave signals (anterior and posterior placement) [14], and **b** the f wave reconstruction error (anterior placement) [9]

Two other modified 12-lead ECG systems for improving AF analysis have been derived using an iterative lead selection principle. Only those leads were chosen which increase the information contained in each selected lead set [20]. Similarly to the previously described lead systems, only four precordial electrodes were repositioned based on the criteria for selecting leads with additional atrial information [9]. Depending on the constraint that either V_1 and V_2 or V_1 and V_4 should be kept at their conventional sites, two electrode placements were derived of which one had the two additional electrodes V_{RI} (right inferior) and V_W (waist), see Fig. 2.3b. Both lead systems were associated with similar f wave reconstruction error, about 10% lower than that obtained with the standard 12-lead ECG, although the electrodes were placed on different parts of the body. A relatively small improvement in reconstruction error implies that modifications of the 12-lead ECG do not result in markedly increased atrial information content. Considering the increased complexity of the electrode placement, it is doubtful whether such modifications will be adopted clinically.

2.2.3 Reduced Lead Systems

So far, no specialized lead system is used in clinical routine when ambulatory monitoring is prescribed in patients with AF. Therefore, five-electrode, standard ambulatory monitoring is typically applied, capable of recording six limb leads, i.e., I, II, III, aVR, aVL, aVF, in combination with a single precordial lead, e.g., V_1 [21]. It is well-known that standard ambulatory monitors lead to reduced quality of life and have lower patient compliance [2, 22]. For this reason, single-lead monitors are considered as a promising alternative for long-term ambulatory monitoring of AF [22]. To facilitate AF detection in reduced-lead ECGs, it is desirable to employ electrode placement optimized for f wave analysis.

A reduced lead system for atrial activity enhancement was proposed already in the very first book on electrocardiography, authored by Sir Thomas Lewis and published in 1913 [23]. The Lewis lead system consists of five leads, where two, L_1 and L_2, are derived for the purpose of enhancing the f waves, see Fig. 2.4a. The bipolar chest lead L_I is obtained by placing electrode 1 over the upper end of the sternum, and electrode 2 on the right side of the sternum at the second intercostal space. Lead L_2 is the voltage between electrodes 2 and 3, placed on the right side of the sternum at the fourth intercostal space.

The Lewis lead system has two out of six electrodes placed directly on the pectoral muscle, where artifacts due to arm movement are likely to occur. In order to avoid leads on the chest muscles, a modified Lewis lead was proposed which is more immune to noise and with good projection of the f waves [24]. The modified Lewis lead L_M is obtained by removing electrode 2 of the Lewis lead system, and moving electrode 3 one intercostal space downwards, i.e., from the fourth to the fifth, to improve the immunity to arm movement artifacts, see Fig. 2.4b. A comparative study showed that L_I and L_2 exhibit a high atrial-to-ventricular amplitude ratio [24]. However, f wave enhancement is achieved at the expense of a much reduced ventricular amplitude, rather than increased f wave amplitude. Despite the fact that L_I and L_2 are proximal, L_I was found to be twice as susceptible to electromyographic noise as L_2, and, therefore, L_2 is considered the preferred lead. For long-term monitoring, where high noise levels are often encountered, L_M may be more advantageous since it has larger f wave amplitude than L_2. Hence, L_2 offers better immunity to electromyographic noise [24].

Recent results have shown that the largest P wave amplitude is obtained when the distance between electrodes is 12–18 cm [25]. Depending on torso size, the distance between the electrodes of the modified Lewis lead L_M is 14–20 cm, whereas the distance for L_I and L_2 is less than half. Since L_I, L_2, and L_M are roughly along the same axis with respect to the heart's electrical vector, the longer distance between electrodes is probably the primary reason for a larger f wave amplitude in L_M than in L_I and L_2.

A short distance between the bipolar electrodes is desirable since the electrodes can then be integrated into a single recording device, and, consequently, increase patient compliance [22, 27]. However, a reduced distance between the electrodes will also reduce P and f wave amplitudes, which in turn may reduce the performance

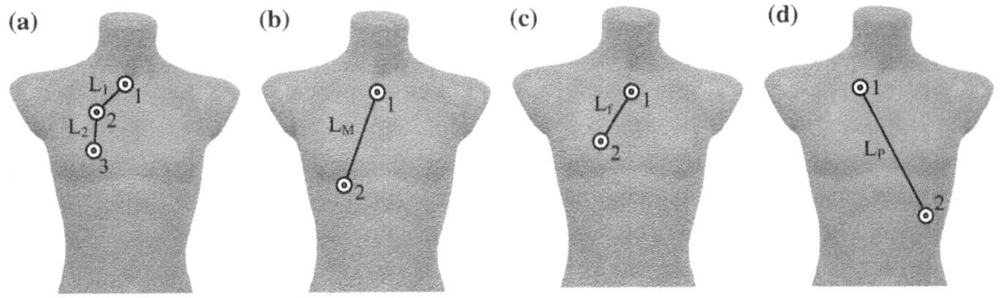

Fig. 2.4 Reduced lead ECG systems suitable for ambulatory AF monitoring. **a** Original Lewis leads [23], **b** modified Lewis lead [24], **c** f-lead [25], and **d** P-lead [26]. Note that all lead systems involve bipolar leads, although only the leads enhancing atrial activity are shown

of AF detectors exploring the presence of f waves [28, 29]. Thus, another approach to finding the best lead for ambulatory monitoring of f waves is to account for both signal amplitude and distance between the electrodes. Based on the analysis of 120 combinations of bipolar leads, obtained from 16 electrodes placed on the precordial area, the best leads for monitoring of P and f waves were identified [25]. A lead for P wave monitoring is obtained by placing one electrode on the right side of the sternum at the second intercostal space, and another electrode on the sternum in line with the fourth intercostal space. A lead for f wave monitoring (referred to as "f-lead") is obtained by placing one electrode on the sternum at the level of the second intercostal space, and another electrode on the right side of the sternum at the fourth intercostal space, see Fig. 2.4c. Although a distance of 8 cm was found to be optimal for both P and f waves, these leads involve different electrodes, and, thus, a lead can only be optimal for either P or f wave monitoring. Considering that one electrode is placed on the sternum, while the other on the right side of the chest, both leads are susceptible to motion artifacts. Hence, the signal quality during daily activities should be investigated before an AF monitor is designed for these particular leads.

However, this approach is not necessarily optimal with respect to maximized atrial amplitude. Therefore, a bipolar lead (referred to as "P-lead") for maximized P wave amplitude was derived based on the analysis of 117-lead BSPM, recorded from more than 200 healthy individuals [26]. Since only healthy individuals were included, the best lead coincided, not surprisingly, with the electrical axis of the heart, see Fig. 2.4d. To obtain the P-lead, one electrode has to be placed on the right sternal clavicular junction and the other on the midpoint of the left costal margin in line with the seventh intercostal space. The study showed that the P-lead has nearly three times larger P wave amplitude than L_2, and 35% larger amplitude than L_M. While the P-lead is attractive for AF detectors involving P wave analysis, it will not necessarily produce larger f wave amplitude.

Although the EASI lead system was not specifically developed for the analysis of atrial activity, lead ES provides relatively large P wave amplitude [24]. This lead system uses four electrodes placed on the torso, where the electrodes E, A, and I are placed at the same sites as in the Frank lead system [19]. The electrode S is placed over the upper end of the sternum (the manubrium) and the electrode E at the bottom of the sternum at the level of the fifth intercostal space.

The aforementioned reduced lead systems, except EASI, were developed for the purpose of enhancing atrial activity, and, therefore, less suitable for evaluating ventricular beat morphology, except for fundamental information such as the occurrence times of the QRS complexes. Since RR interval irregularity, together with P wave absence and f wave presence, represent the landmark properties of AF, a single-lead system may even provide sufficient information for AF detection. Considering that the electrodes of the reduced lead systems are closely positioned to the right atrium, the spectral content of the f waves is largely related to the right atrium. However, it remains to be shown, for example, how well AFR, determined from the surface ECG, agrees with that measured by an intra-atrial recording. Moreover, there is a lack of studies examining noise immunity of different ECG lead systems.

2.3 Recording Devices

For many years, the resting ECG and 24-h ambulatory monitoring were the only available techniques for analyzing AF. However, the rapid development of electronics and communication technologies has given rise to novel approaches to AF monitoring and screening, ranging from invasive devices, providing a convenient way to continuously monitor arrhythmias for months and years, to short-term screening recorders, see Fig. 2.5 [30]. Contemporary mobile ECG devices (smartphones, smartwatches, smart wristbands, tablets) can immediately transfer data to the physician via wireless communication (Bluetooth, Wi-Fi, GSM networks) and Internet. Together with the rapidly growing cloud-based software, new opportunities are created to collect and analyze large amounts of data. Since most mobile devices can simultaneously acquire other types of information than the ECG, e.g., physical activity, body position, and respiration, such information can be used to provide a more complete picture of the factors initiating and maintaining AF.

Given that there are many devices on the market capable of recording the ECG, this section only provides an overview of the most representative technologies applied to AF detection.

2.3.1 Standard Resting ECG

Since the standard 12-lead ECG is globally recognized, cost-effective, straightforward to record, and easy to interpret for a trained physician/technician, the majority of patients with AF are identified using the 12-lead ECG. Its main disadvantage is the

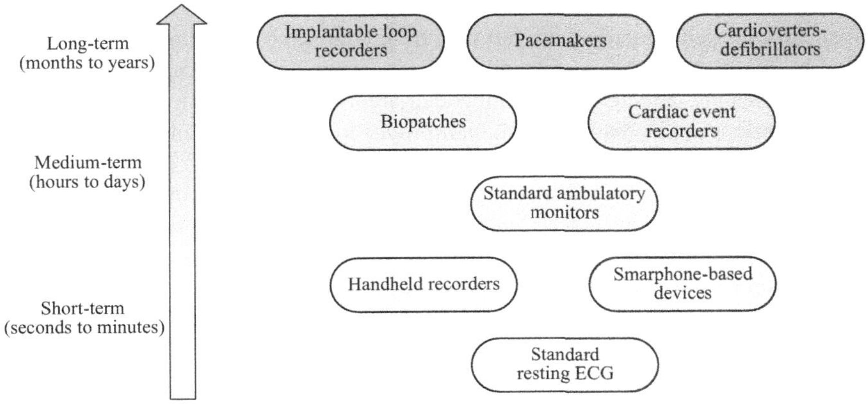

Fig. 2.5 Types of recording devices used for detection of AF, presented according to monitoring duration

short recording time, normally lasting only 10–30 s. Thus, persistent and permanent AF can be detected with the 12-lead ECG, whereas paroxysmal AF, especially in its early stages, is likely to go undetected.

2.3.2 Standard Ambulatory Monitors

Since the ambulatory monitor was introduced by Norman Holter in 1961 [31], it has become widely accepted in clinical applications. The monitor is a portable device capable of recording the ECG continuously for one or two days during normal daily activities, usually with a three-lead configuration. Occasionally, seven-day and even one-month ambulatory monitoring may be prescribed. The ambulatory monitor is a valuable tool not only for arrhythmia detection, including AF, but also to evaluate drug effects on AF recurrence. Once monitoring is finished, the ECG is analyzed offline using commercial software. Although the software for AF detection has improved over the years, the results from automated AF detection still need to be manually reviewed by a physician/technician to ensure that AF episodes are correctly detected.

The major drawbacks of ambulatory monitoring are the adhesive electrodes and the connecting wires, which can be uncomfortable for certain patients and sometimes lead to premature termination of monitoring [32]. Moreover, certain patients are allergic to adhesive electrodes, and therefore other techniques need to be considered.

2.3.3 Cardiac Event Recorders

External cardiac event recorders are portable devices similar to the standard ambulatory monitor, but smaller and lighter since a single-lead ECG is usually recorded. Most cardiac event recorders are not operating continuously, but record when activated by the patient when symptoms occur, or started automatically when rhythm abnormalities are detected by an embedded algorithm. Two main types of cardiac event recorders can be distinguished: *continuous loop recorders* and *symptom event recorders*.

Continuous loop recorders are continuously refreshing, i.e., recording and erasing the data. Data refreshing is terminated when the device is triggered by the patient or an algorithm. In such a way, the ECG signal of the entire event, as well as a few minutes before and after, are stored in the memory. Due to limited storage capacity, only the onset and the end of the episode are saved if the arrhythmia lasts for a longer period of time. Similar to the standard ambulatory monitor, the continuous loop recorder is connected to adhesive electrodes.

Symptom event recorders are, in contrast to continuous loop recorders, not required to be worn at all times, but can be temporarily attached to the body by the patient when arrhythmia symptoms are experienced. However, this device is neither suitable for capturing the very onset of an arrhythmia, nor for detecting nocturnal and asymptomatic events. Moreover, when symptoms are severe, e.g., fainting, it is difficult for the patient to correctly attach the device to the body.

Various studies have demonstrated that cardiac event recorders are prone to false alarms due to ectopic beats, since runs of such beats may resemble AF. For example, a study of the external loop recorder Vitaphone 3100 BT (Vitaphone GmbH, Mannheim, Germany) showed that each patient in sinus rhythm, in average, had more than five false positive ECGs during the 24-h recording period, caused by ventricular and supraventricular premature beats and sinus arrhythmia [33]. Similar performance was reported in another study where nearly 3,000 events were collected, and roughly 1,200 were classified as AF by a proprietary algorithm [34]. However, only 5% were confirmed as AF after manual revision.

2.3.4 *Biopatches*

The biopatch technology provides a comfortable and safe way to monitor health status, by employing a leadless, wearable, single-use device which is designed to record the long-term continuous ECG [35]. In addition to the ECG, this type of device can acquire other physiological parameters, e.g., skin temperature, accelerometer data, and respiration. Due to its minimalistic design, the device can be placed on body areas associated with less motion artifacts, and thus record signals with better quality.

The Zio-Patch device (iRhythm Technologies, San Francisco, CA, USA) is a noninvasive, small size (123 × 53 × 10.7 mm), lightweight (34 g), single-use ECG monitor, capable of recording the ECG up to 14 days [36]. Unlike standard ambulatory monitors, the Zio-Patch can remain attached during showering to ensure continuous monitoring. The device is attached over the left pectoral muscle with skin adhesive. Besides continuous monitoring, symptomatic arrhythmia events can be captured when the patient presses a button on the device. Once monitoring is completed, the device is mailed back to the manufacturer, where the data are analyzed. The detection of AF is performed in a two-step procedure: Arrhythmia episodes are first identified using an algorithm relying on heart rate, rhythm irregularity, and ECG morphology, and then the detected episodes are reviewed by a technician to eliminate false positives.

In a large study by the Zio-Patch, including more than 26,700 patients, the mean wear time was found to be 7.6 $\pm$ 3.6 days, thus being about half the expected monitoring time [22]. Although only 16% reached the maximum monitoring duration of about 13 days, 96% of the patients exceeded the ambulatory monitoring time of 48 h. Given that nearly 25% of all AF cases were detected after 48 h, this finding justifies the Zio-Patch for detection of AF episodes which otherwise would have been missed with standard ambulatory monitoring. In 87% of all patients, the device produced analyzable signal quality during at least 22 h per day.

The NUVANT Mobile Cardiac Telemetry system (Corventis, San Jose, CA, USA) is another biopatch device that offers prolonged monitoring of arrhythmias [37, 38].

In contrast to the Zio-Patch, the NUVANT system analyzes the ECG in real-time. The system consists of a wearable sensor, designed for single-lead recordings, and a portable transmitter which transmits the data to a proprietary monitoring center via a cloud-based application. The device incorporates activation functionality allowing the patient to trigger the device on-demand when symptoms are experienced. The sensor is activated automatically and starts recording the ECG immediately after being attached to the body. The data are transmitted whenever rhythm abnormalities are detected, followed by review of certified technicians. Clinical reports on rhythm trends and AF burden are then prepared. The sensor lasts up to 7.5 days, however, multiple sensors can be employed in sequence to extend the monitoring period up to 30 days. Preliminary studies performed by the Corventis team themselves demonstrated an AF prevalence of 20% among those who used the NUVANT system [37, 38]. However, larger independent clinical studies are needed to establish the clinical usefulness of this device relative to other monitoring technologies.

Compactness, absence of wires, and water resistance of biopatches contribute to better patient compliance and signal quality, which in turn lead to that more cases with AF are detected than with standard ambulatory monitoring. Therefore, further shrinking of device size and extended monitoring duration will likely promote biopatch-based AF monitoring as an alternative to implanted loop recorders. Nevertheless, the clinical implications and the cost effectiveness of biopatches have to be further investigated.

2.3.5 Handheld Recorders

Handheld recorders rely on the single-lead ECG, acquired by placing the hands (thumbs, fingers, palms) on two electrodes during a period from 10 s to several minutes. Handheld recorders have been proposed as an alternative screening tool to pulse palpation. Compared to other ECG modalities, handheld recorders offer certain advantages such as low cost, ease of use, and the absence of adhesive electrodes and connecting wires.

Among the handheld AF screening devices, thumb-ECG recorders are gaining recognition around the world. The Zenicor thumb-ECG recorder (Zenicor Medical Systems AB, Stockholm, Sweden) is used by several hundred clinics in Scandinavia. The device records a 30-s single-lead ECG several times a day, at predetermined times as well as when the patient has AF-related symptoms. The signal is transmitted to a web server via a mobile connection. In such a way, ECGs are stored on the internet, automatically categorized into clinically useful groups, and, if needed, made available to a trained technician for evaluation. The Zenicor device was evaluated on a large population of 7,173 individuals of 75–76 years of age who underwent intermittent screening during two weeks. The results showed that four times more cases with AF were detected than with 24-h standard ambulatory monitoring [39, 40]. Based on information from the thumb-ECG, indicating new onset AF, anticoagulant treatment was initiated in 93% of all patients.

The AfibAlert AF monitor (Lohman Technologies, Sussex, WI., USA) is another commercially available thumb-ECG recorder. The AfibAlert acquires the ECG in two different ways: either by pressing thumbs on the electrodes or using wrist electrodes. The signal is recorded for 45 s, and then analyzed for AF. The preliminary decision is immediately reported by an LED indicator. If AF is suspected, the patient has to transmit the ECG to a physician for confirmation of the preliminary decision by uploading the data through a USB connection. The company website declares an AF detection accuracy of 94%, however, no clinical study has been published which supports this figure.

The MyDiagnostick recorder (Applied Biomedical Systems BV, Maastricht, The Netherlands) is designed to record a palm ECG. The device has the form of a stick with metallic handles at both ends, serving as electrodes. In order to acquire data for arrhythmia detection, the user has to hold the metallic handles for one minute. To reduce the number of false positives, the procedure is repeated twice. The recorded ECG is then analyzed using an embedded, proprietary AF detection algorithm. The patient is informed about the outcome of the analysis via an LED indicator. The MyDiagnostick device was tested on 181 patients, where the majority had confirmed AF. Thus, a highly exaggerated AF prevalence of 53% was reported when the ECGs were acquired [41]. Sensitivity of 94% and specificity of 93% resulted when the recommended protocol of three subsequent measurements was followed (see Sect. 4.5 for the definition of different detection performance measures). Since most patients had AF, the influence of non-AF ECGs with ectopic beats on the false positive rate remains to be established.

In summary, handheld devices offer a simple and fast means for detecting AF, since measurements can be performed whenever arrhythmia symptoms are experienced. Moreover, such devices may be used by a physician/technician to check whether the patient needs a standard 12-lead ECG for confirmation of AF. On the other hand, the huge amount of data to be manually reviewed is a significant problem for some of the handheld ECG recorders. In addition, poor signal quality due to large electrode–skin impedance and motion artifacts, rapid changes in the ECG signal due to lost electrode contact, and low f wave amplitude are obstacles which make the analysis particularly challenging [42].

2.3.6 Smartphone-Based Devices

Smartphone-based devices are emerging tools for screening of general health status [43, 44]. In 2016, there were 2.3 billion smartphone users around the world, and more than 259,000 mobile health applications available on app stores for personal use. It is highly likely that smartphones incorporating healthcare technologies will occupy a large part of the medical screening devices in the future. Considering that smartphones are well-suited for data acquisition, storage, and processing, as well as for display and transmission of analysis results, they represent an inexpensive means for mass screening of AF [45, 46].

Smartphones are used to acquire the ECG between the fingers of the left and right hands, placed on the electrodes at the back of the smartphone case. AliveCor Heart Monitor (AliveCor, San Francisco, CA, USA) and CardiacDesigns ECG Check (CardiacDesigns, Park City, UT, USA) have cases with integrated dry electrodes for acquiring a single-lead ECG. Before a recording is made, special instructions on arm relaxation are provided with the aim of reducing the noise level and the amount of artifacts. The ECG is recorded for about one minute and transmitted to the microphone of the iPhone, using a modulated ultrasound signal. The ECG is sent to a cardiologist for review when an abnormal rhythm has been identified by a proprietary algorithm.

Although studies are underway to assess the suitability of this technology for mass screening [47], its performance remains unclear. For example, the original study reported excellent performance of the AliveCor Heart Monitor with sensitivity of 98% and specificity of 97% [48]. However, a subsequent study on two different groups, cardiac patients and geriatric patients, revealed much lower sensitivities of 55% and 79%, respectively [49]. This dramatic reduction in sensitivity was explained by errors in the software and the decision taken by the company to favor specificity over sensitivity [50]. The motivation behind this decision was to minimize the number of false positives, since the device is sold to patients who will not necessarily seek ECG revision by certified technicians.

2.3.7 Implantable Devices

Implantable loop recorders are invasive leadless devices used exclusively for diagnostic purposes. Such recorders have proved to be useful for diagnosing recurrent syncope events when the patient temporarily has lost consciousness and then recovers spontaneously [51]. In case of AF, the implantable loop recorder plays a special role in certain situations, e.g., when evaluating the success of AF treatment procedures such as radiofrequency or cryoablation, assessing the efficacy of rate control therapy, or detecting asymptomatic paroxysmal AF episodes after cryptogenic ischemic stroke.

Several implantable loop recorders with embedded AF detection are available for clinical use, e.g., Reveal XT (Medronic, Minneapolis, MN, USA), SJM Confirm (St, Jude Medical, St. Paul, MN, USA), and Sleuth (Transoma Medical, St. Paul, MN, USA). These devices include two built-in electrodes suitable for recording the bipolar electrogram, have a thickness of several millimeters, weigh less than 20 g, and are inserted subcutaneously. Similarly to the external, continuous loop recorder earlier mentioned, the implantable equivalent involves looping memory, and can operate either in automated self-activation mode or patient-activated mode using a handheld control device when symptoms are experienced.

The Reveal XT implantable loop recorder identifies atrial tachycardia and AF on the basis of Poincaré plot analysis [52], see also Sect. 4.2.2. In addition, the device can be programmed to detect arrhythmia episodes with a user-defined minimum duration.

In a study with 247 patients using a minimum AF episode duration of two minutes, the sensitivity and the positive predictive value were found to be 88.2% and 73.5%, respectively. However, these figures increased to 92.1 and 79.6% when the minimum duration was increased to as much as six minutes [53]. As a result, six minutes has usually been preferred in clinical studies. Manual review of simultaneously recorded electrograms showed that false detection due to ectopic beats represented the most common problem [53, 54]. A more detailed review of the causes showed that 35% of the false positives were due to activity of the pectoral muscle, 15% due to atrial and ventricular premature beats, 4% due to false QRS detection, and 1.5% due to T wave oversensing [55].

Many clinical studies have demonstrated that continuous AF monitoring using implantable loop recorders is superior to noninvasive techniques. Hence, implantable cardiac monitors are gaining in popularity, although the false positive rate is usually high. However, mass implantation of invasive devices is unrealistic due to the high cost associated with the device and the required surgical procedure, as well as the potential risk of infection. Another notable drawback is that the device has to be replaced after 2–3 years of usage, although emerging, energy-effective hardware and software solutions may, at least in theory, extend the operation time of the device up to 10 years [56].

Implantable devices, such as the pacemaker, the cardioverter–defibrillator, and the biventricular pacemaker for cardiac resynchronization therapy are used for therapeutic purposes. However, they can also be programmed to detect arrhythmias such as AF. This type of device can record the intra-atrial electrogram directly in the heart via an atrial lead. In contrast to the surface ECG, the intra-atrial electrogram mostly reflects atrial activations, whereas the ventricular activity usually has lower amplitude (Fig. 2.6). Thus, a device with an atrial electrode not only makes it possible to

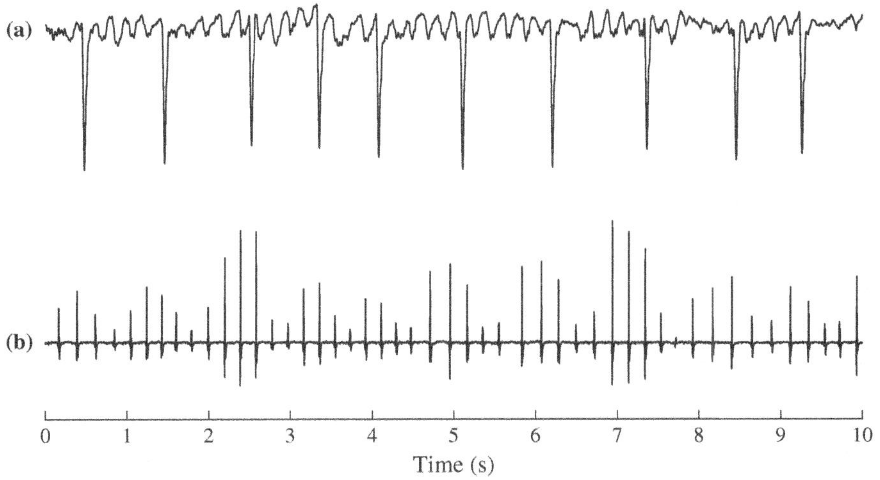

Fig. 2.6 Simultaneously recorded **a** ECG and **b** intra-atrial electrogram during AF

detect episodes of rapid atrial rate, but also to characterize individual episodes with respect to atrial rate.

Since the dual-chamber cardioverter–defibrillator has an electrode placed in the right ventricle, both the intra-atrial and intra-ventricular signals are involved in the detection of atrial tachyarrhythmias. A small study showed that 98% of 132 AF episodes were correctly detected by a dual-chamber cardioverter–defibrillator Medtronic Jewel AF (Medtronic, Minneapolis, MN, USA) [57]. All false positives were due to over-sensing of far-field ventricular activity. It should be noted that all falsely detected episodes were shorter than five minutes, with a mean duration of 2.6 ± 2.0 min for atrial tachycardia and 3.2 ± 1.6 min for AF. This result suggests that the false positive rate increases for arrhythmic events of shorter duration. Although AF may be discriminated from atrial tachycardia or atrial flutter using information on AFR and irregularity of atrial events, some AF episodes may be incorrectly classified as atrial tachycardia.

As mentioned above, incorrect AF detection in a dual-chamber cardioverter–defibrillator may be due to far-field ventricular activity in the intra-atrial electrogram. In some patients, especially when the atrial electrode is positioned outside the right atrial appendage, the amplitude of the ventricular activity is very large. Therefore, to avoid over-sensing of the far-field ventricular activity, the detection sensitivity of atrial waves has either to be reduced or the post-ventricular atrial blanking prolonged [58]. For this reason, in many clinical studies, an episode is flagged as atrial tachyarrhythmia when the atrial rate exceeds 190 beats per minute for at least six minutes. Moreover, a rapid atrial rate may be due to other supraventricular tachycardias, such as atrial flutter or atrial tachycardia, or even to bursts of atrial premature beats. Thus, manual review of intra-atrial signals may not be enough to distinguish between AF and other arrhythmias.

2.3.8 Non-ECG Devices

Emerging technologies for signal acquisition provide interesting means for recording physiological signals in a less obtrusive way, without the need for disposable electrodes. Several photoplethysmographic approaches to AF detection have been proposed, including the built-in camera of an iPhone [59], a web camera [60], an earlobe sensor [61], and a smart wristband [62] to acquire a pulsatile signal, see Fig. 2.7. Since AF detection based on the RR interval series of the ECG has been found useful (Sect. 4.2), it can be anticipated that AF detection based on the PP interval series of the photoplethysmogram (PPG) should be equally useful, although the PP intervals do not always match the RR intervals [63, 64]. In fact, the RR-based approach dominates in PPG-based AF detection, just as it does in ECG-based AF detection, because information on pulse morphology is more vulnerable to noise and artifacts.

Since a camera is available in any smartphone, the least expensive alternative to mass AF screening is based on the smartphone. The main idea is to produce a pulsatile PPG signal from the video obtained by placing the fingertip directly on the

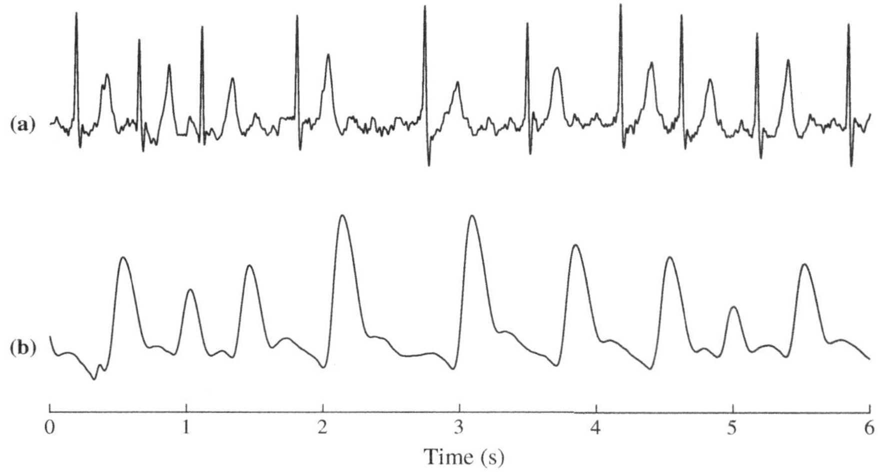

Fig. 2.7 Simultaneously recorded **a** ECG and **b** photoplethysmogram during AF

camera lens, with the LED flash illuminating the fingertip [46]. The resulting PPG signal is then processed with respect to AF detection. An iPhone-based prototype was validated in 76 patients with persistent AF undergoing electrical cardioversion, based on the pulsatile signal acquired before and after cardioversion [65]. Detection performance was expressed in terms of sensitivity and specificity, found to be 96% and 97%, respectively. The Cardiio Rhythm smartphone application (Cardiio, Cambridge, MA, USA) exhibited similar performance, with sensitivity and specificity of 93% and 98%, respectively [45]. In that study, the signal was acquired under the instruction of a trained observer. However, when acquired without supervision, the signal quality will most likely deteriorate, leading to reduced performance.

The commercially available smart wristbands facilitate unobtrusive AF monitoring, since the PPG can be acquired intermittently or even continuously for several days. Using the Empatica E4 smart wristband (Empatica, Milan, Italy), the PPG signal was acquired at rest for 10 min in 31 patients with persistent AF, 29 with sinus rhythm, and 9 with non-AF arrhythmias [62]. The sensitivity and specificity of AF detection were found to be 75% and 96%, respectively.

Given that the oscillometric principle of self-screening blood pressure monitors involves the analysis of a pulsatile signal, which in turn represents pressure oscillations in the sphygmomanometer cuff, the same signal can be employed for evaluation of pulse rhythm irregularity. Such an approach is especially attractive for mass screening, since home blood pressure monitors are widespread among hypertensive patients at high risk for developing AF. The Microlife BP A200 (Microlife AG, Widnau, Switzerland) and the Omron M6 (Omron Healthcare, Kyoto, Japan) are widely used blood pressure monitors with an integrated function for AF detection. In both monitors, detection is performed during cuff deflation by calculating the mean and standard deviation of 10 consecutive pulse intervals. Then, the irregularity index is computed by dividing the standard deviation by the mean, i.e., the coefficient of variation. Since these measures are especially sensitive to the presence of ectopic beats, intervals 25% shorter and 25% longer than the mean are removed from the series before computation of the irregularity index. The only notable difference between these two blood pressure monitors lies in their respective recommendations: three consecutive measurements should be performed using the Microlife BP A200, whereas only one for the Omron M6.

Several studies have investigated the feasibility of the Microlife BP A200 blood pressure monitor to detect AF. The performance differed slightly between the studies, depending primarily on the number of consecutive measurements taken for decision-making. Rather high sensitivity of 92–100% and specificity of 89–97% were obtained when three consecutive measurements were performed [66–69]. However, such monitors are prone to false alarms due to the presence of ectopic beats or highly variable pulse rates. For example, respiratory sinus arrhythmia is very common in the younger population, thus it is not surprising that 18% of the measurements were false positives for individuals of 13–18 years of age [70]. Hence, it is essential to evaluate the usefulness of this technique for AF detection, requiring large-scale studies where the ECG is simultaneously recorded.

Although the idea to detect AF using the PPG is promising, artifacts tend to play an important role when recordings are made at home, without the supervision of trained staff. Therefore, reliable artifact detection is necessary to ensure that the workload, as well as the expenses, generated by many false detections can be held to a minimum. Another problem arises in situations when the patient has impaired blood flow in the fingers—a problem commonly encountered in patients suffering from diabetes. So far, no guidelines exist on how to interpret the PPG signal, and, therefore, the ECG still needs to be recorded to confirm the presence of AF. As a result, PPG-based devices are suitable for AF screening, but not for diagnostic purposes.

2.3.9 Monitoring Strategies

At an early stage of arrhythmia progression, AF detection is particularly challenging because the episodes may be asymptomatic, brief, and infrequent. Therefore, AF is usually identified during planned examinations of health status or by pronounced symptoms. Today, pulse palpation followed by a 12-lead ECG or 24-h ambulatory monitoring is the standard procedure for AF screening in individuals over 65 years [1]. However, ambulatory monitoring, let alone the 12-lead ECG, is usually insufficient for detecting paroxysmal AF. Hence, there is an ongoing debate on the monitoring strategies which are better suited for specific tasks, such as, monitoring of AF recurrence after catheter ablation, cryptogenic ischemic stroke, and coronary artery bypass grafting [71, 72]. Moreover, when selecting the most appropriate strategy for AF detection, factors such as cost effectiveness and patient compliance should be considered as well.

A comprehensive study was conducted in which various strategies of intermittent AF monitoring (24-h, 7-, 14-, and 30-day standard ambulatory monitoring) were investigated in terms of the likelihood to detect at least one AF episode during a one-year period [72]. Since the study excluded AF episodes shorter than five minutes, the chance to detect even a single AF episode during the monitoring period increases if a high-performing algorithm for brief AF detection is applied. The study was based on mathematical simulations using data from invasive, continuous monitoring of 647 patients. The simulation results showed that in order to identify paroxysmal AF in half of the monitored patients, four random tests of 24-h ambulatory monitoring should, in average, be prescribed. To reach a sensitivity of 80%, at least three random tests of 30-day ambulatory monitoring, five tests of 14 days, or seven tests of 7 days are needed.

An essential point to be made regarding the temporal occurrence pattern of AF episodes is that it is considerably more challenging to detect AF with intermittent monitoring when AF episodes are highly aggregated in time, see Fig. 2.8 [72]. In such cases 24-h ambulatory monitoring can turn out to be completely ineffective, requiring extended monitoring to improve the detection rate.

A number of studies have been conducted which compare the standard monitoring strategy, i.e., the 12-lead ECG or 24-h ambulatory monitoring, to potentially more advantageous AF detection strategies [72]. For example, the above-mentioned large scale population study, involving individuals of 75–76 years of age, showed that short-term intermittent screening, using the handheld Zenicor ECG recorder for at least twice a day over two weeks, detected new onset AF in 7.4% of all patients [73].

Another problem is the selection of effective strategy for AF detection after cryptogenic ischemic stroke; both intermittent screening and continuous monitoring are considered. For example, patients having suffered from a stroke or a transient ischemic attack were screened for one month, using a patient-activated event recorder [74]. Only patients with a negative outcome of the initial 24-h ambulatory monitoring were prescribed with screening (one ECG recorded per day of about

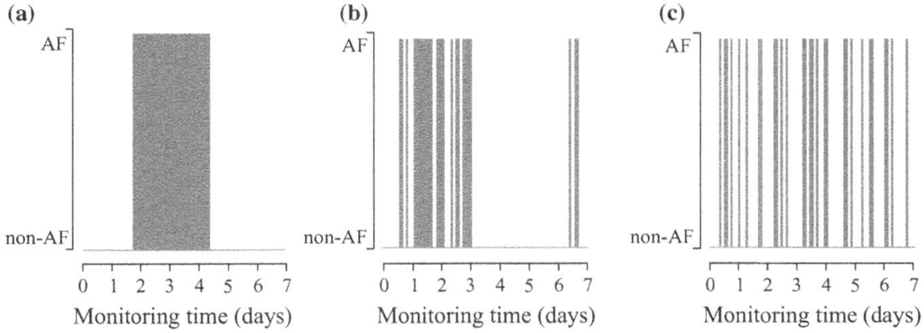

Fig. 2.8 Different temporal occurrence pattern of AF episodes, all three patterns having identical AF burden (equal to 0.3). **a** One single episode, **b** highly aggregated episodes, and **c** numerous episodes spread out over the monitoring period. Episodes with AF are indicated with dark areas

30-s duration). Surprisingly, as many as 9.2% of all cases with new onset AF were identified using such a simplistic screening approach. It was emphasized in [74] that the cost of transtelephonic ECG monitoring was almost one third lower than that of 24-h ambulatory monitoring.

Since stroke patients are predisposed to have brief AF episodes, intermittent screening may result in lower than actual AF detection rates. Hence, continuous monitoring, using either invasive or noninvasive technologies, has been applied in several studies. For example, one-month monitoring after ischemic stroke, using non-invasive event-triggered loop recorder, improved the detection rate of paroxysmal AF by more than five times compared to 24-h ambulatory monitoring [75]. In another study, patients having suffered from cryptogenic ischemic stroke were prescribed with continuous AF monitoring using an insertable loop recorder, where half a year of monitoring yielded up to a sixfold higher AF detection rate compared to 24-h ambulatory monitoring [76].

These findings clearly demonstrate that 24-h ambulatory monitoring is not particularly efficient for detecting AF, despite the fact that it represents the standard procedure in most countries. However, a unified agreement on how each different situation (opportunistic AF screening, evaluation of cardioversion/catheter ablation success, AF detection after cryptogenic stroke, monitoring of drug effect) should be handled remains to be established in order to achieve the highest efficiency of AF detection.

References

1. P. Kirchhof, S. Benussi, D. Kotecha, A. Ahlsson, D. Atar, B. Casadei, M. Castella, H.C. Diener, H. Heidbuchel, J. Hendriks, G. Hindricks, A.S. Manolis, J. Oldgren, B.A. Popescu, U. Schotten, B. Van Putte, P. Vardas, S. Agewall, J. Camm, G. Baron Esquivias, W. Budts, S. Carerj, F. Casselman, A. Coca, R. De Caterina, S. Deftereos, D. Dobrev, J.M. Ferro, G. Filippatos, D. Fitzsimons, B. Gorenek, M. Guenoun, S.H. Hohnloser, P. Kolh, G.Y. Lip, A. Manolis, J. McMurray, P. Ponikowski, R. Rosenhek, F. Ruschitzka, I. Savelieva, S. Sharma, P. Suwalski, J.L. Tamargo, C.J. Taylor, I.C. Van Gelder, A.A. Voors, S. Windecker, J.L. Zamorano, K. Zeppenfeld, 2016 ESC guidelines for the management of atrial fibrillation developed in collaboration with EACTS. Eur. Heart J. **37**, 2893–2962 (2016)
2. L. Roten, M. Schilling, A. Häberlin, J. Seiler, N.G. Schwick, J. Fuhrer, E. Delacrétaz, H. Tanner, Is 7-day event triggered ECG recording equivalent to 7-day Holter ECG recording for atrial fibrillation screening? Heart **98**, 645–649 (2012)
3. M.S. Guillem, A.M. Climent, F. Castells, D. Husser, J. Millet, A. Arya, C. Piorkowski, A. Bollmann, Noninvasive mapping of human atrial fibrillation. J. Cardiovasc. Electrophysiol. **20**, 507–513 (2009)
4. K. Harris, D. Edwards, J. Mant, How can we best detect atrial fibrillation? J. R. Coll. Physicians. Edinb. **42**, 5–22 (2012)
5. J. Mant, D.A. Fitzmaurice, F.D.R. Hobbs, S. Jowett, E.T. Murray, R. Holder, M. Davies, G.Y.H. Lip, Accuracy of diagnosing atrial fibrillation on electrocardiogram by primary care practitioners and interpretative diagnostic software: analysis of data from screening for atrial fibrillation in the elderly (safe) trial. Br. Med. J. **335**, 1–6 (2007)

6. A.L. Bakker, G. Nijkerk, B.E. Groenemeijer, R.A. Waalewijn, E.M. Koomen, R.L. Braam, H.J. Wellens, The Lewis lead: making recognition of P waves easy during wide QRS complex tachycardia. Circulation **119**, e592–e593 (2009)
7. A. Mizuno, K. Masuda, K. Niwa, Usefulness of Lewis lead for visualizing P-wave. Circ. J. **78**, 2774–2775 (2014)
8. P.G. Platonov, V.D.A. Corino, M. Seifert, F. Holmqvist, L. Sörnmo, Atrial fibrillatory rate in the clinical context: natural course and prediction of intervention outcome. Europace **16**, iv110–iv119 (2014)
9. M.S. Guillem, A. Bollmann, A.M. Climent, D. Husser, J. Millet, F. Castells, How many leads are necessary for a reliable reconstruction of surface potentials during atrial fibrillation? IEEE Trans. Inf. Technol. Biomed. **13**, 330–340 (2009)
10. M.S. Guillem, A.V. Sahakian, S. Swiryn, Derivation of orthogonal leads from the 12-lead electrocardiogram. Performance of an atrial-based transform for the derivation of P loops. J. Electrocardiol. **41**, 19–25 (2008)
11. M. Haissaguerre, M. Hocini, A.J. Shah, N. Derval, F. Sacher, P. Jaïs, R. Dubois, Noninvasive panoramic mapping of human atrial fibrillation mechanisms: a feasibility report. J. Cardiovasc. Electrophysiol. **24**, 711–717 (2013)
12. M.S. Guillem, A.M. Climent, J. Millet, Á. Arenal, F. Fernández-Avilés, J. Jalife, F. Atienza, O. Berenfeld, Noninvasive localization of maximal frequency sites of atrial fibrillation by body surface potential mapping. Circ. Arrhythm. Electrophysiol. **6**, 294–301 (2013)
13. Z. Ihara, V. Jacquemet, J.M. Vesin, A. van Oosterom, Adaption of the standard 12-lead ECG system focusing on atrial electrical activity, in Proceedings of Computers in Cardiology vol. **32**, 203–205 (2005)
14. Z. Ihara, A. van Oosterom, V. Jacquemet, R. Hoekema, Adaptation of the 12-lead electrocardiogram system dedicated to the analysis of atrial fibrillation. J. Electrocardiol. **40**, 68.e1–68.e8 (2007)
15. D. Husser, M. Stridh, L. Sörnmo, I. Toepffer, H.U. Klein, S.B. Olsson, A. Bollmann, Electroatriography–time-frequency analysis of atrial fibrillation from modified 12-lead ECG configurations for improved diagnosis and therapy. Med. Hypotheses **68**, 568–573 (2007)
16. W. Wenger, P. Kligfield, Variability of precordial electrode placement during routine electrocardiography. J. Electrocardiol. **29**, 179–184 (1996)
17. A. Bollmann, N. Kanuru, K. McTeague, P. Walter, D.B. DeLurgio, J. Langberg, Frequency analysis of human atrial fibrillation using the surface electrocardiogram and its response to ibutilide. Am. J. Cardiol. **81**, 1439–1445 (1998)
18. S. Petrutiu, A.V. Sahakian, W.B. Fisher, S. Swiryn, Manifestation of left atrial events in the surface electrocardiogram, in Proceedings of Computers in Cardiology vol. **33**, 1–4 (2006)
19. G.E. Dower, A. Yakush, S.B. Nazzal, R.V. Jutzy, C.E. Ruiz, Deriving the 12-lead electrocardiogram from four (EASI) electrodes. J. Electrocardiol. **21**(Suppl.), S182–187 (1988)
20. R.L. Lux, C.R. Smith, R.F. Wyatt, J.A. Abildskov, Limited lead selection for estimation of body surface potential maps in electrocardiography. IEEE Trans. Biomed. Eng. **25**, 270–276 (1978)
21. B.J. Drew, R.M. Califf, M. Funk, E.S. Kaufman, M.W. Krucoff, M.M. Laks, P.W. Macfarlane, C. Sommargren, S. Swiryn, G.F. Van Hare, Practice standards for electrocardiographic monitoring in hospital settings. Circulation **110**, 2721–2746 (2004)
22. M.P. Turakhia, D.D. Hoang, P. Zimetbaum, J.D. Miller, V.F. Froelicher, U.N. Kumar, X. Xu, F. Yang, P.A. Heidenreich, Diagnostic utility of a novel leadless arrhythmia monitoring device. Am. J. Cardiol. **112**, 520–524 (2013)
23. T. Lewis, *Clinical Electrocardiography* (Shaw & Sons, New York, 1913)
24. A. Petrėnas, V. Marozas, G. Jaruševičius, L. Sörnmo, A modified Lewis ECG lead system for ambulatory monitoring of atrial arrhythmias. J. Electrocardiol. **48**, 157–163 (2015)
25. S. Nedios, I. Romero, J.-H. Gerds-Li, E. Fleck, C. Kriatselis, Precordial electrode placement for optimal ECG monitoring: implications for ambulatory monitor devices and event recorders. J. Electrocardiol. **47**, 669–676 (2014)

26. A. Kennedy, D.D. Finlay, D. Guldenring, R.R. Bond, J. McLaughlin, Detecting the elusive P-wave: a new ECG lead to improve the recording of atrial activity. IEEE Trans. Biomed. Eng. **63**, 243–249 (2016)
27. M.A. Rosenberg, M. Samuel, A. Thosani, P.J. Zimetbaum, Use of a noninvasive continuous monitoring device in the management of atrial fibrillation: a pilot study. Pacing Clin. Electrophysiol. **36**, 328–333 (2013)
28. S. Babaeizadeh, R.E. Gregg, E.D. Helfenbein, J.M. Lindauer, S.H. Zhou, Improvements in atrial fibrillation detection for real-time monitoring. J. Electrocardiol. **42**, 522–526 (2009)
29. S. Asgari, A. Mehrnia, M. Moussavi, Automatic detection of atrial fibrillation using stationary wavelet transform and support vector machine. Comput. Biol. Med. **60**, 132–142 (2015)
30. J.S. Steinberg, N. Varma, I. Cygankiewicz, P. Aziz, P. Balsam, A. Baranchuk, D.J. Cantillon, P. Dilaveris, S.J. Dubner, N. El-Sherif, J. Krol, M. Kurpesa, M.T. La Rovere, S.S. Lobodzinski, E.T. Locati, S. Mittal, B. Olshansky, E. Piotrowicz, L. Saxon, P.H. Stone, L. Tereshchenko, G. Turitto, N.J. Wimmer, R.L. Verrier, W. Zareba, R. Piotrowicz, 2017 ISHNE-HRS expert consensus statement on ambulatory ECG and external cardiacmonitoring/telemetry. Heart Rhythm **17**, e55–e96 (2017)
31. N.J. Holter, New method for heart studies: continuous electrocardiography of active subjects over long period is now practical. Science **134**, 1214 (1961)
32. H.T. Tu, S. Spence, J.M. Kalman, S.M. Davis, Twenty-eight day Holter monitoring is poorly tolerated and insensitive for paroxysmal atrial fibrillation detection in cryptogenic stroke. Int. Med. J. **44**, 505–508 (2014)
33. A. Müller, W. Scharner, T. Borchardt, W. Och, H. Korb, Reliability of an external loop recorder for automatic recognition and transtelephonic ECG transmission of atrial fibrillation. J. Telemed. Telecare **15**, 391–396 (2009)
34. B.O. Velthuis, J. Bos, K. Kraaier, J. Stevenhagen, J.M. van Opstal, J. van der Palen, M.F. Scholten, Performance of an external transtelephonic loop recorder for automated detection of paroxysmal atrial fibrillation. Ann. Noninvasive Electrocardiol. **18**, 564–570 (2013)
35. S.S. Lobodzinski, ECG patch monitors for assessment of cardiac rhythm abnormalities. Prog. Cardiovasc. Dis. **56**, 224–229 (2013)
36. P.M. Barrett, R. Komatireddy, S. Haaser, S. Topol, J. Sheard, J. Encinas, A.J. Fought, E.J. Topol, Comparison of 24-hour Holter monitoring with 14-day novel adhesive patch electrocardiographic monitoring. Am. J. Med. **127**, 95.e11–95.e17 (2014)
37. J. M. Engel, V. Mehta, R. Fogoros, A. Chavan, Study of arrhythmia prevalence in NUVANT mobile cardiac telemetry system patients, in *Proceedings Conference of the IEEE Engineering in Medicine and Biology Society* (2012), pp. 2440–2443
38. J.M. Engel, N. Chakravarthy, G. Nosbush, M. Merkert, M.D.R. Fogoros, A. Chavan, Comparison of arrhythmia prevalence in NUVANT Mobile Cardiac Telemetry System patients in the US and India, in *Proceedings Conference of the IEEE Engineering in Medicine and Biology Society* (2014), pp. 2730–2733
39. L. Friberg, J. Engdahl, V. Frykman, E. Svennberg, L.-Å. Levin, M. Rosenqvist, Population screening of 75- and 76-year-old men and women for silent atrial fibrillation (STROKESTOP). Europace **15**, 135–140 (2012)
40. E. Svennberg, J. Engdahl, F. Al-Khalili, L. Friberg, V. Frykman, M. Rosenqvist, Mass screening for untreated atrial fibrillation: the STROKESTOP study. Circulation **131**, 2176–2184 (2015)
41. B. Vaes, S. Stalpaert, K. Tavernier, B. Thaels, D. Lapeire, W. Mullens, J. Degryse, The diagnostic accuracy of the MyDiagnostick to detect atrial fibrillation in primary care. BMC Fam. Pract. **15**, 113 (2014)
42. M. Stridh, M. Rosenqvist, Automatic screening of atrial fibrillation in thumb-ECG recordings, in Proceedings of Computing in Cardiology vol. **39**, 193–196 (2012)
43. E. Agu, P. Pedersen, D. Strong, B. Tulu, Q. He, L. Wang, Y. Li, The smartphone as a medical device: assessing enablers, benefits and challenges, in *10th Annual IEEE Communications Society Conference on Sensor, Mesh and Ad Hoc Communications and Networks (SECON)* (2013), pp. 76–80
44. A.R.J. Mitchell, P. Le Page, Living with the handheld ECG. BMJ Innov. **2** (2015)

45. P.-H. Chan, C.-K. Wong, Y. C. Poh, L. Pun, W.W.-C. Leung, Y.-F. Wong, M.M.-Y. Wong, M.-Z. Poh, D. W.-S. Chu, C.-W. Siu, Diagnostic performance of a smartphone-based photoplethysmographic application for atrial fibrillation screening in a primary care setting. J. Am. Heart Assoc. **5** (2016)
46. B. Freedman, Screening for atrial fibrillation using a smartphone: Is there an app for that? J. Am. Heart Assoc. **5** (2016)
47. N. Lowres, S.B. Freedman, J. Redfern, A. McLachlan, I. Krass, A. Bennett, T. Briffa, A. Bauman, L. Neubeck, Screening education and recognition in community pharmacies of atrial fibrillation to prevent stroke in an ambulant population aged $\geq$65 years (SEARCH-AF stroke prevention study): a cross-sectional study protocol. Br. Med. J. Open **2** (2012)
48. J.K. Lau, N. Lowres, L. Neubeck, D.B. Brieger, R.W. Sy, C.D. Galloway, D.E. Albert, S.B. Freedman, iPhone ECG application for community screening to detect silent atrial fibrillation: a novel technology to prevent stroke. Int. J. Cardiol. **165**, 193–194 (2013)
49. L. Desteghe, Z. Raymaekers, M. Lutin, J. Vijgen, D. Dilling-Boer, P. Koopman, J. Schurmans, P. Vanduynhoven, P. Dendale, H. Heidbuchel, Performance of handheld electrocardiogram devices to detect atrial fibrillation in a cardiology and geriatric ward setting. Europace **19**, 29–39 (2017)
50. D.E. Albert, Letter on Desteghe et al., Performance of handheld electrocardiogram devices to detect atrial fibrillation in a cardiology and geriatric ward setting. Europace, euw218 (2016)
51. K. Kanjwal, V.M. Figueredo, B. Karabin, B. Grubb, The implantable loop recorder: current uses, future directions. J. Innov. Cardiac Rhythm Manag. **2**, 215–222 (2011)
52. S. Sarkar, D. Ritscher, R. Mehra, A detector for a chronic implantable atrial tachyarrhythmia monitor. IEEE Trans. Biomed. Eng. **55**, 1219–1224 (2008)
53. G. Hindricks, E. Pokushalov, L. Urban, M. Taborsky, K.-H. Kuck, D. Lebedev, G. Rieger, H. Pürerfellner, and on behalf of the XPECT Trial Investigators, Performance of a new leadless implantable cardiac monitor in detecting and quantifying atrial fibrillation results of the XPECT trial. Circ. Arrhythm. Electrophysiol. **3**, 141–147 (2010)
54. A.S. Montenero, A. Quayyum, P. Franciosa, D. Mangiameli, A. Antonelli, L. Barbieri, N. Bruno, F. Zumbo, M. Vimercati, Implantable loop recorders: a novel method to judge patient perception of atrial fibrillation. Preliminary results from a pilot study. J. Interv. Card. Electrophysiol. **10**, 211–220 (2004)
55. C. Eitel, D. Husser, G. Hindricks, M. Frühauf, S. Hilbert, A. Arya, T. Gaspar, U. Wetzel, A. Bollmann, C. Piorkowski, Performance of an implantable automatic atrial fibrillation detection device: impact of software adjustments and relevance of manual episode analysis. Europace **13**, 480–485 (2011)
56. O. Andersson, K.H. Chon, L. Sörnmo, J.N. Rodrigues, A 290 mV sub-V_T ASIC for real-time atrial fibrillation detection. IEEE Trans. Biomed. Circuits Syst. **9**, 377–386 (2015)
57. C.D. Swerdlow, W. Schsls, B. Dijkman, W. Jung, N.V. Sheth, W.H. Olson, B.D. Gunderson, Detection of atrial fibrillation and flutter by a dual-chamber implantable cardioverter-defibrillator. For the Worldwide Jewel AF Investigators. Circulation **101**, 878–885 (2000)
58. J.W. Fung, J. Sperzel, C.M. Yu, J.Y. Chan, R.N. Gelder, M.X. Yang, R. Rooke, P. Boileau, G. Fröhlig, Multicenter clinical experience with an atrial lead designed to minimize far-field R-wave sensing. Europace **11**, 618–624 (2009)
59. J. Lee, B.A. Reyes, D.D. McManus, O. Mathias, K.H. Chon, Atrial fibrillation detection using an iPhone 4S. IEEE Trans. Biomed. Eng. **60**, 203–206 (2013)
60. J.P. Couderc, S. Kyal, L. Mestha, B. Xu, D. Peterson, X. Xia, B. Hall, Detection of atrial fibrillation using contactless facial video monitoring. Heart Rhythm **12**, 195–201 (2015)
61. T. Conroy, J.H. Guzman, B. Hall, G. Tsouri, J.-P. Couderc, Detection of atrial fibrillation using an earlobe photoplethysmographic sensor. Physiol. Meas. **38**, 1906–1918 (2017)
62. V.D.A. Corino, R. Laureanti, L. Ferranti, G. Scarpini, F. Lombardi, L.T. Mainardi, Detection of atrial fibrillation episodes using a wristband device. Physiol. Meas. **38**, 787–799 (2017)
63. G. Lu, F. Yang, J.A. Taylor, J.F. Stein, A comparison of photoplethysmography and ECG recording to analyse heart rate variability in healthy subjects. J. Med. Eng. Technol. **33**, 634–641 (2009)

64. E. Gil, M. Orini, R. Bailón, J.M. Vergara, L. Mainardi, P. Laguna, Photoplethysmography pulse rate variability as a surrogate measurement of heart rate variability during non-stationary conditions. Physiol. Meas. **31**, 1271 (2010)
65. D.D. McManus, J. Lee, O. Maitas, N. Esa, R. Pidikiti, A. Carlucci, J. Harrington, E. Mick, K.H. Chon, A novel application for the detection of an irregular pulse using an iPhone 4S in patients with atrial fibrillation. Heart Rhythm **10**, 315–319 (2013)
66. J. Wiesel, L. Fitzig, Y. Herschman, F.C. Messineo, Detection of atrial fibrillation using a modified Microlife blood pressure monitor. Am. J. Hypertens. **22**, 848–852 (2009)
67. G.S. Stergiou, N. Karpettas, A. Protogerou, E.G. Nasothimiou, M. Kyriakidis, Diagnostic accuracy of a home blood pressure monitor to detect atrial fibrillation. J. Hum. Hypertens. **23**, 654–658 (2009)
68. G. Marazzi, F. Iellamo, M. Volterrani, M. Lombardo, F. Pelliccia, D. Righi, F. Grieco, L. Cacciotti, L. Iaia, G. Caminiti, G. Rosano, Comparison of Microlife BP A200 Plus and Omron M6 blood pressure monitors to detect atrial fibrillation in hypertensive patients. Adv. Ther. **29**, 64–70 (2012)
69. J. Wiesel, B. Arbesfeld, D. Schechter, Comparison of the Microlife blood pressure monitor with the Omron blood pressure monitor for detecting atrial fibrillation. Am. J. Cardiol. **114**, 1046–1048 (2014)
70. A.J. Cheung, B.M.Y. Cheung, False detection of atrial fibrillation in children by a blood pressure monitor with atrial fibrillation detection function. Br. Med. J. Case Rep. 1–2 (2015)
71. G. Hindricks, C. Piorkowski, Atrial fibrillation monitoring: mathematics meets real life. Circulation **126**, 791–792 (2012)
72. E.I. Charitos, U. Stierle, P.D. Ziegler, M. Baldewig, D.R. Robinson, H. Sievers, T. Hanke, A comprehensive evaluation of rhythm monitoring strategies for the detection of atrial fibrillation recurrence: insights from 647 continuously monitored patients and implications for monitoring after therapeutic interventions. Circulation **126**, 806–814 (2012)
73. J. Engdahl, L. Andersson, M. Mirskaya, M. Rosenqvist, Stepwise screening of atrial fibrillation in a 75-year-old population: Implications for stroke prevention. Circulation **127**, 930–937 (2013)
74. N. Gaillard, S. Deltour, B. Vilotijevic, A. Hornych, S. Crozier, A. Leger, R. Frank, Y. Samson, Detection of paroxysmal atrial fibrillation with transtelephonic EKG in TIA or stroke patients. Neurology **74**, 1666–1670 (2010)
75. D.J. Gladstone, M. Spring, P. Dorian, V. Panzov, K.E. Thorpe, J. Hall, H. Vaid, M. O'Donnell, A. Laupacis, R. Côté, M. Sharma, J.A. Blakely, A. Shuaib, V. Hachinski, S.B. Coutts, D.J. Sahlas, P. Teal, S. Yip, J.D. Spence, B. Buck, S. Verreault, L.K. Casaubon, A. Penn, D. Selchen, A. Jin, D. Howse, M. Mehdiratta, K. Boyle, R. Aviv, M.K. Kapral, M. Mamdani, Atrial fibrillation in patients with cryptogenic stroke. N. Engl. J. Med. **370**, 2467–2477 (2014)
76. T. Sanna, H.-C. Diener, R.S. Passman, V. Di Lazzaro, R.A. Bernstein, C.A. Morillo, M.M. Rymer, V. Thijs, T. Rogers, F. Beckers, K. Lindborg, J. Brachmann, Cryptogenic stroke and underlying atrial fibrillation. N. Engl. J. Med. **370**, 2478–2486 (2014)

Chapter 3
Databases and Simulation

Leif Sörnmo, Andrius Petrėnas and Vaidotas Marozas

3.1 Public ECG Databases

The availability of public databases is essential as it enables researchers to establish whether a novel method performs better than the existing ones. Many of the public ECG databases relevant to engineering-oriented research on atrial fibrillation (AF) are available for download at PhysioNet (www.physionet.org), a free web resource with a huge collection of physiological signals and software [1]. The Physionet databases have played, and continue to play, a crucial role in the development of AF detectors and the evaluation of their performance (Chap. 4), whereas they hardly play any role in the development of methods for f wave extraction (Chap. 5) and f wave characterization (Chap. 6).

The PhysioNet databases include beat-based annotations such as occurrence time and type of beat, but often also arrhythmia-based annotations such as type and onset/end of arrhythmia. Annotations on beat occurrence time may be automated and provided by a well-performing QRS detector, whereas arrhythmia-based annotations are usually provided by one or several experts, implying a considerable work effort to annotate a database consisting of long-term continuous ECG recordings. Unfortunately, information on the annotation process is usually scarce, and details are almost invariably missing on the number of annotators involved, the level of expertise among the annotators, and how consensus was reached in cases of disagreement.

L. Sörnmo (✉)
Department of Biomedical Engineering and Center for Integrative Electrocardiology, Lund University, Lund, Sweden
e-mail: leif.sornmo@bme.lth.se

A. Petrėnas · V. Marozas
Biomedical Engineering Institute, Kaunas University of Technology, Kaunas, Lithuania
e-mail: andrius.petrenas@ktu.lt

V. Marozas
e-mail: vaidotas.marozas@ktu.lt

L. Sörnmo (ed.), *Atrial Fibrillation from an Engineering Perspective*,
Series in BioEngineering, https://doi.org/10.1007/978-3-319-68515-1_3

Considering that some ECG databases have evolved into virtually becoming standards, information on the annotation process should, preferably, be transparent to the user.

In the following, the most popular public databases employed in engineering-oriented research are briefly described.

The MIT–BIH Atrial Fibrillation Database (AFDB) consists of 25 10-h, two-lead ambulatory ECG recordings of patients with AF, mostly paroxysmal [2]. The signals were acquired using an analog device with a bandwidth of approximately 0.1–40 Hz, sampled at a rate of 250 Hz, and quantized with 12-bit resolution over a range of ± 10 mV. Two of the 25 recordings contain only the RR interval series, but no ECG signal, and can therefore only be used in RR-based analysis. Information on lead placement is missing.

The database was manually annotated with respect to type of beat, type and onset/end of arrhythmia, resulting in a total of 297 AF episodes with durations ranging from as few as 3 beats to tens of thousands of beats.

The distributions of AF episode duration and RR intervals provide interesting information on the properties of AFDB. Figure 3.1a presents the histogram of episode duration, with an exponential-like decay, except that 29 episodes have durations exceeding 2000 beats. Together, these 29 episodes account for as much as 82% of the total time the patients are in AF; when computed in individual patients, this percentage is commonly referred to as "AF burden." The fact that a small number of episodes can dominate the total time a patient is in AF highlights an important limitation of the commonly used detection performance measures, to be further discussed in Sect. 4.5.

Figure 3.1b presents the histogram of all RR intervals in AFDB, with most RR intervals ranging from 0.3 to 1.5 s. As many as 25% of all RR intervals are shorter than 0.5 s, thus imposing an important constraint on methods exploring f waves in the TQ interval; this constraint applies especially to methods for f wave extraction, see Chap. 5. For an RR interval of 500 ms and a QT interval with a typical length of 350 ms, the TQ interval is only 150 ms, which for a dominant atrial frequency (DAF) of 5 Hz implies that less than one f wave is contained in the TQ interval.

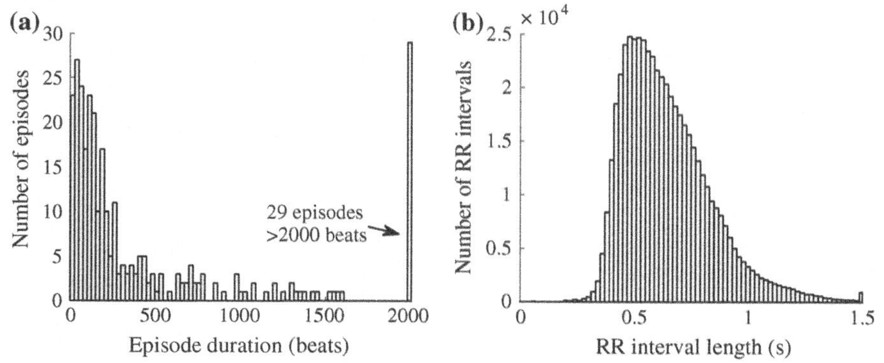

Fig. 3.1 Histograms of **a** AF episode duration and **b** RR intervals in AF, determined from the MIT–BIH AF Database

The Long-Term AF Database (LTAFDB) consists of 84 two-lead ambulatory ECG recordings obtained in patients with paroxysmal or persistent AF, lasting from 24 to 25 h [3]. The signals were sampled at a rate of 128 Hz and quantized with 12-bit resolution over a range of ± 10 mV. Information on bandwidth and lead placement is missing.

The beat-based annotations were automated, whereas the arrhythmia-based annotations resulted from manual review of the output of a commercial system for ECG analysis. More than 7000 AF episodes are contained in LTAFDB, and therefore it is the public database with the largest number of episodes.

The temporal occurrence pattern of AF episodes is presented in Fig. 3.2 for four different patients; the onset and end of an episode are given by manual annotations. These four examples illustrate that the temporal occurrence pattern can differ dramatically between patients.

The AF Termination Database (AFTDB) is a subset of LTAFDB composed of 80 1-min excerpts from patients with spontaneously terminating or persistent AF [4]. The database was compiled for the purpose of predicting spontaneous termination of AF. The 80 records are divided into a training set with 30 records and two test sets with 30 and 20 records, respectively.

The Short Single-Lead AF Database (SSAFDB) consists of 12,186 single-lead ECG recordings obtained from a smartphone-based device, lasting from 9 to 60 s [5]. The signals were sampled at a rate of 300 Hz, quantized with 16-bit resolution over

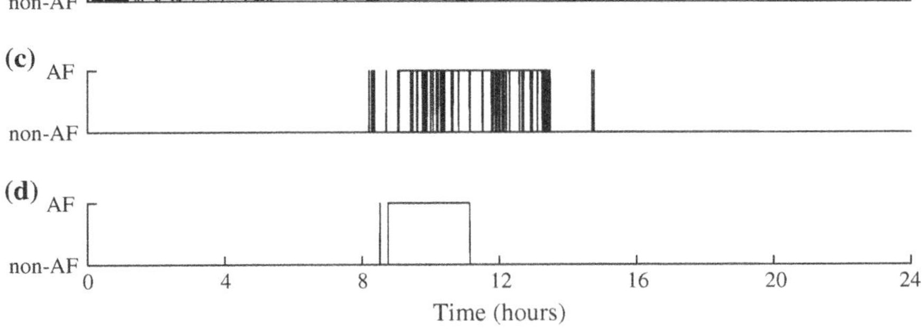

Fig. 3.2 Examples of temporal occurrence patterns of episodes in paroxysmal AF, obtained from four patients monitored over a 24-h period, being part of the Long-Term AF Database. **a** A few long episodes which together extend virtually the entire monitoring period, **b** numerous, often short episodes which together extend virtually the entire monitoring period, **c** many short episodes aggregated in a 5-h period, and **d** a short episode followed by a much longer 3-h episode

a range of ±5 mV, with a bandwidth from 0.5 to 40 Hz. Although the lead is not specified, the vast majority is lead I since it is the simplest to record with the device.

The database is divided into a training set with 8,528 recordings and a test set with 3,658 recordings. Each recording is manually annotated using the following four categories: 1. Normal sinus rhythm, 2. AF, 3. other rhythm, and 4. too noisy to classify, with 5076, 758, 2415, and 279 recordings in each of the categories of the training set. A category applies to the entire ECG recording, even if an arrhythmia is only partially present. No beat-based annotations are provided.

Since the smartphone-based device is used for home-based screening, and thus operated by the patient, the quality of the recording is generally much lower than, for example, in long-term continuous recordings. In addition, f wave amplitude is generally lower in lead I than in lead V_1, which is the preferred lead for f wave analysis. Signal quality can be quantified using an index which determines the suitability of analyzing f waves in 5-s signal segments [6], see also Sect. 6.5 for a brief description. The signal quality index is normalized to the interval [0, 1], where 1 represents the highest quality; a suitable cut-off value for acceptable signal quality is 0.25. Figure 3.3a presents the histogram of the signal quality index, computed in nonoverlapping, 5-s segments of all recordings of SSAFDB annotated as AF. Using 0.25 as the cut-off value, 83% of all recordings in SSAFDB have a signal quality which is too low for f wave analysis.

The original purpose of compiling SSAFDB was to evaluate the performance of classifiers designed to handle short ECG segments, whereas long-term ambulatory ECG databases such as AFDB and LTAFDB have primarily been used to evaluate performance in terms of how accurately AF episodes can be detected. Thus, different types of algorithms are evaluated on SSAFDB and AFDB/LTAFDB.

The MIT–BIH Arrhythmia Database (MITDB) contains 48 half-hour excerpts of two-channel ambulatory ECG recordings, obtained from 47 subjects [7]. The signals

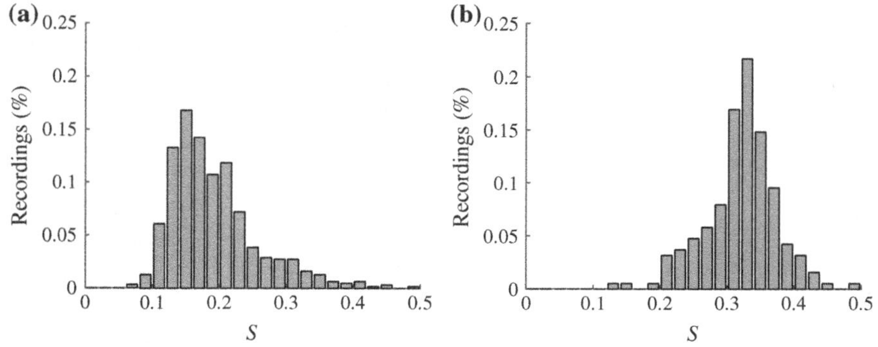

Fig. 3.3 Signal quality assessed on all AF recordings in **a** the Short Single-Lead AF Database and **b** the Lund AF Database (lead V_1), using an index (S) which determines the suitability of analyzing f waves [6]. The results are presented as relative histograms

were sampled at a rate of 360 Hz and quantized with 11-bit resolution over a range of ±10 mV. Information on bandwidth and lead placement is missing.

Since only eight recordings contain AF, with a total of 105 episodes, the main value of this database is to investigate detection performance in the presence of non-AF arrhythmias such as atrial flutter, bigeminy, and trigeminy.

The MIT–BIH Normal Sinus Rhythm Database (NSRDB) includes 18 long-term ECG recordings of subjects without significant arrhythmias. Hence, only the specificity of an AF detector can be investigated with this database, for example, in the presence of respiratory sinus arrhythmia.

3.2 Non-public ECG Databases

Although public databases have eliminated much of the time-consuming work involved with data collection, the need to collect databases which are well-matched to a particular research problem nevertheless remains. This will ensure that methods development and performance evaluation are carried out on relevant data. For example, the development of methods for f wave characterization calls for databases obtained with ECG leads which are more relevant than those of the above-mentioned public databases. In fact, the collection of matched databases promotes diversity in research in a way which public databases historically have not done. Although most matched databases are non-public at the outset, either proprietary or available at a cost, it can be hoped that they sooner or later become public to benefit a larger group of researchers.

Considering that many public databases were collected using old recording technology, where MITDB is one of the oldest, dating to 1982, another important motivation for collecting databases is to benefit from modern recording technology, offering higher sampling rate, larger bandwidth, lower noise level, more leads, and longer acquisition period.

The Lund AF Database exemplifies the numerous non-public databases collected over the years, with the purpose of developing and evaluating methods for f wave characterization [8]. The database contains 211 12-lead extended ECG recordings obtained at rest from patients with AF, mostly persistent (in some studies, a 1-min segment was extracted from each patient in this database to ensure AF presence throughout the segment). The signals were sampled at a rate of 1000 Hz, quantized with 16-bit resolution over a range of ±10 mV, with a bandwidth from 0.1 to 300 Hz. No annotations are provided.

Figure 3.4a presents the RR interval histogram of the Lund AF Database, resembling the RR interval histogram of AFDB shown in Fig. 3.1b. Since the histogram in Fig. 3.4a is obtained from signals recorded at rest, it would likely have been shifted leftwards towards shorter intervals had the database been recorded during physical activity, with implications on the length of the TQ interval and related analysis.

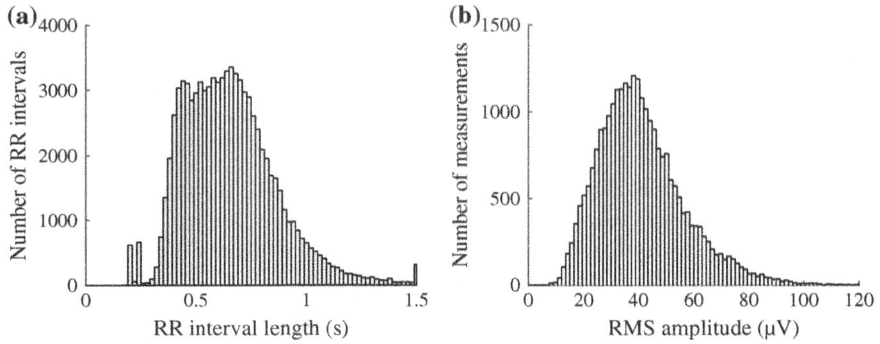

Fig. 3.4 Histograms of **a** RR intervals and **b** f wave amplitude in lead V_1, determined from 1-min segments of the Lund AF Database

The histogram of f wave amplitude in lead V_1 is presented in Fig. 3.4b. Here, amplitude is defined as the root mean square (RMS) value of the samples contained in the TQ interval, beginning 350 ms after a QRS complex and ending 50 ms before the preceding QRS complex; no amplitude measurement was made in TQ intervals shorter than 250 ms. Section 6.2 provides an overview of different approaches to measuring f wave amplitude.

Figure 3.3b quantifies that the signal quality in lead V_1 of the Lund AF database is superior to that of SSAFDB. This result is, of course, expected since the former database was recorded during rest, under the supervision of a technician who made sure that the electrodes were properly attached. Using a cut-off value of 0.25, 11% of all recordings have signal quality which is too poor for f wave analysis, to be contrasted with the above-mentioned 83% of SSAFDB.

3.3 Simulation of Atrial Fibrillation

Although databases with ECG signals are central to methodological development and evaluation, model-based simulation offers certain advantages such as the possibility to investigate conditions which are difficult to deal with experimentally and the possibility to control the properties of the simulated signal by a set of parameters. As a result, the agreement between simulated and estimated signals can be quantitatively assessed and expressed in terms of suitable performance measures. If desired, these measures can be computed for simulated signals with different signal-to-noise ratios (SNRs). The simulation advantages were first exploited in the context of f wave extraction, since none of the public ECG databases lend themselves well to performance evaluation, and later in the context of detection of brief AF episodes, since annotated ECG database with such episodes are largely missing.

Three f wave simulation models with widely different complexity are briefly described below. Since none of these models produce a signal with ventricular activity, the simulated f wave signal is usually added to ECG signals obtained from subjects in normal sinus rhythm, provided that the P waves have been first cancelled. In doing so, the inherent variation in QRS morphology, e.g., due to respiration, is transferred from the recorded to the simulated ECG signal—a transfer which is important in f wave extraction since morphologic variation can have substantial influence on performance. The RR intervals of normal sinus rhythm are also transferred to the simulated ECG signal—a transfer which may be acceptable when the simulated ECG signal is investigated for f wave extraction, but clearly unacceptable for AF detection.

The *f wave sawtooth model* is widely used in algorithmic development, first introduced in [9] and later employed in, e.g., [10–14]. This signal model is defined by a sum of K amplitude- and frequency-modulated sinusoids with harmonically related frequencies,

$$d(n) = \sum_{k=1}^{K} a_k(n) \sin\left(k\omega_0 n + \frac{\Delta f}{f_f} \sin(\omega_f n)\right), \quad n = 0, \ldots, N-1, \tag{3.1}$$

where $\omega_0 = 2\pi f_0$ is the fundamental frequency, i.e., the model counterpart to the DAF. The fundamental frequency ω_0 is modulated by $\omega_f = 2\pi f_f$ with a maximum deviation of Δf. The time-varying amplitude $a_k(n)$ is defined so that $d(n)$ exhibits a sawtooth characteristic,

$$a_k(n) = \frac{2}{k\pi} \left(a + \Delta a \sin(\omega_a n)\right), \tag{3.2}$$

where a is the sawtooth amplitude, Δa is the maximum modulation amplitude, and $\omega_a = 2\pi f_a$ is the modulation frequency of the amplitude. The model in (3.1) offers certain flexibility since both f wave amplitude and frequency are modulated.

An important limitation of the sawtooth model was brought to light when the problem of f wave extraction was addressed using an artificial neural network [15]: the network could learn the predictable changes in amplitude and frequency of the simulated f wave signal, leading to exaggerated performance figures.

The *f wave replication model* produces a signal based on the observed samples of the TQ intervals [16]; no mathematical modeling is involved. Interpolation between two successive TQ intervals fills in the intermediate QT interval with f wave samples, using the approach originally described in [9]. The f waves of the first TQ interval are replicated in the QT interval and subjected to linear weighting, and the f waves in the second, subsequent TQ interval are replicated in the same way, but time-reversed. The interpolated samples of the intervening QT interval result from summation of the two replicated and weighted signals. Other techniques for TQ-based interpolation are described in Sect. 5.3.

While the f wave replication model can produce realistic signals, neither the repetition rate nor the amplitude of f waves can be controlled. Another major limitation is that the length of the TQ intervals decreases as the heart rate increases, implying that the risk of producing unrealistic f wave signals becomes increasingly higher at higher heart rates.

A much more sophisticated approach to simulating f wave signals is based on a *biophysical model* of the atria [17], see also [18, 19]. The model is based on anatomical information derived from magnetic resonance imaging, accounting for the entries and exits of the vessels, the locations of the valves connecting the atria to the ventricles, as well as several other aspects. The electrical activity of the atria is modeled in terms of membrane kinetics, where the presence of heterogeneities in action potential duration creates the substrate for sustained AF. Volume conduction theory is employed to describe the propagation of currents from the electrical sources of the atria through the passive body tissues to the body surface, influencing the amplitude and morphology of the simulated multi-lead f wave signals.

Since none of the three above-mentioned simulation models account for switching between non-AF rhythms and AF, they cannot be used when addressing the problem of detecting AF. To fill this void, a model of paroxysmal AF has been proposed [20], including not only rhythm switching but also the possibility to chose whether the simulated signal should be composed of synthetic or real components, described in Sects. 3.4 and 3.5, respectively.

3.4 Simulation of Paroxysmal AF Using Synthetic Components

The simulation of multi-lead ECGs in paroxysmal AF is based on phenomenological, mathematical modeling of ventricular rhythm, ventricular morphology, atrial morphology, and rhythm switching, whereas the noise added to the simulated signal derives from a public database, see Fig. 3.5. Thus, the resulting signal is composed of synthetic components whose properties are controlled by a set of parameters defining, e.g., episode duration, variability of the RR interval series in sinus rhythm and AF, f and P wave morphology, QRST complex morphology, and percentage of atrial

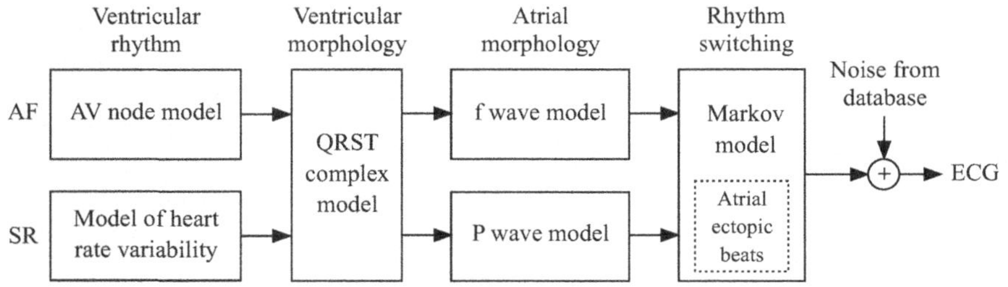

Fig. 3.5 Simulation of ECG signals using synthetic components. The same model of QRST complexes is employed in sinus rhythm (SR) and AF

premature beats (APBs). For each new realization of the simulated signal, the model parameters are generated randomly from uniform distributions in predefined ranges so that realistic ECG signals with unique intersubject morphologies can be produced.

The simulation model assumes a vectorcardiogram (VCG) lead system initially, consisting of the orthogonal leads X, Y, and Z. Once suitably processed, these leads are transformed to the standard 12-lead ECG system. A detailed description of the simulation model is found in [20], together with a list of the default model parameter values.

3.4.1 Atrial Fibrillation

Ventricular rhythm. A statistical model of the atrioventricular (AV) node with dual pathways is used to generate RR intervals in AF [21]. In this model, the ventricles are assumed to be activated by atrial impulses arriving to the AV node according to a Poisson process with mean arrival rate λ_a, which is closely related to the DAF. The joint probability density function (PDF) of the consecutive RR intervals $x_0, x_1, ..., x_{N-1}$ is given by

$$p_x(x_0, x_1, ..., x_{N-1}) = \prod_{n=0}^{N-1} (\epsilon p_{x,s}(x_n) + (1-\epsilon) p_{x,f}(x_n)), \tag{3.3}$$

where ϵ is the probability of an atrial impulse conducted through the slow pathway, whose refractory period is defined by a deterministic part τ_s and a stochastic part $\tau_{s,p}$. Hence, the probability of an atrial impulse to take the fast pathway, whose refractory period is defined by τ_f and $\tau_{f,p}$, is $(1-\epsilon)$. For an atrial impulse taking the slow pathway, the interval x between two successive ventricular activations, i.e., the RR interval, is described by the following PDF [21]:

$$p_{x,s}(x) = \begin{cases} 0, & 0 < x < \tau_s, \\ \dfrac{\lambda_a(x-\tau_s)}{\tau_{s,p}} \exp\left[-\dfrac{\lambda_a(x-\tau_s)^2}{2\tau_{s,p}}\right], & \tau_s \le x < \tau_s + \tau_{s,p}, \\ \lambda_a \exp\left[-\dfrac{\lambda_a \tau_{s,p}}{2} - \lambda_a(x-\tau_s-\tau_{s,p})\right], & x \ge \tau_s + \tau_{s,p}. \end{cases} \tag{3.4}$$

The PDF of the fast pathway is described by $p_{x,f}(x)$, being identical to (3.4) except that τ_s is replaced with τ_f and $\tau_{s,p}$ with $\tau_{f,p}$. Chapter 7 provides a comprehensive overview of AV node models for simulation of RR intervals in AF, including the statistical AV node model in [21].

f waves. The f wave sawtooth model in (3.1) is supplemented with a stochastic component so that more complex, less predictable f waves can be produced [15].

Using, for convenience, a continuous-time framework, the f wave model signal $f_l(t)$ of the l-th vectorcardiographic lead is composed of two components,

$$f_l(t) = d_l(t) + s_l(t), \quad l \in \{X,Y,Z\}, \tag{3.5}$$

where $d_l(t)$ is defined similarly to (3.1),

$$d_l(t) = \sum_{k=1}^{K} a_{l,k}(t) \sin\left(k\Omega_{l,0}t + \frac{\Delta F}{F_k}\sin(2\pi F_k t)\right), \tag{3.6}$$

but with the difference that lead dependence is introduced, i.e., $\Omega_{l,0} = 2\pi F_{l,0}$ and

$$a_{l,k}(n) = \frac{2}{k\pi}\left(a_l + \Delta a_l \sin(\Omega_{a,l}n)\right), \quad k = 1, \ldots, K. \tag{3.7}$$

In paroxysmal AF, the DAF (corresponding to $F_{l,0}$) is typically contained in the interval 3–7 Hz [3], while, in persistent and permanent AF, it is typically higher and contained in the interval 5–12 Hz. Moreover, it is well-known that the DAF depends on anatomical location [22], which in the model is accounted for by setting $F_{X,0}$ to a value 5% larger than $F_{Y,0}$, and $F_{Z,0}$ to a value 5% smaller than $F_{Y,0}$. The mean arrival rate λ_a of atrial impulses arriving to the AV node is taken as the average of the frequencies $F_{X,0}$, $F_{Y,0}$, and $F_{Z,0}$.

The stochastic f wave component $s_l(t)$ results from multi-bandpass filtering of white noise, with two passbands symmetrically related to $F_{l,0}$ by $[0.65F_{l,0}, 0.95F_{l,0}]$ and $[1.05F_{l,0}, 1.35F_{l,0}]$. The variance of the input white noise $\sigma^2_{l,s}$ is taken as a fraction of the sawtooth amplitude a_l in (3.7).

The first minutes after AF onset and the last minute before AF termination are associated with more organized f waves and a lower DAF [23–25], which in the model is accounted for by using bandpass filters with narrower passbands for the first three minutes and the last minute of the episode. A set of bandpass filters is used with gradually wider passbands, starting at $[0.8F_{l,0}, 0.95F_{l,0}]$ and $[1.2F_{l,0}, 1.35F_{l,0}]$ and ending at $[0.65F_{l,0}, 0.95F_{l,0}]$ and $[1.05F_{l,0}, 1.35F_{l,0}]$, respectively. To account for the lower DAF, $F_{l,0}$ is multiplied with a factor which increases linearly from 0.8 to 1 during the first three minutes of an AF episode. Conversely, $F_{l,0}$ is multiplied with a factor which decreases linearly from 1 to 0.8 during the last minute of an AF episode. Figure 3.6 illustrates simulated f waves at the onset, the midpoint, and the end of an AF episode.

A further generalization of the sawtooth model, to make the f wave signal even less regular, is to employ an adaptive non-harmonic model in which amplitude and frequency modulation is described by a random walk whose steps are sampled from a zero-mean Gaussian distribution [26].

QRST complexes. The three-dimensional, single-dipole ECG model proposed in [27] is used for simulating QRST complexes, building on the dynamical model based on three coupled, ordinary differential equations [28]. The three orthogonal

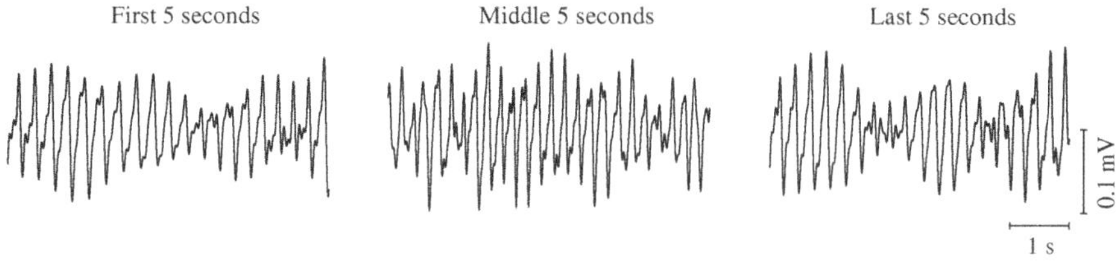

Fig. 3.6 Simulated f waves at the onset, the midpoint, and the end of an AF episode, produced by the sawtooth-based model in (3.5)

leads are obtained by projecting the dipole vector onto the recorded leads. The dipole vector, defined by $q_X(t)$, $q_Y(t)$, and $q_Z(t)$, is modeled as a summation of P different Gaussian functions,

$$q_l(t) = \sum_{p=1}^{P} \alpha_{l,p} \exp\left[-\frac{(t-\mu_{l,p})^2}{2\sigma_{l,p}^2}\right], \quad l \in \{\mathrm{X,Y,Z}\}, \tag{3.8}$$

where each Gaussian is appropriately scaled in amplitude and time with $\alpha_{l,p}$ and $\sigma_{l,p}$, respectively, and shifted in time with $\mu_{l,p}$. To allow for a wide variety of QRST morphologies, $\alpha_{l,p}$, $\sigma_{l,p}$, and $\mu_{l,p}$ are assigned uniform distributions [20]. In contrast to the models in [27, 28], where the aim was to simulate a signal with recurrent heartbeats, the aim of the paroxysmal AF simulation model is to produce a single QRST complex, and, therefore, the VCG loop defined by the orthogonal leads $q_X(t)$, $q_Y(t)$, and $q_Z(t)$ is traversed only once. Amplitude variation is introduced by letting $\alpha_{l,p}$ vary according to a sinusoidal function whose frequency is randomly chosen in the interval [0.05, 0.15] Hz to mimic Mayer waves.

The resulting three-lead QRST complex $q_X(t)$, $q_Y(t)$, and $q_Z(t)$ is placed at the occurrence time produced by the AV node model, accompanied by resampling of the T wave to ensure that the duration fits into the current RR interval. Since the QT interval is usually shorter in AF than in sinus rhythm, it is set to a fixed value (360 ms) based on observations reported in [29, 30].

3.4.2 Sinus Rhythm

Ventricular rhythm. The RR intervals in sinus rhythm are simulated according to the technique described in [28], where parasympathetic stimulation (respiratory sinus arrhythmia) and baroreflex regulation are modeled by a bimodal power spectrum of the RR interval series, defined by two Gaussian functions

$$S_{\mathrm{RR}}(\Omega) = \frac{P_1}{\sqrt{2\pi\sigma^2_{\mathrm{RR},1}}}\exp\left[-\frac{(\Omega-\Omega_1)^2}{2\sigma^2_{\mathrm{RR},1}}\right] + \frac{P_2}{\sqrt{2\pi\sigma^2_{\mathrm{RR},2}}}\exp\left[-\frac{(\Omega-\Omega_2)^2}{2\sigma^2_{\mathrm{RR},2}}\right], \tag{3.9}$$

where Ω_1 and Ω_2 ($\Omega_1 < \Omega_2$) are the mean frequencies with related "variance" $\sigma^2_{\mathrm{RR},1}$ and $\sigma^2_{\mathrm{RR},2}$ and spectral power P_1 and P_2, respectively. The low- to high-frequency power ratio is determined by P_1/P_2. The higher frequency Ω_2 is usually related to the respiratory rate.

The resulting RR interval series is obtained by computing the inverse Fourier transform of the spectrum $S_{\mathrm{RR}}(\Omega)$. The desired heart rate and heart rate variability are set by scaling the RR interval series and adding an offset value. Very low frequency oscillations are modeled by a zero-mean component added to the output of the model in [28]. This component is produced by a third-order autoregressive model, identified from a lowpass filtered (cut-off frequency 0.001 Hz) RR interval series taken from NSRDB [20].

P waves. A linear combination of Hermite functions is used to model P waves in the orthogonal leads,

$$p_l(t) = \sum_{i=1}^{3} w_{l,i}\phi_i(t), \quad l \in \{\mathrm{X,Y,Z}\}, \tag{3.10}$$

where $w_{l,i}$ are lead-dependent weights. The first three Hermite functions are defined by

$$\phi_1(t) = \frac{1}{\sqrt{\sigma_{\mathrm{P},1}\sqrt{\pi}}}\cdot\exp\left[-\frac{t^2}{2\sigma^2_{\mathrm{P},1}}\right], \tag{3.11}$$

$$\phi_2(t) = -\frac{\sqrt{2}}{\sqrt{\sigma_{\mathrm{P},2}\sqrt{\pi}}}\frac{t}{\sigma_{\mathrm{P},2}}\cdot\exp\left[-\frac{t^2}{2\sigma^2_{\mathrm{P},2}}\right], \tag{3.12}$$

$$\phi_3(t) = \frac{1}{\sqrt{2\sigma_{\mathrm{P},3}\sqrt{\pi}}}\left(\frac{2t^2}{\sigma^2_{\mathrm{P},3}}-1\right)\cdot\exp\left[-\frac{t^2}{2\sigma^2_{\mathrm{P},3}}\right], \tag{3.13}$$

with mono-, bi-, and triphasic morphology, respectively. The width of $\phi_i(t)$ is determined by $\sigma_{\mathrm{P},i}$, which is treated as a lead-independent parameter. The Hermite functions were originally proposed in [31] for modeling of QRS complex morphology, and later explored for different purposes in ECG analysis, see, e.g., [32–35].

Depending on polarity and morphology, P waves may be classified into three different types [36], of which P waves of Type 2 are the ones which are considered for simulation, characterized by positive, monophasic morphology in leads X and Y, and biphasic morphology in lead Z with a transition from negative to positive polarity. This type of P wave is predominant in patients with paroxysmal AF [36, 37]. Since P waves are monophasic in leads X and Y, larger values are assigned to $w_{\mathrm{X},1}$ and $w_{\mathrm{Y},1}$, whereas a larger value is assigned to $w_{\mathrm{Z},2}$ to emphasize the biphasic morphology in lead Z. To account for the fact that P wave morphology varies over time, $w_{l,i}$ and $\sigma_{P,i}$ vary according to a sinusoidal function whose frequency is randomly chosen in the interval [0.05, 0.15] Hz.

QRST complexes. The technique used for simulating QRST complexes in AF is also used in sinus rhythm. Resampling of the T wave is based on the well-known Bazett's formula, setting the corrected QT interval to 420 ms [38]. Immediately after AF termination, T wave duration increases linearly over the next seven beats to produce a smooth QT interval transition from AF to sinus rhythm. The choice of a seven-beat transition is ad hoc, since the QT interval transition in AF has not been much investigated in the literature.

3.4.3 Atrial Premature Beats

Since APBs are frequent in AF patients [39–42], it is important to account for their presence in the simulation model. Using a simple two-state Markov chain, a certain percentage of APBs is introduced, chosen from the following four types of unifocal APBs [43]:

1. APBs with reset of the sinus node. The sum of the length of the preceding and the subsequent RR intervals is less than twice the normal RR interval, simulated by 20% shortening of the preceding RR interval and by leaving the subsequent RR interval unchanged.
2. Interpolated APBs occur in between two adjacent sinus beats, simulated by splitting an RR interval into two intervals with 60/40 proportions.
3. APBs with delayed reset of the sinus node, simulated by 20% shortening of the preceding RR interval and 20% prolongation of the subsequent RR interval.
4. APBs with full compensatory pause, simulated by 20% shortening of the preceding RR interval, and subtracting the shortened RR interval from twice the normal RR interval to obtain the subsequent RR interval.

The likelihood of generating consecutive APBs, i.e., couplets, triplets, and short runs, is increased by setting the percentage of APBs to a large value. To account for the fact that P waves associated with APBs often deviate in amplitude and morphology from normal P waves in sinus rhythm, a new set of parameter values is generated and used to simulate P waves preceding APBs. The QRST complexes are generated in the same way as is done in sinus rhythm. Figure 3.7 illustrates simulated ECGs with different types of APBs.

3.4.4 Respiration

To account for the fact that respiration influences QRST morphology through changes in the electrical axis of the heart, the simulated VCG signal is transformed by a

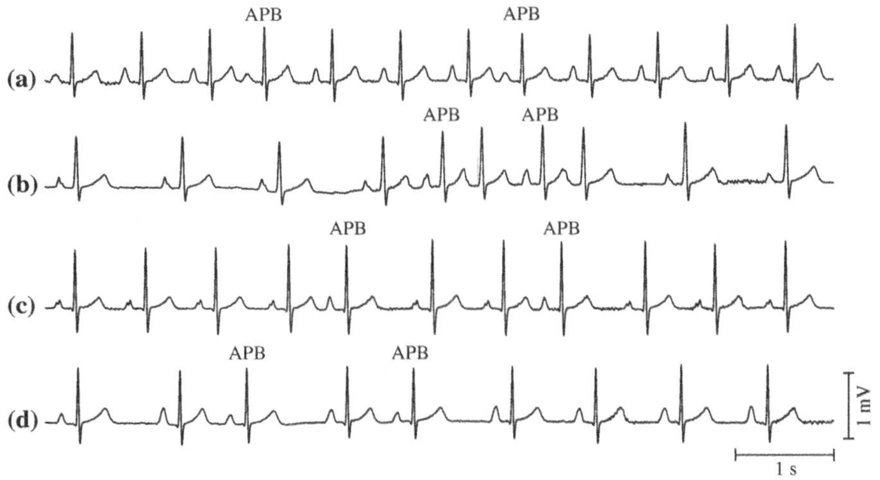

Fig. 3.7 Simulated ECGs containing **a** atrial premature beats (APBs) with reset of the sinus node (type 1), **b** interpolated APBs (type 2), **c** APBs with delayed reset of the sinus node (type 3), and **d** APBs with full compensatory pause (type 4)

rotation matrix $\mathbf{Q}(t)$, composed of three successive rotations around each of the axes [44],

$$\mathbf{Q}(t) = \mathbf{Q}_{\mathrm{X}}(t)\mathbf{Q}_{\mathrm{Y}}(t)\mathbf{Q}_{\mathrm{Z}}(t). \tag{3.14}$$

The three rotation matrices are defined by the time-varying angles $\varphi_{\mathrm{X}}(t)$, $\varphi_{\mathrm{Y}}(t)$, and $\varphi_{\mathrm{Z}}(t)$,

$$\mathbf{Q}_{\mathrm{X}}(t) = \begin{bmatrix} 1 & 0 & 0 \\ 0 & \cos\varphi_{\mathrm{X}}(t) & \sin\varphi_{\mathrm{X}}(t) \\ 0 & -\sin\varphi_{\mathrm{X}}(t) & \cos\varphi_{\mathrm{X}}(t) \end{bmatrix}, \tag{3.15}$$

$$\mathbf{Q}_{\mathrm{Y}}(t) = \begin{bmatrix} \cos\varphi_{\mathrm{Y}}(t) & 0 & \sin\varphi_{\mathrm{Y}}(t) \\ 0 & 1 & 0 \\ -\sin\varphi_{\mathrm{Y}}(t) & 0 & \cos\varphi_{\mathrm{Y}}(t) \end{bmatrix}, \tag{3.16}$$

$$\mathbf{Q}_{\mathrm{Z}}(t) = \begin{bmatrix} \cos\varphi_{\mathrm{Z}}(t) & \sin\varphi_{\mathrm{Z}}(t) & 0 \\ -\sin\varphi_{\mathrm{Z}}(t) & \cos\varphi_{\mathrm{Z}}(t) & 0 \\ 0 & 0 & 1 \end{bmatrix}. \tag{3.17}$$

It is assumed that angular variation is proportional to the amount of air in the lungs during a respiratory cycle, a property modeled as the product of two sigmoidal functions reflecting inspiration and expiration,

$$\psi(t) = \frac{1}{1 + e^{-\gamma_{\mathrm{in}} t}} \frac{1}{1 + e^{\gamma_{\mathrm{ex}}(t-\delta)}}, \tag{3.18}$$

where γ_{in} and γ_{ex} define the duration of inspiration and expiration, respectively, and δ defines the delay between inspiration and expiration. In lead X, the angular variation across successive respiratory cycles is defined by

$$\varphi_{\text{X}}(t) = \sum_{i=0}^{\infty} \xi_{\text{X}} \psi(t - iT_r), \tag{3.19}$$

where T_r is the duration of a respiratory cycle (inversely related to the fixed respiratory frequency, i.e., $T_r = 2\pi/\Omega_r$), and ξ_{X} is the maximum angular variation. The angular variation in leads Y and Z is determined in a similar way, defined by ξ_{Y} and ξ_{Z}, respectively. The choice of realistic model parameter values is discussed in [45], as well as an extension of the model in (3.19) so that a time-varying respiratory frequency can be accounted for.

In sinus rhythm, the respiratory frequency Ω_2 in (3.9), influencing the ventricular rhythm through the autonomic system, should, preferably, be set to Ω_r. In AF, the autonomic influence of respiration on ventricular rhythm is not modeled since the cardiorespiratory interaction is negligible [46].

3.4.5 Additive Noise

Three types of noise frequently encountered in ambulatory recordings—baseline wander, muscle noise, and electrode motion artifacts—can be added to the simulated ECG. These types of noise are extracted from the MIT–BIH Noise Stress Test Database, composed of a number of 30-min recordings which predominantly contain baseline wander, electromyographic noise, and electrode motion artifacts [47]. The two leads of the recordings in this database are labeled leads X and Y, whereas the noise in lead Z is constructed by computing the square root of the sum of squares of leads X and Y (an offset value is added before squaring, and the mean is subtracted after taking the square root).

3.4.6 Transformation from VCG to 12-Lead ECG

Different transformation matrices are applied to f waves, P waves, and QRST complexes when computing the standard 12-lead ECG from the VCG. The f wave transformation is based on the inverse of the P wave optimized transformation matrix [48], multiplied with a diagonal scaling matrix determining the tendency of f wave amplitude in the 12-lead ECG [20]. The diagonal matrix accounts for the fact that f wave amplitude is typically largest in V_1 and then gradually decreases as the leads move away from the atria. The decrease in amplitude can be explained by a much more scattered electrical vector in AF than in sinus rhythm, combined with increased distance to the electrode site. The resulting simulated 12-lead ECG with f waves, but not QRST complexes, is illustrated in Fig. 3.8a, and a real 12-lead ECG, whose f waves resemble the simulated ones, is illustrated in Fig. 3.8b.

The inverse of the P wave optimized transformation matrix in [48] is used to reconstruct P waves in the 12-lead ECG, see Fig. 3.9.

The Dower matrix [49, 50] is used to compute the QRST complexes, as well as the noise, in the 12-lead ECG. However, the transformation of the QRST complexes and the noise is done separately so that the noise can be scaled in each lead to the desired RMS value before being added to the 12-lead signal composed of both atrial and ventricular activity.

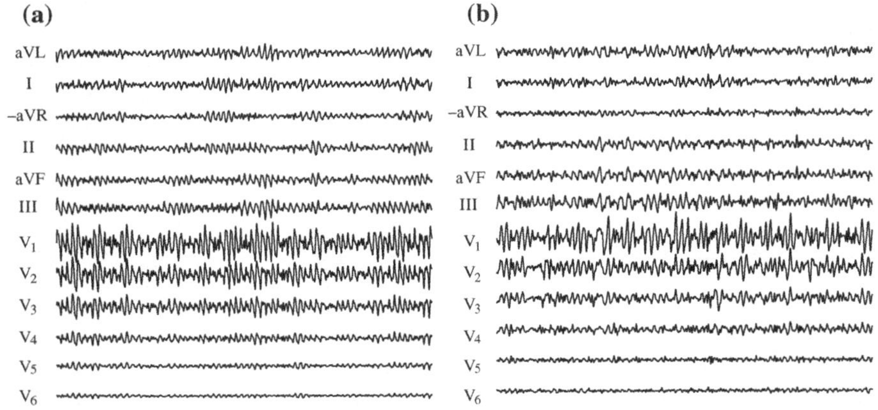

Fig. 3.8 **a** Simulated f waves produced by the model in (3.5), and **b** f waves extracted from a real ECG using an echo state network [15]

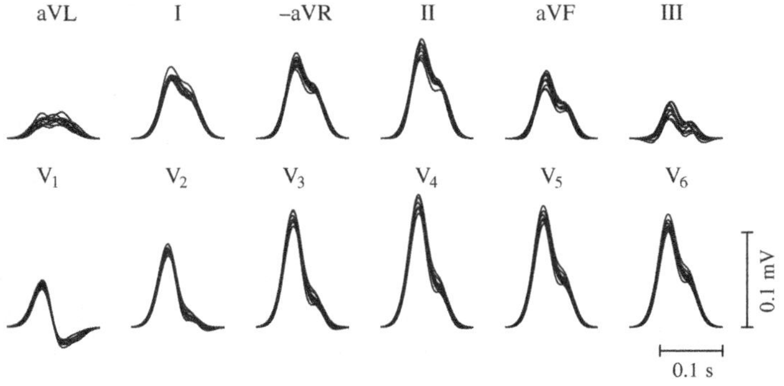

Fig. 3.9 Ten superimposed realizations of P waves in the standard 12-lead ECG, modeled as a linear combination of the first three Hermite functions using randomly generated weights

3.4.7 Switching Between Atrial Fibrillation and Sinus Rhythm

The switching between sinus rhythm and AF is modeled by a two-state continuous-time Markov chain, where the time d spent in a state, also referred to as episode duration, is determined by the exponential PDF

$$p(d) = \begin{cases} \beta_d e^{-\beta_d d}, & d \geq 0, \\ 0, & d < 0. \end{cases} \quad (3.20)$$

The parameter β_d defines the rate of episodes. The median duration of an AF episode is given by

$$\bar{d}_{\mathrm{AF}} = \frac{\ln 2}{\beta_{\mathrm{AF}}}, \quad (3.21)$$

where β_{AF} denotes the rate of AF episodes, cf. (3.4). The median duration of an episode with sinus rhythm is assumed to be given by

$$\bar{d}_{\mathrm{SR}} = \frac{B}{(1-B)} \cdot \bar{d}_{\mathrm{AF}}, \quad (3.22)$$

where B $(0 < B < 1)$ determines the total time AF is present, and thus B can be viewed as a descriptor of mean AF burden. The sole parameter controlling episode duration is $\bar{d}_{\mathrm{AF}}$, and no minimum episode duration is specified.

A more advanced, non-Markovian switching model has been proposed which account for aspects of AF progression related to genetic disposition, age-, and AF history-related remodeling [51]. The model can simulate individual AF episodes as well as the natural progression of AF in patients over a period of decades.

The possibility to generate episodes with varying duration is valuable when simulating arrhythmia progression. Evidence shows that brief episodes progress to longer episodes [52, 53], implying that it is of interest to evaluate detection performance as a function of episode duration. Moreover, brief but rare episodes have been observed in patients after cryptogenic stroke and transient ischemic attack [54–57]. Such signals can be simulated with the model described in this section, using, for example, a median episode duration of 30 beats and a low AF burden of 0.001.

3.5 Simulation of Paroxysmal AF Using Real Components

Alternatively, the simulator can produce signals based on real ECG components, randomly selected from the three databases which are used to characterize ventricular rhythm, atrial activity (f or P waves), and QRST complexes, see Fig. 3.10.

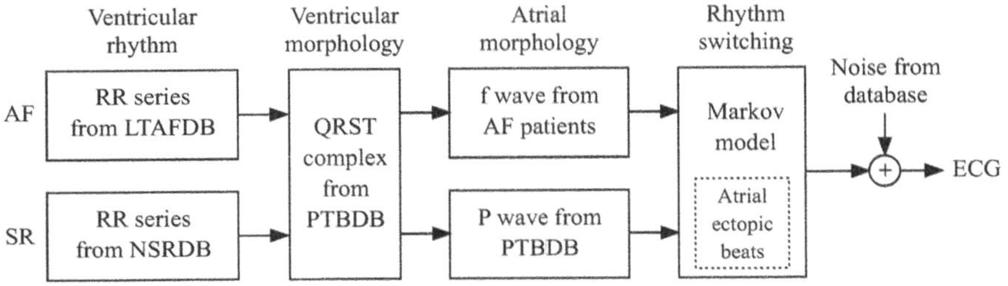

Fig. 3.10 Simulation of ECG signals using real components, taken from the Long-Term AF Database (LTAFDB), the MIT–BIH Normal Sinus Rhythm Database (NSRDB), and the PTB Diagnostic ECG Database (PTBDB)

These components, together with the above-described noise types, are added to produce the standard 12-lead ECG.

Ventricular rhythm. The Long Term Atrial Fibrillation Database was used for creating a set of AF rhythms. A total of 69 different RR interval series were extracted from the 84 long-term ECG recordings; the 15 remaining recordings were excluded due to their relatively short duration with AF (<5000 beats). Similarly, the entire NSRDB, consisting of 18 long-term ECG recordings, was used to create a set of sinus rhythms. Switching between paroxysmal AF and sinus rhythm is modeled in the same way as for synthetic components, cf. Sect. 3.4.7.

For each simulated signal, the RR interval series of the prevailing rhythm is randomly selected from the proper rhythm set, and repeated by concatenation until the desired length is attained. While heart rate is often higher in AF than in sinus rhythm, this may not be the case when concatenating randomly selected RR intervals in sinus rhythm and AF. Therefore, whenever the mean RR interval is shorter in sinus rhythm than in AF, the mean RR interval in sinus rhythm is adjusted to become identical to the mean RR interval in AF.

It should be noted that when simulating ECGs using real components, the atrial and ventricular rates are unrelated since the f waves and the RR interval series are extracted from different databases.

f and P waves. A set of 20 segments with real, multi-lead f waves is extracted from the Lund AF database with 12-lead ECGs, acquired from patients with persistent AF [8]. An echo state network was applied for f wave extraction [15], see also Sect. 5.5.3. Lead V_6 was used as reference lead when extracting f waves in the remaining 11 ECG leads, whereas lead V_5 was used when extracting f waves in lead V_6, see Fig. 3.8b.

In sinus rhythm, the original, real P wave, along with the subsequent QRST complex, is retained, while, in AF, only the QRST complex is retained and a continuous f wave signal added.

QRST complexes. A set of 100 15-lead ECGs (12 standard leads plus Frank leads) with sinus rhythm, selected from the Physikalisch–Technische Bundesanstalt Database, serves as the basis for modeling QRST complexes. Following baseline removal and QRST delineation [58], the original T wave is resampled to have a fixed width and then adjusted to the prevailing heart rate according to the procedure described in Sect. 3.4.1. Since the ECGs of this database last for only about two minutes, the QRST complexes are repeated by concatenation until the desired duration is achieved. The TQ interval is interpolated using cubic spline interpolation. All other steps required to generate QRST complexes are similar to those described in Sect. 3.4.1.

Simulated signals composed of either synthetic or real ECG components are illustrated in Fig. 3.11.

3.6 Relevance of Simulated Signals

The question whether a simulation model produces realistic signals is not easily answered since the term "realistic" is difficult to quantify. Historically, this question

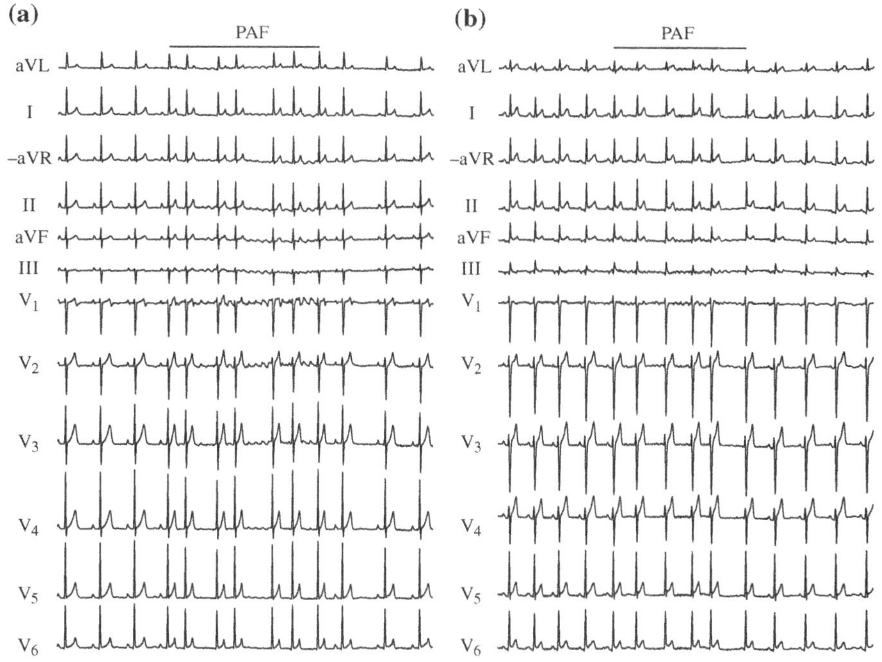

Fig. 3.11 Simulated 12-lead ECGs containing a brief AF episode, composed of **a** synthetic components and **b** real components. Using synthetic components, the 12-lead ECG is obtained from the simulated signals in leads X, Y, and Z, following linear transformations. Using real components, the original 12-lead ECG is taken from the Lund AF database, followed by removal of P waves and addition of extracted f waves

has not received much, if any, attention in papers describing simulation models of the ECG, see, e.g., [28, 31, 59, 60], although the models have turned out to be most valuable in the development of signal processing algorithms—an observation which applies particularly to the simulation model in [28]. To provide a quantitative answer, the idea to let expert cardiologists assess blindly the realism of simulated ECG signals was first materialized in [20], involving not only the simulated ECG signals produced by the model in Sect. 3.4, but also real ECG signals [20]. The results showed that the simulated signals were, for the most part, realistic, but they also showed that the approach to modeling of the QT interval in AF needed improvement. To make the outcome of expert assessment more powerful, it would have been desirable with more than two cardiologists so that more far-reaching conclusions could have been drawn.

In the context of AF detection, an indirect approach to evaluating signal realism is to analyze simulated signals using some suitable detector, and then compare the obtained results with those obtained using the same detector on an existing database containing real ECGs [20]. Neither this approach has been considered in the past, although it may provide valuable insight into whether the simulated signals are too "doctored" to be used for the development of AF detectors.

The degree of sophistication of a simulation model is another way to judge model relevance, hinted at in [17] where the f wave replication model was labeled as "primitive" and the above-mentioned model of normal sinus rhythm [28] as "simple," whereas the biophysical model proposed by the authors themselves was labeled as "more sophisticated" in producing ECG signals. Considering that the biophysical model accounts for detailed electroanatomical information, whereas the other two models do not, such labeling seems reasonable. But does a higher degree of sophistication imply that the model is better suited for the development of signal processing algorithms and performance evaluation? The fact that biophysical models have hardly been considered at all for such purposes provides an answer to this question, with implementational and computational complexity, difficulty to control basic signal characteristics such as f wave amplitude and repetition rate, and the lack of rhythm switching models as probable reasons. From an algorithmic viewpoint, it is not obvious why biophysical models necessarily produce ECG signals which are more relevant than those of phenomenological models, such as the ones described in Sects. 3.4 and 3.5.

References

1. A.L. Goldberger, L.A. Amaral, L. Glass, J.M. Hausdorff, P.C. Ivanov, R.G. Mark, J.E. Mietus, G.B. Moody, C.K. Peng, H.E. Stanley, PhysioBank, PhysioToolkit, and PhysioNet: components of a new research resource for complex physiologic signals. Circulation **101**, E215–220 (2000)
2. G.B. Moody, R.G. Mark, A new method for detecting atrial fibrillation using R-R intervals, in Proceedings of Computers in Cardiology vol. **10**, 227–230 (1983)
3. S. Petrutiu, A.V. Sahakian, S. Swiryn, Abrupt changes in fibrillatory wave characteristics at the termination of paroxysmal atrial fibrillation in humans. Europace **9**, 466–470 (2007)

4. G.B. Moody, Spontaneous termination of atrial fibrillation: a challenge from PhysioNet and Computers in Cardiology 2004, in Proceedings of Computers in Cardiology vol. **31**, 101–104 (2004)
5. G.D. Clifford, C. Liu, B. Moody, L.-W.H. Lehman, I. Silva, Q. Li, A. Johnson, R.G. Mark, AF classification from a short single lead ECG recording: the PhysioNet Computing in Cardiology Challenge 2017, in Proceedings of Computing in Cardiology vol. **44**, 1 (2017)
6. M. Henriksson, A. Petrėnas, V. Marozas, F. Sandberg, L. Sörnmo, Model-based assessment of f-wave signal quality in patients with atrial fibrillation. IEEE Trans. Biomed. Eng. (2018, accepted)
7. R.G. Mark, P.S. Schluter, G.B. Moody, P.H. Devlin, D. Chernoff, An annotated ECG database for evaluating arrhythmia detectors. Proc. IEEE Front. Eng. Health Care, 205–210 (1982)
8. M. Stridh, L. Sörnmo, C.J. Meurling, S.B. Olsson, Sequential characterization of atrial tachyarrhythmias based on ECG time-frequency analysis. IEEE Trans. Biomed. Eng. **51**, 100–114 (2004)
9. M. Stridh, L. Sörnmo, Spatiotemporal QRST cancellation techniques for analysis of atrial fibrillation. IEEE Trans. Biomed. Eng. **48**, 105–111 (2001)
10. F. Sandberg, M. Stridh, L. Sörnmo, Robust time-frequency analysis of atrial fibrillation using hidden Markov models. IEEE Trans. Biomed. Eng. **55**, 502–511 (2008)
11. V.D.A. Corino, L.T. Mainardi, M. Stridh, L. Sörmno, Improved time-frequency analysis of atrial fibrillation signals using spectral modelling. IEEE Trans. Biomed. Eng. **56**, 2723–2730 (2008)
12. R. Alcaraz, J.J. Rieta, Surface ECG organization analysis to predict paroxysmal atrial fibrillation termination. Comput. Biol. Med. **39**, 697–706 (2009)
13. R. Sassi, V.D.A. Corino, L.T. Mainardi, Analysis of surface atrial signals: time series with missing data? Ann. Biomed. Eng. **37**, 2082–2092 (2009)
14. H. Dai, S. Jiang, Y. Li, Atrial activity extraction from single lead ECG recordings: evaluation of two novel methods. Comput. Biol. Med. **43**, 176–183 (2013)
15. A. Petrėnas, V. Marozas, L. Sörnmo, A. Lukoševičius, An echo state neural network for QRST cancellation during atrial fibrillation. IEEE Trans. Biomed. Eng. **59**, 2950–2957 (2012)
16. F. Castells, J.J. Rieta, J. Millet, V. Zarzoso, Spatiotemporal blind source separation approach to atrial activity estimation in atrial tachyarrhythmias. IEEE Trans. Biomed. Eng. **52**, 258–267 (2005)
17. V. Jacquemet, A. van Oosterom, J.-M. Vesin, L. Kappenberger, Analysis of electrocardiograms during atrial fibrillation: a biophysical approach. IEEE Med. Biol. Eng. Mag. **25**, 79–88 (2006)
18. O. Blanc, N. Virag, J.-M. Vesin, L. Kappenberger, A computer model of human atria with reasonable computation load and realistic anatomical properties. IEEE Trans. Biomed. Eng. **48**, 1229–1237 (2001)
19. N. Virag, V. Jacquemet, C.S. Henriquez, S. Zozor, O. Blanc, J.-M. Vesin, E. Pruvot, L. Kappenberger, Study of atrial arrhythmias in a computer model based on magnetic resonance images of human atria. Chaos **12**, 754–763 (2002)
20. A. Petrėnas, V. Marozas, A. Sološenko, R. Kubilius, J. Skibarkienė, J. Oster, L. Sörnmo, Electrocardiogram modeling during paroxysmal atrial fibrillation: application to the detection of brief episodes. Physiol. Meas. **38**, 2058–2080 (2017)
21. V.D.A. Corino, F. Sandberg, L.T. Mainardi, L. Sörnmo, An atrioventricular node model for analysis of the ventricular response during atrial fibrillation. IEEE Trans. Biomed. Eng. **58**, 3386–3395 (2011)
22. M.S. Guillem, A.M. Climent, J. Millet, Á. Arenal, F. Fernández-Avilés, J. Jalife, F. Atienza, O. Berenfeld, Noninvasive localization of maximal frequency sites of atrial fibrillation by body surface potential mapping. Circ. Arrhythm. Electrophysiol. **6**, 294–301 (2013)
23. F. Ravelli, M. Masè, M.D. Greco, L. Faes, M. Disertori, Deterioration of organization in the first minutes of atrial fibrillation: a beat-to-beat analysis of cycle length and wave similarity. J. Cardiovasc. Electrophysiol. **18**, 60–65 (2007)
24. R. Alcaraz, J.J. Rieta, Non-invasive organization variation assessment in the onset and termination of paroxysmal atrial fibrillation. Comput. Methods Programs Biomed. **93**, 148–154 (2009)

25. M. Masè, M. Marini, M. Disertori, F. Ravelli, Dynamics of AV coupling during human atrial fibrillation: role of atrial rate. Am. J. Physiol. Heart Circ. Physiol. **309**, H198–H205 (2015)
26. J. Malik, N. Reed, C.-L. Wang, H.-T. Wu, Single-lead f-wave extraction using diffusion geometry. Physiol. Meas. **38**, 1310–1334 (2017)
27. R. Sameni, G.D. Clifford, C. Jutten, M.B. Shamsollahi, Multichannel ECG and noise modeling: application to maternal and fetal ECG signals. J. Adv. Signal Process., 1–14 (2007)
28. P.E. McSharry, G.D. Clifford, L. Tarassenko, L.A. Smith, A dynamical model for generating synthetic electrocardiogram signals. IEEE Trans. Biomed. Eng. **50**, 289–294 (2003)
29. G.R. Pai, J.M. Rawles, The QT interval in atrial fibrillation. Brit. Heart J. **61**, 510–513 (1989)
30. D.L. Musat, M. Adhaduk, M.W. Preminger, A. Arshad, T. Sichrovsky, J.S. Steinberg, S. Mittal, Correlation of QT interval correction methods during atrial fibrillation and sinus rhythm. Am. J. Cardiol. **112**, 1379–1383 (2013)
31. L. Sörnmo, P.O. Börjesson, M.E. Nygårds, O. Pahlm, A method for evaluation of QRS shape features using a mathematical model for the ECG. IEEE Trans. Biomed. Eng. **28**, 713–717 (1981)
32. P. Laguna, R. Jané, S. Olmos, N.V. Thakor, H. Rix, P. Caminal, Adaptive estimation of QRS complex by the Hermite model for classification and ectopic beat detection. Med. Biol. Eng. Comput **34**, 58–68 (1996)
33. T.H. Linh, S. Osowski, M. Stodolski, On-line heart beat recognition using Hermite polynomials and neuro-fuzzy network. IEEE Trans. Instrum. Measure. **52**, 1224–1231 (2003)
34. H. Haraldsson, L. Edenbrandt, M. Ohlsson, Detecting acute myocardial infarction in the 12-lead ECG using Hermite expansions and neural networks. Artif. Intell. Med. **32**, 127–136 (2004)
35. A. Sandryhaila, S. Saba, M. Puschel, J. Kovacevic, Efficient compression of QRS complexes using Hermite expansion. IEEE Trans. Signal Process. **60**, 947–955 (2012)
36. R. Havmöller, J. Carlson, F. Holmqvist, A. Herreros, C. Meurling, S.B. Olsson, P.G. Platonov, Age-related changes in P wave morphology in healthy subjects. BMC Cardiovasc. Disord. **7**, 22 (2007)
37. F. Holmqvist, M.S. Olesen, A. Tveit, S. Enger, J. Tapanainen, R. Jurkko, R. Havmöller, S. Haunsø, J. Carlson, J.H. Svendsen, P.G. Platonov, Abnormal atrial activation in young patients with lone atrial fibrillation. Europace **13**, 188–192 (2011)
38. H.C. Bazett, An analysis of the time relations of electrocardiograms. Heart **7**, 353–370 (1920)
39. S.-A. Chen, M.-H. Hsieh, C.-T. Tai, C.-F. Tsai, V.S. Prakash, W.-C. Yu, T.-L. Hsu, Y.-A. Ding, M.-S. Chang, Initiation of atrial fibrillation by ectopic beats originating from the pulmonary veins: electrophysiological characteristics, pharmacological responses, and effects of radiofrequency ablation. Circulation **100**, 1879–1886 (1999)
40. D. Wallmann, D. Tüller, K. Wustmann, P. Meier, J. Isenegger, M. Arnold, H.P. Mattle, E. Delacrétaz, Frequent atrial premature beats predict paroxysmal atrial fibrillation in stroke patients: an opportunity for a new diagnostic strategy. Stroke **38**, 2292–2294 (2007)
41. M. Weber-Krüger, K. Gröschel, M. Mende, J. Seegers, R. Lahno, B. Haase, C.-F. Niehaus, F. Edelmann, G. Hasenfuß, R. Wachter, R. Stahrenberg, Excessive supraventricular ectopic activity is indicative of paroxysmal atrial fibrillation in patients with cerebral ischemia. PLoS ONE **8**, e67602 (2013)
42. D.J. Gladstone, P. Dorian, M. Spring, V. Panzov, M. Mamdani, J.S. Healey, K.E. Thorpe, for EMBRACE Steering Committee and Investigators, Atrial premature beats predict atrial fibrillation in cryptogenic stroke: results from the EMBRACE trial. Stroke **46**, 936–941 (2015)
43. T. Thong, J. McNames, M. Aboy, B. Goldstein, Prediction of paroxysmal atrial fibrillation by analysis of atrial premature complexes. IEEE Trans. Biomed. Eng. **4**, 561–569 (2004)
44. M. Åström, E. Carro, L. Sörnmo, P. Laguna, B. Wohlfart, Vectorcardiographic loop alignment and the measurement of morphologic beat-to-beat variability in noisy signals. IEEE Trans. Biomed. Eng. **47**, 497–506 (2000)
45. R. Bailón, L. Sörnmo, P. Laguna, A robust method for ECG-based estimation of the respiratory frequency during stress testing. IEEE Trans. Biomed. Eng. **53**, 1273–1285 (2006)
46. M.M. Platiša, T. Bojić, S.U. Pavlović, N.N. Radovanović, A. Kalauzi, Uncoupling of cardiac and respiratory rhythm in atrial fibrillation. Biomed. Tech. (Berlin) **61**, 657–663 (2016)

47. G.B. Moody, W.K. Muldrow, R.G. Mark, A noise stress test for arrhythmia detectors. Proc. Comput. Cardiol. **11**, 381–384 (1984)
48. M.S. Guillem, A.V. Sahakian, S. Swiryn, Derivation of orthogonal leads from the 12-lead electrocardiogram. Performance of an atrial-based transform for the derivation of P loops. J. Electrocardiol. **41**, 19–25 (2008)
49. G.E. Dower, A lead synthesizer for the Frank system to simulate the standard 12-lead electrocardiogram. J. Electrocardiol. **1**, 101–116 (1968)
50. G.E. Dower, H.B. Machado, J.A. Osborne, On deriving the electrocardiogram from vectorcardiographic leads. Clin. Cardiol. **3**, 87–95 (1980)
51. E.T.Y. Chang, Y.T. Lin, T. Galla, R.H. Clayton, J. Eatock, A stochastic individual-based model of the progression of atrial fibrillation in individuals and populations. PLoS ONE **11**, e0152349 (2016)
52. M.C. Wijffels, C.J. Kirchhof, R. Dorland, M.A. Allessie, Atrial fibrillation begets atrial fibrillation. A study in awake chronically instrumented goats. Circulation **92**, 1954–1968 (1995)
53. C.R. Kerr, K.H. Humphries, M. Talajic, G.J. Klein, S.J. Connolly, M. Green, J. Boone, R. Sheldon, P. Dorian, D. Newman, Progression to chronic atrial fibrillation after the initial diagnosis of paroxysmal atrial fibrillation: results from the Canadian Registry of Atrial Fibrillation. Am. Heart J. **149**, 489–496 (2005)
54. A.H. Tayal, M. Tian, K.M. Kelly, S.C. Jones, D.G. Wright, D. Singh, J. Jarouse, J. Brillman, S. Murali, R. Gupta, Atrial fibrillation detected by mobile cardiac outpatient telemetry in cryptogenic TIA or stroke. Neurology **71**, 1696–1701 (2008)
55. C.G. Favilla, E. Ingala, J. Jara, E. Fessler, B. Cucchiara, S.R. Messé, M.T. Mullen, A. Prasad, J. Siegler, M.D. Hutchinson, S.E. Kasner, Predictors of finding occult atrial fibrillation after cryptogenic stroke. Stroke **46**, 1210–1215 (2015)
56. J.W. Keach, S.M. Bradley, M.P. Turakhia, T.M. Maddox, Early detection of occult atrial fibrillation and stroke prevention. Heart **101**, 1097–1102 (2015)
57. D.J. Miller, K. Shah, S. Modi, A. Mahajan, S. Zahoor, M. Affan, The evolution and application of cardiac monitoring for occult atrial fibrillation in cryptogenic stroke and TIA. Curr. Treat. Options Neurol. **18**, 17 (2016)
58. P. Laguna, R. Jané, P. Caminal, Automatic detection of wave boundaries in multilead ECG signals: validation with the CSE database. Comput. Biomed. Res. **27**, 45–60 (1994)
59. A. van Oosterom, T.F. Oostendorp, ECGSIM: an interactive tool for studying the genesis of QRST waveforms. Heart **90**, 165–168 (2004)
60. J. Behar, F. Andreotti, S. Zaunseder, Q. Li, J. Oster, G.D. Clifford, An ECG simulator for generating maternal-foetal activity mixtures on abdominal ECG recordings. Physiol. Meas. **35**, 1537–1550 (2014)

Chapter 4
Detection of Atrial Fibrillation

Leif Sörnmo, Andrius Petrėnas and Vaidotas Marozas

4.1 Introduction

The detection of episodes of atrial fibrillation (AF) has been dealt with for more than three decades in research, and yet the challenge remains to develop a detector fully capable of handling all the problems associated with the analysis of continuous long-term ECG recordings as well as of recordings acquired by handheld devices for AF screening. Unacceptably high false alarm rates have been reported, mostly due to the presence of ectopic beats and noisy signal segments, but also due to non-AF arrhythmias manifested by rhythms patterns resembling those of AF, see, e.g., [1]. For the human reader, the following three properties are essential when detecting AF episodes:

1. the presence of a highly irregular rhythm,
2. the absence of P waves, and
3. the presence of f waves.

These properties are, to various extents, explored when developing algorithms for AF detection.

L. Sörnmo (✉)
Department of Biomedical Engineering and Center for Integrative Electrocardiology, Lund University, Lund, Sweden
e-mail: leif.sornmo@bme.lth.se

A. Petrėnas · V. Marozas
Biomedical Engineering Institute, Kaunas University of Technology, Kaunas, Lithuania
e-mail: andrius.petrenas@ktu.lt

V. Marozas
e-mail: vaidotas.marozas@ktu.lt

L. Sörnmo (ed.), *Atrial Fibrillation from an Engineering Perspective*,
Series in BioEngineering, https://doi.org/10.1007/978-3-319-68515-1_4

Translating "highly irregular rhythm" into a detection parameter is challenging, since not much is known a priori about the features which are best suited for characterizing irregularity in AF. An abundance of detection parameters have been proposed in the literature, many of them reviewed in this chapter, and each parameter is designed to capture some specific feature of rhythm irregularity. An early study on the characterization of irregularity in AF, without also addressing the AF detection problem, posed the fundamental question whether the series of RR intervals in AF is random or deterministic [2]. The results in that study showed that the RR intervals are not entirely unpredictable, as evidenced by the nonzero correlation between the observed and the predicted RR intervals at different correlation lags. However, these findings did not apply to all patients of the analyzed data set, and, therefore, parameters related to prediction/correlation are unlikely to be good candidates for AF detection. In another study, spectral analysis demonstrated that the RR interval series during AF has a white noise-like spectrum when analyzed on a minute-by-minute scale [3].

Heart rate may be considered in AF detection as it tends to be higher in AF episodes than in sinus rhythm. Although it is obvious that heart rate alone cannot be used for detection, the power of a detection parameter describing rhythm irregularity may still be boosted by integrating information on heart rate into the definition of a parameter. Heart rate is usually characterized by the mean of the RR intervals contained in a detection window.

The detection of AF is compounded by the fact that certain arrhythmias are manifested by RR interval patterns closely resembling those observed in AF. This problem is particularly pronounced when all detection parameters describe rhythm characteristics. Hence, it is highly desirable that the detector can recognize the characteristics of confounding non-AF rhythm patterns so that the number of false alarms is minimized. Runs of ventricular premature beats (VPBs), frequent atrial premature beats (APBs), and atrial flutter, as well as bigeminy and trigeminy, are all important sources to false alarms; representative RR interval series for some of these confounding rhythms are displayed in Fig. 4.1. Another source of false alarms is inaccurate QRS detection, e.g., caused by muscle noise, motion artifacts, or large-amplitude T waves. Moreover, the risk of detecting non-AF rhythm patterns becomes increasingly higher as the detection window becomes increasingly shorter, which is required to detect short AF episodes.

When information on P waves and/or f waves is considered in AF detection, it should be paired with information on signal quality, indicating to what degree wave measurements can be trusted. Otherwise, garbage measurements may completely disrupt detection performance. Given that many clinical studies explore information derived from continuous long-term ECG recordings, often characterized by a substantial variation in noise level, information on signal quality should be an integral part of the decision-making process.

An AF episode of at least 30 s duration is considered clinically significant—a definition which was published in the ACC/AHA/ESC 2006 guidelines for management of AF patients [4], and in widespread use among clinicians. The motivation behind 30 s as minimum duration was not clearly stated, although the guidelines pointed

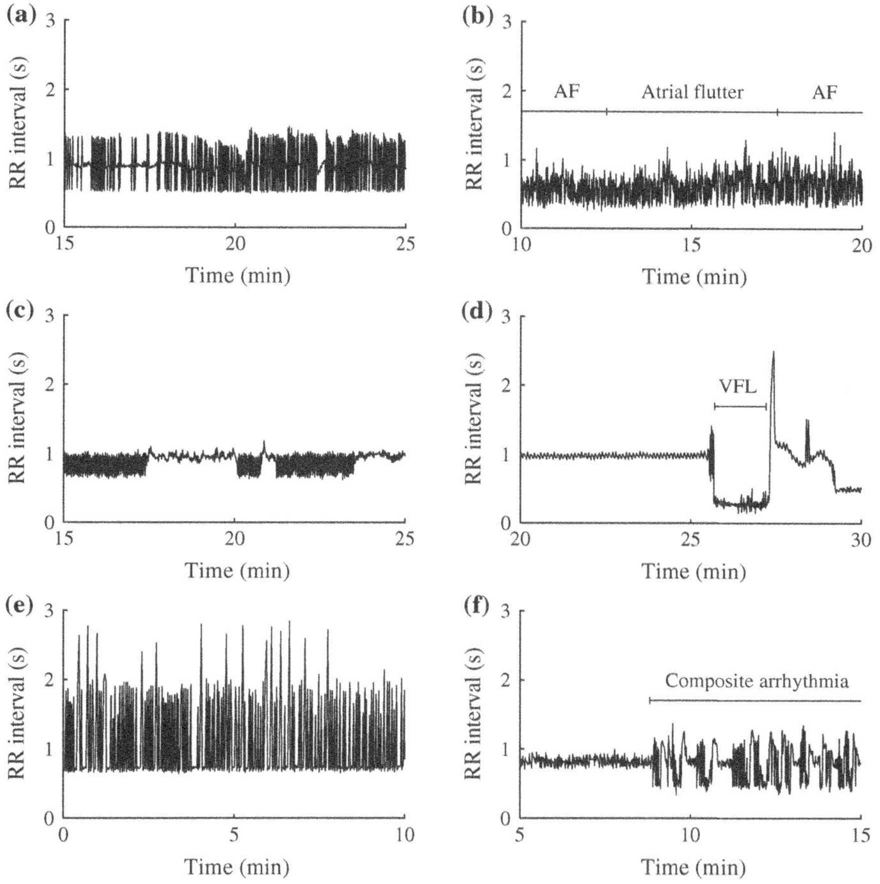

Fig. 4.1 Illustration of RR interval patterns which may confound detection of AF episodes. **a** Multiple ventricular premature beats, including bigeminy and trigeminy, **b** atrial flutter surrounded by AF, **c** second degree atrioventricular block, **d** episode of ventricular flutter (VFL), **e** sinus bradycardia, and **f** episode of a composite arrhythmia including AF, atrial flutter, atrial bigeminy, supraventricular tachycardia, atrioventricular junctional rhythm, and atrial premature beats. All examples are taken from the MIT–BIH Arrhythmia Database

out that AF episodes briefer than 30 s may be relevant in "certain clinical situations involving symptomatic patients, pre-excitation or in assessing the effectiveness of therapeutic interventions." Interestingly, the more recent guidelines published in 2014 [5] did not mention anything about minimum episode duration, whereas the 2016 guidelines [6] brought back the 30 s minimum duration previously published in 2006.

In recent years, the significance of AF episodes briefer than 30 s has received increasing attention in clinical research, especially concerning issues related to the future risk of stroke and its prevention.[1] It has been suggested that such brief episodes are directly coupled to the formation of atrial thrombus, and, therefore, may be viewed as biomarkers of prolonged episodes occurring outside of the monitoring period [12–14]. When monitoring is performed during a month-long period, a patient with numerous brief episodes can very well have a higher AF burden than a patient with a few episodes which all exceed 30 s, meaning a higher thromboembolic risk for the patient with brief episodes [15], see also [16] and page 43. The concept "AF burden" is defined as the proportion of the total recording time a patient is in AF. The minimum duration of an episode which still convey clinically significant information remains to be established.

Long-term AF monitoring requires automated event detection for efficient and practical handling. Thus, the properties of the detector play a central role as they impose a lower limit on how brief an episode can be and still be detected. Most detectors described in the literature have a design that precludes the detection of episodes briefer than 30 s due to the principle adopted for detection. For example, AF detection based on RR interval histogram analysis requires a large number of RR intervals to ensure that the histogram is reasonably reliable. Indeed, some ECG-based detectors are blind to episodes briefer than two minutes, whereas, in implantable devices, a minimum episode duration of as much as six minutes has been used [17, 18]. Clinical studies reporting results on the presence of episodes briefer than 30 s have relied on commercial detectors, implementing proprietary algorithms whose detection performance have not been published [12, 13, 19, 20]. Therefore, manual review of possible AF events briefer than 30 s has been required to carry through the study [12]. Consequently, it is of substantial interest to design and evaluate AF detectors which facilitate the investigation of the clinical significance of brief episodes.

The duration of an AF episode is highly variable, extending from less than 30 s up to seven days; episodes extending beyond seven days are designated as persistent AF [4]. Similar to the problem of detecting QRS complexes, where a least informative approach is often recommended with respect to assumptions on signal properties [21], an AF detector should not involve firm assumptions on episode duration, nor on the minimum distance between two subsequent AF episodes. By merging two detected episodes, even if separated by just a few seconds, clinically relevant information could be excluded.

With the advent of handheld and smartphone-based devices for AF screening comes new possibilities to identify previously undetected AF [22–29], cf. Sects. 2.3.5 and 2.3.6, but also new challenges related to the signal quality of such patient-operated devices which, in general, is poorer than the quality associated with the clinical modalities, see Fig. 4.2 for an illustration of poor signal quality. Since handheld and smartphone-based devices are designed to record a single lead, not necessarily reflecting atrial activity, rhythm-based detection is the typical mode of operation, with information on f and P wave morphology as a bonus.

---

[1] Paroxysmal AF manifested by episodes briefer than 30 s is sometimes referred to as *occult paroxysmal AF*, especially when asymptomatic or undetected by conventional methods [7–11].

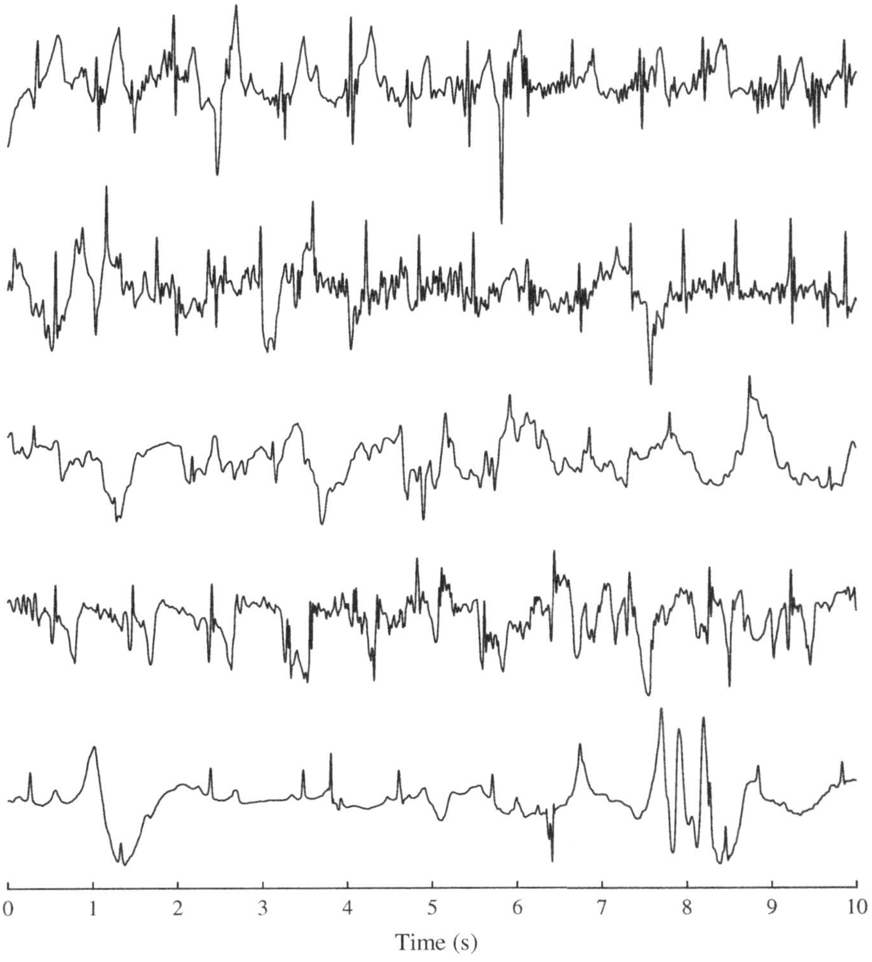

Fig. 4.2 Five examples of poor-quality ECGs recorded using a smartphone-based device. The signals are part of the database made available for the PhysioNet/Computing in Cardiology Challenge 2017 [30]

In this chapter, the main design principles used in AF detection are reviewed, either exploring rhythm information only, i.e., the RR interval series, (Sect. 4.2) or information on both rhythm and atrial wave morphology (Sect. 4.3). Aspects on detector implementation are briefly considered in Sect. 4.4, and different performance measures used in AF detection are described in Sect. 4.5. Although several reflections on performance are interspersed throughout the chapter, Sect. 4.6 has detection performance as its main theme, with a discussion on aspects which need to be considered when evaluating performance. The chapter ends with a discussion on different types of ECG-derived information which may be explored to improve detection performance (Sect. 4.7).

4.2 Rhythm-Based AF Detection

Since reliable information on the absence/presence of P and f waves is difficult to extract at low signal-to-noise ratios (SNRs), the vast majority of AF detectors rely entirely on parameters quantifying RR interval irregularity, e.g., the degree of randomness, variability, and complexity. Another important explanation to the dominance of rhythm-based detectors is that their implementation in hardware requires far less energy than do detectors which also involve morphologic information. The RR interval series constitutes the sole input data to most detectors implemented in an implantable device, since morphologic information is difficult to extract from invasive recordings.

Over the years, detector design has been based on *ad hoc principles*, involving one or a few parameters which are fed to a simple classifier, while neither model-based statistical detection nor physiological considerations have played a significant role in the design. Nonetheless, it is obvious from the results listed in Table 4.1 that ad hoc principles have helped to push the limits of detection performance as both sensitivity and specificity have improved; for a definition of these two performance measures, see Sect. 4.5.[2] Still, further improvement of detector performance is warranted so that, for example, the problem of false alarms due to frequent ectopic beats, together masquerading as an AF episode, non-AF arrhythmias, or noisy signals can be adequately addressed.

Apart from using the RR interval series $x(0), \ldots, x(N-1)$ itself as detector input, the first difference,

$$\Delta x(n) = x(n) - x(n-1), \quad n = 1, \ldots, N-1, \tag{4.1}$$

sometimes also serves as input, where N is the number of RR intervals and n is the interval index (and thus not ECG sample index). Unless the ECG recording is very short, i.e., on the order of 10–20 s, the input data is usually processed using a *sliding time window* approach in which the detection parameters are repeatedly computed as the window slides forward in time. Sliding by one RR interval at a time offers the best time resolution of episode onset and end; however, it may be necessary to take larger "slides" to reduce the amount of computations, for example, 50 intervals at a time [36].

The main principles explored for rhythm-based AF detection are described in the following. To simplify the description, detection parameters are assumed to be

[2]Several other detectors have been proposed besides those listed in Table 4.1. However, for various reasons, their respective performance was not evaluated on AFDB.

Table 4.1 Performance of rhythm-based AF detectors expressed in terms of sensitivity (Se) and specificity (Sp), using the MIT–BIH Atrial Fibrillation Database (AFDB) for evaluation, see Sect. 3.1. The subset $AFDB_1$ is identical to AFDB, except that records 4936 and 5091 are excluded for reasons of incorrect annotations. The detectors are ordered with respect to their year of publication

Method by	Year	Database	Se (%)	Sp (%)
Tateno and Glass [31]	2001	AFDB	94.4	97.2
Dash et al. [32]	2009	$AFDB_1$	94.4	95.1
Lian et al. [33]	2011	AFDB	95.8	96.4
Lake and Moorman [34]	2011	AFDB	91	94
Huang et al. [35]	2011	AFDB	96.1	98.1
Shouldice et al. [36]	2012	AFDB	92	96
Lee et al. [37]	2013	$AFDB_1$	98.2	97.7
Zhou et al. [38]	2014	AFDB	96.9	98.3
Asgari et al. [39]	2015	AFDB	97.0	97.1
Petrėnas et al. [40]	2015	AFDB	97.1	98.3
Zhou et al. [41]	2015	AFDB	97.4	98.4

computed in a fixed window, however, it is straightforward to replace it with a sliding window. The interested reader may want to follow up with some other rhythm-based detectors proposed over years [42–47].

4.2.1 *Irregularity Parameters*

Table 4.2 presents a list of parameters considered in the design of AF detectors, grouped into five categories, namely statistical dispersion, entropy, parameters based on symbolic dynamics, parameters based on the Poincaré plot, and parameters based on the time-varying coherence function. Of these categories, statistical parameters reflecting dispersion, e.g., the root mean square of successive differences, the mean of absolute successive differences, and the coefficient of variation, are the most commonly used. Some detectors base their decisions on just one parameter, combined with simple thresholding, whereas other detectors rely on a combination of parameters as input to the classifier. Certain parameters are intimately related to a statistical test, for example, the number of turning points, and, therefore, the test is described together with the parameter, instead of in Sect. 4.2.6 where different types of classifier are described.

Statistical Dispersion Parameters

The coefficient of variation (CV) of $x(n)$ has been used in AF detection [31, 48], defined by

Table 4.2 List of parameters used in rhythm-based AF detection, grouped into five different categories: statistical dispersion, entropy, symbolic dynamics, Poincaré plot-based, and time-varying coherence function

Detection parameter	Publication
Coefficient of variation	[31, 48]
Root mean square of successive differences	[32, 37]
Normalized mean of absolute successive differences	[48]
Number of turning points	[32]
Histogram-based parameters	[31, 35]
Shannon entropy	[32, 37, 38]
Sample entropy	[34, 49]
Simplified sample entropy	[40]
Symbolic dynamics	[38, 41]
Poincaré plot of $x(n)$ versus $x(n-1)$ + bin count	[50]
Poincaré plot of $\Delta x(n)$ versus $\Delta x(n-1)$ + bin count	[51]
Poincaré plot of $x(n)$ versus $\Delta x(n-1)$ + bin count	[33]
Time-varying coherence function	[37]

$$P_{\mathrm{CV}} = \frac{\sigma_x}{m_x}, \tag{4.2}$$

where m_x and σ_x denote the mean and the standard deviation of $x(n)$, respectively. The parameter P_{CV} describes dispersion but also reflects changes in heart rate since RR interval shortening, often occurring in an AF episode, is related to a smaller m_x. Using $\Delta x(n)$ instead of $x(n)$ in (4.2), the resulting mean $m_{\Delta x}$ becomes close to zero, and, therefore, to avoid division with zero, as well as to maintain the dependence on changes in heart rate, it is substituted by m_x. The performance of two single-parameter detectors, both based on P_{CV} but computed either from $x(n)$ or $\Delta x(n)$, were studied in [31]; the two detectors were found to have about the same performance.

The root mean square of successive differences (RMSSD) is defined by

$$P_{\mathrm{RMSSD}} = \sqrt{\frac{1}{N-1}\sum_{n=1}^{N-1} \Delta x^2(n)}. \tag{4.3}$$

Since this parameter does not reflect changes in heart rate, a heart rate dependent detection threshold can be introduced to implicitly handle such changes [32]. Accordingly, P_{RMSSD} can alternatively be interpreted as a heart rate normalized parameter applied to a fixed threshold test. Thus, the test involving a heart rate normalized P_{RMSSD} is identical to P_{CV}, with the mean and standard deviation of $\Delta x(n)$ inserted in (4.2).

Yet another dispersion parameter is the normalized mean of absolute successive differences (NMASD) [48], defined by

$$P_{\text{NMASD}} = \frac{\dfrac{1}{N-1}\displaystyle\sum_{n=1}^{N-1} |\Delta x(n)|}{m_x}. \tag{4.4}$$

The motivation for using P_{NMASD} instead of P_{CV}, when based on $\Delta x(n)$, is unclear as the former parameter represents an approximation of the latter. Therefore, it is not surprising that the detection performance of P_{NMASD} was found to be almost the same as that of P_{CV} [48].

Thus, it may be concluded that the three dispersion parameters in (4.2)–(4.4) convey similar information. As shown below, yet another detection parameter conveys information on RR interval dispersion, though developed in the context of the Poincaré plot.

Number of Turning Points

The turning point test is a nonparametric, statistical test to determine whether the samples of a time series can be modeled by independent and identically distributed random variables. In a completely random series, any three successive samples are equally likely to occur in any of the six possible orders. In four of the orders, a turning point exists if the middle sample is a local maximum or a local minimum. Thus, the probability of a turning point in a three-sample series is $2/3$.

For a series with N samples, the number of turning points N_{TP} can be counted and compared to the expected number of turning points m_{TP} of a completely random series. If N_{TP} is too many standard deviations σ_{TP} away from m_{TP}, the series cannot be considered as completely random. Making use of the result that the mean and the standard deviation of N_{TP} are given by [52]

$$m_{\text{TP}} = \frac{2(N-2)}{3}, \tag{4.5}$$

$$\sigma_{\text{TP}} = \sqrt{\frac{16N-29}{90}}, \tag{4.6}$$

respectively, and that N_{TP} obeys an asymptotically normal distribution for a sufficiently large N, a two-sided statistical test can be used. When the number of observed turning points falls outside the 95% confidence limits, defined by $m_{\text{TP}} \pm 1.96\sigma_{\text{TP}}$, the hypothesis stating that the series is completely random can be rejected.

In AF detection, the number of observed turning points, together with other parameters, is employed for characterizing RR interval irregularity in AF [32]. Rather than using a statistical test with 95% confidence limits, the limits are determined to optimize detection performance with respect to sensitivity and specificity. When the number of turning points falls outside the optimized limits, the RR interval series is likely to exhibit periodicity, for example, due to respiratory-modulated sinus rhythm. Since it has been shown that RR intervals in AF may exhibit certain correlation [2], the turning point test loses some of its power in detecting random RR interval series. Moreover, the turning point information is likely to cause false alarms in the presence of ectopic beats and rapid changes in rhythm, and, therefore, it is less suitable for AF detection.

Histogram-Based Parameters

Since RR interval histograms determined in sinus rhythm or AF exhibit considerable differences in shape, their shapes have been explored for AF detection. However, to make the histogram approach work, the bins must be sufficiently well-populated so that a histogram can be produced which is representative of the prevailing rhythm. This requirement implies that a large number of RR intervals has to be used for histogram construction—100 beats appears to be a minimum number [31, 35]—which, on the other hand, implies lower accuracy of the estimated onset and end times of an AF episode. If fewer and wider bins are used to allow a shorter window, the histogram becomes increasingly inadequate for discrimination between different types of cardiac rhythms. Therefore, an inherent limitation of histogram-based detection is the need of a long window, which thus precludes the detection of brief episodes.

A straightforward approach to histogram-based detection is to define a set of heuristic features which characterize the histogram, e.g., the height and the number of nonempty bins. Since a histogram in AF is usually much broader in shape than a histogram in sinus rhythm, AF is characterized by a lower height and fewer nonempty bins. If a change in heart rate occurs within the detection window, a histogram in sinus rhythm will broaden and become increasingly similar to the shape of an AF histogram. To some extent, however, this transitional problem can be avoided using the ΔRR interval histogram, since differencing not only removes slow trends present in the RR interval series, but it also makes the histogram span over a smaller range of values.

A more sophisticated approach to histogram-based detection is to compare the RR interval histogram of the detection window with a set of template histograms, stratified according to the mean RR interval length [31].[3] Each template histogram is constructed from all the RR intervals contained in (nonoverlapping) windows with a mean RR interval length falling inside an interval with predefined limits, ranging, for example, from 350–399 to 1100–1149 ms in steps of 50 ms [31]. Windows whose mean length falls outside any of the predefined intervals are discarded from further analysis. The template histograms are constructed prior to detection, preferably from a huge AF database to ensure that the histograms are sufficiently representative of the underlying probability density function (PDF); the same procedure applies to ΔRR intervals.

---

[3]The relationship between histogram shape and mean heart rate has previously been investigated in noninvasive studies on atrioventricular node physiology in AF, leading to the concept of heart rate stratified histograms [53, 54].

In AF detection, the observed RR interval histogram is computed in a sliding detection window and compared to each of the template histograms [31]. For this comparison, the nonparametric Kolmogorov–Smirnov test can be used since it measures the probability of the observed RR intervals being drawn from the same population as the fixed data set, i.e., the RR intervals used for constructing the template histograms [55]. This test involves a statistic defined by the largest distance between the cumulative histogram of the observed data set and the cumulative template histogram, assessing whether the two cumulative histograms are different, see Fig. 4.3. The Kolmogorov–Smirnov test is suitable to use when two cumulative probability distributions differ in a global fashion near the center, but less suitable when the two distributions differ with respect to the number of peaks. For example, the largest distance between a bimodal and a unimodal cumulative probability distribution, both determined in AF, may not be large enough to show that the two data sets come from different populations. In such cases, the Anderson–Darling test is a better choice since it makes use of a weighted sum of the squared deviations between the two cumulative probability distributions, rather than just the largest distance at one single point [55].

Poor performance was reported when the RR series was used as input to the Kolmogorov–Smirnov test, with sensitivity and specificity of 66.3% and 99.0%, respectively [31]. Using instead ΔRR intervals as input, the sensitivity improved dramatically to 94.4%, whereas the decrease in specificity to 97.2% was relatively modest. While the authors did not provide any specific explanation to this improvement, it may be that the use of ΔRR intervals leads to better performance since the related histogram is more unimodal than that of the RR intervals, and therefore better suited for use with the Kolmogorov–Smirnov test.

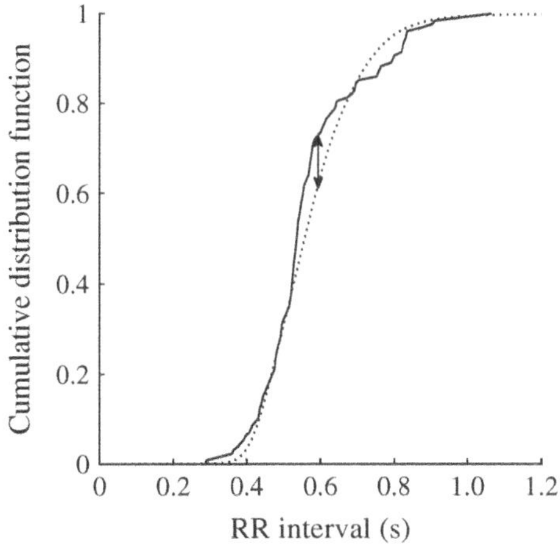

Fig. 4.3 The Kolmogorov–Smirnov test requires that the largest distance between two cumulative histograms is determined. In this example, both histograms belong to RR intervals in AF. The largest distance is marked with an arrow

The multi-template histogram approach offers the advantage of providing a much more detailed characterization of the shape of the RR interval distribution than does the single template histogram in which all RR intervals are merged. On the other hand, it is well-known that the shape of RR interval histograms exhibits considerable intra- as well as inter-patient variability, and unimodal as well as bimodal shapes are often observed in AF [56–58]. Consequently, an AF detector relying on a set of template histograms is likely to perform less satisfactory when these types of variability are pronounced.

Another approach to histogram-based AF detection is to compare two ΔRR interval histograms determined from the first and the last part of the detection window [35], thus replacing the above-mentioned comparison to template histograms. The sum of the squared difference between the corresponding bin counts of the two histograms is used as a detection parameter: this difference remains small as long as the same rhythm persists, but increases when a transition from sinus rhythm to AF occurs, or vice versa. Since the information carried by the squared difference turned out to be insufficient for achieving satisfactory detection performance, the number of nonempty bins, the height of the histogram, and the standard deviation of the ΔRR intervals were also used as detection parameters to improve discrimination between sinus rhythm and AF.

Shannon Entropy

The Shannon entropy quantifies the uncertainty (unpredictability) of the information content of a "message" such as the RR interval series [59]. In statistical terms, the entropy increases as the PDF becomes increasingly uniform, and decreases when the PDF becomes increasingly concentrated around a certain value. In other words, large entropy indicates low predictability of the information content, and vice versa. The Shannon entropy (ShEn) is defined by

$$I_{\mathrm{ShEn}} = -\sum_{i=1}^{B} p(x_i)\log_2(p(x_i)), \tag{4.7}$$

where the message is synonymous to the outcome of a random variable x assuming B different values, i.e., $(x_1, \ldots, x_B)$; the probability of each value is given by $p(x_i)$. Since I_{ShEn} ranges from 0 to $\log_2(B)$, the right hand side of (4.7) is sometimes normalized with $\log_2(B)$ to facilitate interpretation. In practice, the probability $p(x_i)$ is estimated from the message itself, usually by computing the histogram. The probability of the i-th bin is estimated by

$$\hat{p}(x_i) = \frac{N(i)}{N}, \tag{4.8}$$

where $N(i)$ denotes the count of the i-th bin.

The Shannon entropy I_{ShEn} has been considered in AF detection since it is typically much larger in AF than in sinus rhythm [32]. The computation of I_{ShEn} is based on a modified RR interval series in which the longest and the shortest RR intervals are first removed to reduce the influence of outlier values. The histogram is constructed from the remaining RR intervals, with the bins equally spaced over an interval defined by the shortest and longest RR intervals of the modified series. The authors concluded that at least 16 bins should be used to obtain I_{ShEn} with reasonable accuracy.

It has been found that I_{ShEn} is associated with a degradation in performance at higher heart rates, i.e., from about 90 beats per minute (bpm) and higher [49]. This finding can be explained by noting that the probability distribution $\hat{p}(x_i)$ becomes increasingly narrower as the heart rate increases, illustrated by the following example where the variation in heart rate, set to 5 bpm, is identical at different heart rates. For a heart rate of 60 bpm, the RR intervals corresponding to 55 and 65 bpm have the lengths 1090 and 923 ms, respectively, and thus the difference in length is 167 ms. On the other hand, for a heart rate of 120 bpm, the RR intervals corresponding to 115 and 125 bpm have the lengths 521 and 480 ms, respectively, i.e., the difference in length has shrunk to 41 ms. Since I_{ShEn} is computed from the RR intervals, and not from the instantaneous heart rate, it is obvious that the power of I_{ShEn} to discriminate AF from sinus rhythm becomes increasingly worse as the heart rate becomes increasingly higher.

Rather than computing I_{ShEn} directly from the RR interval series, the ΔRR interval series can be mapped to a symbolic series, defined by an alphabet, containing only 10 symbols, which is used for computation of I_{ShEn} [38]. The mapping function quantizes the changes present in the RR interval series by relating the changes to a "reference RR series" resulting from lowpass filtering of the RR interval series. The quantization grid is dynamic in the sense that it is defined by the properties of another, even more lowpass filtered version of the RR interval series; linear, time-invariant lowpass filters are employed, where the lowpass filters are obtained by ad hoc design. The results suggested that the use of symbolic dynamics provides a path to better performance, probably explained by the quantization operation which helps to improve the separation between normal beats and beats in AF when described by I_{ShEn}.

In a subsequent study, bearing considerable resemblance to the one in [38], the authors delved further into the use of symbolic series and Shannon entropy [41]. The main difference between the two detectors is that the instantaneous heart rate is employed, rather than the RR interval series, for generating a symbol series, using a quantization grid with fixed steps. While the authors do not provide any explanation to why the instantaneous heart rate leads to slightly better detection performance, this result seems plausible since the above-mentioned limitation, i.e., when I_{ShEn} is computed from the RR interval series at different heart rates [49], is then sidestepped.

Sample Entropy

While the Shannon entropy is based on the probability of a certain RR interval length to occur, the sample entropy (SampEn) reflects self-similarity of a signal, and therefore used as a measure of signal complexity [60, 61]. The sample entropy is defined as the negative natural logarithm of the conditional probability of a signal repeating itself for m samples within the tolerance r will also repeat itself for $m+1$ samples, where self-matches are excluded [60],

$$I_{\text{SampEn}} = -\ln\left(\frac{B(m+1,r)}{B(m,r)}\right), \tag{4.9}$$

where $B(m,r)$ is the probability of pairs of sequences which match for m samples. A small value of I_{SampEn} indicates that the signal repeats itself and therefore is regular, whereas a large value indicates a complex (irregular) signal. In terms of AF detection, this means that a transition from sinus rhythm to AF is manifested by a considerably increase in I_{SampEn}, and vice versa.

To estimate the probability $B(m,r)$, the RR interval series $x(0),\ldots,x(N-1)$ is first divided into m-length subsequences, described by the vectors

$$\mathbf{x}(i) = \begin{bmatrix} x(i) \\ \vdots \\ x(i+m-1) \end{bmatrix}, \quad i = 0,\ldots,N-m-1. \tag{4.10}$$

Similarity between two subsequences, beginning at i and j, respectively, is measured by the maximum norm, defined by

$$\|\mathbf{x}(i) - \mathbf{x}(j)\|_\infty = \max_{k=0,\ldots,m-1} |x(i+k) - x(j+k)|, \quad i,j = 0,\ldots,N-m-1. \tag{4.11}$$

Two subsequences are considered similar when $\|\mathbf{x}(i) - \mathbf{x}(j)\|_\infty$ is within a fixed tolerance r. Accordingly, the average number of similar subsequences is given by

$$\hat{B}_i(m,r) = \frac{1}{N-m-1} \sum_{j=0, j\neq i}^{N-m-1} H(r - \|\mathbf{x}(i) - \mathbf{x}(j)\|_\infty), \tag{4.12}$$

where self-matches are excluded. The maximum number of similar subsequences is equal to $N-m-1$. The Heaviside step function $H(z)$ is defined by

$$H(z) = \begin{cases} 1, & z \geq 0, \\ 0, & z < 0. \end{cases} \tag{4.13}$$

The probability of two m-length subsequences being similar is estimated by

$$\hat{B}(m, r) = \frac{1}{N-m} \sum_{i=0}^{N-m-1} \hat{B}_i(m, r)$$
$$= \frac{1}{(N-m)(N-m-1)} \sum_{i=0}^{N-m-1} \sum_{j=0, j \neq i}^{N-m-1} H(r - \|\mathbf{x}(i) - \mathbf{x}(j)\|_\infty). \quad (4.14)$$

Since an estimate of $B(m+1, r)$ is required before I_{SampEn} can be computed, (4.11)–(4.14) are also evaluated for $m+1$.

When computing I_{SampEn} in a short window, required for detection of brief AF episodes, the likelihood that none of the few subsequences match is high, especially for a small r. Accordingly, the denominator $\hat{B}(m, r)$ in (4.9) may become zero, leading to that I_{SampEn} is undefined. In order to address this problem, the probabilities in (4.9) can be converted to probability densities by division of the volume of the matching regions [62],

$$-\ln\left(\frac{B(m+1, r)}{(2r)^{m+1}}\right) + \ln\left(\frac{B(m, r)}{(2r)^m}\right) = -\ln\left(\frac{B(m+1, r)}{B(m, r)}\right) + \ln(2r). \quad (4.15)$$

This conversion, serving as a normalization, allows direct comparison of sample entropies computed for different values of r. As a result, the standard approach to selecting r, i.e., an r taken as a fraction of the standard deviation of the input data [60], may be replaced by an approach in which r is data-dependent. The operating value of r is then determined by incrementing r until $B(m, r)$ becomes nonzero; in AF analysis 30 ms has been used as initial value of r, after which r is incremented in steps of 5 ms.

Based on statistical analysis of different RR interval series in AF, it has been observed that the mean RR interval length $\bar{m}_x$ provides predictive information on AF independently of I_{SampEn} [34]. In AF detection, this observation can be accounted for by simply subtracting the logarithm of $\bar{m}_x$ from the expression on the right hand side of (4.15), leading to a new entropy measure, labeled the *coefficient of sample entropy* (CSampEn) and defined by [34]

$$I_{\text{CSampEn}} = I_{\text{SampEn}} + \ln(2r) - \ln(\bar{m}_x). \quad (4.16)$$

The inclusion of $\bar{m}_x$ implies that I_{CSampEn}, as desired, increases in AF when the heart rate is usually higher, whereas it decreases in sinus rhythm when the heart rate is usually lower.

Before I_{CSampEn} can be computed, the subsequence length m needs to be determined. Use of the shortest possible subsequence, i.e., $m = 1$, may be motivated by the observation that the autocorrelation function of RR intervals in AF is essentially zero for nonzero lags [3]. Another, more straightforward motivation is that better detection performance is obtained for $m = 1$ than for a larger m [34]; for additional aspects on the choice of m and r, see Sect. 6.4.4.

It should be pointed out that I_{SampEn} was preceded chronologically by the *approximate entropy* I_{ApEn} [63], defined in the same way as I_{SampEn} except that self-matches are included in (4.12). However, it has been shown that I_{ApEn} is biased, heavily dependent on the number of samples N, and lacks relative consistency [60], and therefore less used than I_{SampEn}.

A variation on I_{SampEn} is the *fuzzy entropy* where the Heaviside function $H(z)$ in (4.13) is replaced by a function which fuzzifies the samples and thereby avoids that similarity of subsequences is treated as either/or [64]. The use of fuzzy entropy has found its way into the analysis of heart rate variability [65] and f wave characterization [66], whereas it remains to be shown whether it can provide better performance in AF detection.

Probability of Pairs of Matching RR Interval Subsequences

A simpler approach to entropy-based AF detection is to only consider the probability $B(m, r)$, forming part of the definition of I_{SampEn} in (4.9) [40, 67]. This approach is advantageous from an implementation viewpoint since $B(m + 1, r)$ is not needed, and neither the ratio of probabilities nor the natural logarithm have to be computed. Another advantage is that the problem of an undefined I_{SampEn} is circumvented. In this approach, the maximum norm in (4.12) is replaced with the Euclidean norm between two m-length subsequences. The following expression is used in place of $\hat{B}(m, r)$ [67],

$$\hat{C}(m, r) = \frac{2}{(N - m)(N - m - 1)} \sum_{i=0}^{N-m-1} \sum_{j=i+1}^{N-m} H(r - \|\mathbf{x}(i) - \mathbf{x}(j)\|), \quad (4.17)$$

where the Euclidean norm is denoted $\|\cdot\|$ and the normalization factor is given by the maximum value of the double sum. The estimator $\hat{C}(m, r)$ differs from $\hat{B}(m, r)$ with respect to the difference between $\mathbf{x}(i)$ and $\mathbf{x}(j)$ which is only counted once in $\hat{C}(m, r)$; self-matches are avoided in both estimators.[4]

An AF detector based on $B(m = 1, r)$ has been proposed in [40], offering the additional implementation advantages that neither the maximization in (4.11) nor the Euclidean distance in (4.17) need to be performed. The probability of two RR intervals differing less than r is estimated by

$$\hat{B}(m = 1, r) = \frac{2}{(N - 1)(N - 2)} \sum_{i=0}^{N-2} \sum_{j=i+1}^{N-1} H(r - |x(i) - x(j)|). \quad (4.18)$$

Before application of a detection threshold, the probability $\hat{B}(m = 1, r)$ is divided by an estimate of the mean length of the RR intervals contained in the detection window

[4]It should be noted that $\hat{C}(m, r)$ constitutes an essential part of the *correlation dimension*, a measure introduced to describe the dimensionality of the space occupied by a set of random samples [68].

to emphasize that AF is usually accompanied by a higher heart rate [40]. Thus, the resulting detection parameter, denoted the *simplified sample entropy* (SSampEn), is defined by

$$I_{\text{SSampEn}} = \frac{\hat{B}(m = 1, r)}{\bar{m}_x}, \tag{4.19}$$

where $\bar{m}_x$ is obtained from exponential averaging of the RR intervals, excluding the RR intervals related to ectopic beats which previously have been flagged by a simple algorithm, see Sect. 4.2.5. The ratio in (4.19) bears considerable resemblance to the coefficient of variation in (4.2), since the numerator is a dispersion measure (though thresholded and therefore not changing in the same continuous way as does the standard deviation in (4.2)) and the denominator is given by the mean of the RR intervals.

Another possible approach to accounting for information on heart rate in (4.18) is to replace the fixed tolerance r with a tolerance defined as a function of the heart rate in the detection window, i.e., $r \to r(\bar{m}_x)$. If a fixed r is still preferred, it can, as already mentioned, be taken as a fraction of the standard deviation determined from a huge data set [60].

4.2.2 *Poincaré-Based Parameters*

The scatter plot of successive pairs of RR intervals $(x(n), x(n+1))$, known as the Poincaré plot, is a simple technique for characterizing different types of cardiac rhythms. This type of plot was introduced for analyzing nonlinear aspects of heart rate variability, constructed from a series of RR intervals spanning over a long time period, i.e., up to several days [69–72]. The Poincaré plot has also served as the guiding design principle when developing AF detectors, but then a much shorter time period determined by the detection window is subject to analysis, i.e., typically ranging from 60 to 120 s. Since the Poincaré plot constructed from the RR intervals in AF is much more scattered than the plot constructed from normal sinus rhythm and atrial or ventricular ectopic rhythms, illustrated in Fig. 4.4, the challenge to be addressed is one of translating the scattering observed in AF to a set of detection parameters. The following two approaches have been pursued:

1. parameters reflecting the density of points in different regions of the Poincaré plot [33, 51, 73], and
2. parameters providing a geometrical characterization of the points in the Poincaré plot [50].

In addition to relying on $(x(n), x(n+1))$ as the basis for producing a Poincaré plot, these two approaches can alternatively rely on $(\Delta x(n), \Delta x(n+1))$ or $(x(n), \Delta x(n))$ which also convey information on beat-to-beat irregularity in the RR interval series.[5]

[5]Yet another approach proposed for characterizing the Poincaré plot is the complex correlation measure, quantifying the point-to-point (temporal) variation of the RR series [74], see also [75].

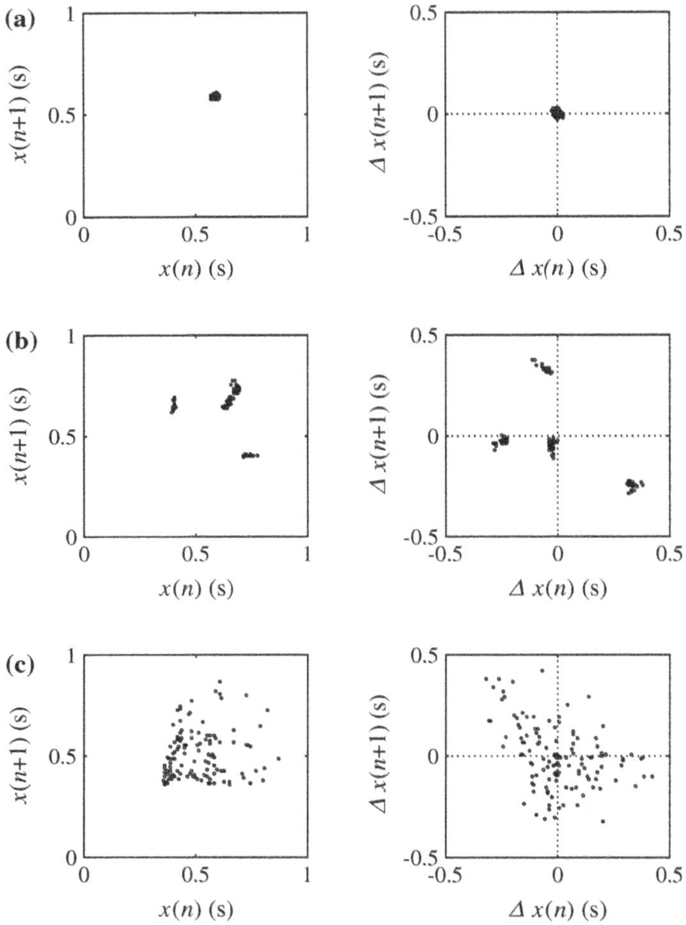

Fig. 4.4 Poincaré plots defined by $(x(n), x(n+1))$ and $(\Delta x(n), \Delta x(n+1))$ (left and right column, respectively), resulting from **a** normal sinus rhythm, **b** sinus rhythm with ectopic beats, and **c** AF. All plots are based on 128 RR intervals

The first AF detector to explore the point density of a Poincaré plot was defined by $(\Delta x(n), \Delta x(n+1))$ [51, 76]. Hence, the proposed analysis is not confined to just the first quadrant, as is the case for $(x(n), x(n+1))$, but covers all four quadrants since $\Delta x(n)$ can assume both positive and negative values. The quadrants are divided into a square grid, where the cells are treated as bins of a two-dimensional histogram; the bin size is a design parameter which should be set to a small value, e.g., 25 ms.

However, this measure has not received any attention in AF detection, probably because it is better suited for discriminating between ectopic rhythms and normal sinus rhythm than between AF and normal sinus rhythm.

Moreover, the Poincaré plot is divided into different regions defined so that their respective populations of points correlate with different rhythms, as manifested by the pattern of the three successive RR intervals required for computing $\Delta x(n)$ and $\Delta x(n+1)$, see Fig. 4.5. First, the total number of bins populated by at least one point ("nonzero bins") is computed for all regions, excluding a circular region enclosing origo which is populated by points related to normal sinus rhythm. Then, the total number of bins is corrected by not only subtracting the number of points in region 0, but also a number reflecting the presence of APBs; APBs tend to cluster in certain regions since they are often accompanied by a compensatory pause. An AF episode is detected whenever the corrected total number of bins exceeds a predefined threshold, provided that the number of points reflecting the presence of atrial tachycardia falls below another predefined threshold. The presence of atrial tachycardia is determined by a heuristic combination of the number of points found in different regions relevant to this particular arrhythmia, see Fig. 4.5; for a detailed description of the algorithm, see [51, 76].

Using the Poincaré plot defined by $(x(n), \Delta x(n))$, a much simpler approach to AF detection has been proposed in [33], particularly well-suited for use in implantable loop recorders. This approach was later applied to AF detection in polysomnographic recordings [73]. In the plot, the first and the fourth quadrants are analyzed since $\Delta x(n)$ can assume both positive and negative values. Again, the two quadrants are divided into a square grid with cells treated as bins. All bins with at least one point are counted, and an AF episode is detected whenever the total count exceeds a predefined, fixed threshold. Obviously, many more bins will be nonzero for an irregular rhythm such as AF than for normal sinus rhythm. In contrast to [51], this approach does not require that the Poincaré plot is divided into different regions, thereby simplifying detector implementation. The count of nonzero bins defines the detection parameter P_{NZPP}.

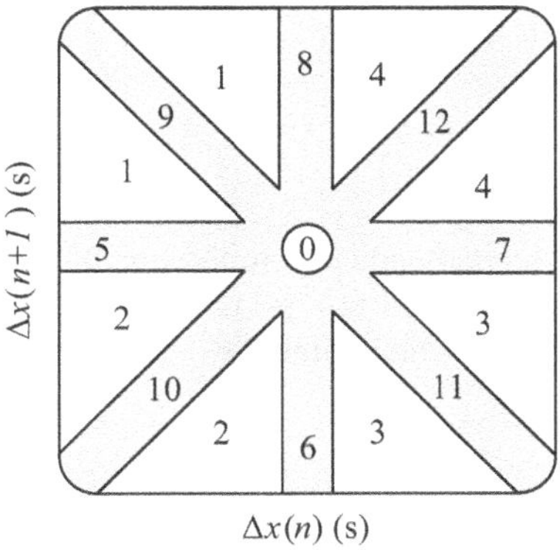

Fig. 4.5 Definition of regions in a Poincaré plot defined by $(\Delta x(n), \Delta x(n+1))$ [51]. Normal sinus rhythm usually populates the circular, origo-centered region 0, whereas AF populates all regions except region 0. Atrial tachycardia usually populates regions 6, 7, 9, and 11, whereas atrial and ventricular premature beats usually populate regions 1–4

As already pointed out, histogram-based detectors suffer from the disadvantage of requiring a large number of RR intervals to achieve adequate performance, especially when a two-dimensional histogram is analyzed. Therefore, it is not surprising that a 2 min detection window is recommended to ensure that the different regions of the Poincaré plot (which may be viewed as a counterpart to histogram bins) are reasonably well-populated [51]. When neither histogram shape nor population size are of importance, a much shorter detection window may be employed, e.g., 64 beats, without having to trade much in performance [33]. The introduction of regions offer, on the other hand, a means to detect other rhythms than AF, e.g., atrial flutter or APBs. It should be pointed out that the relative advantage of using a Poincaré plot defined either by $(x(n), \Delta x(n))$ or $(\Delta x(n), \Delta x(n+1))$, rather than by $(x(n), x(n+1))$, remains to be established.

The second approach to Poincaré-based AF detection involves parameters providing a geometrical characterization of how the points $(x(n), x(n+1))$ populate the plot [50]; see also [77] where some of the original ideas appeared. As will be obvious from the following, detection parameters involving distances in the Poincaré plot are related to the statistical dispersion measures described earlier. Accordingly, the main merit of the Poincaré plot seems to be its use as a conceptual tool for designing parameters, while the plot itself does not provide much novel information. In contrast to the Poincaré-based detector proposed in [51], the geometrical parameters do not treat any particular region of the Poincaré plot as more likely to be populated when AF is present, but simply quantifies certain type of dispersion of the RR interval series.

In normal sinus rhythm, the points of the Poincaré plot are typically dispersed around the line of identity, i.e., $x(n) = x(n+1)$, forming a cluster whose shape resembles an ellipse. One of the axes of the ellipse, usually the major axis, has the same orientation as the line of identity, whereas the other axis is perpendicular. The dispersion of points along these two axes is quantified by first performing a 45° rotation of $(x(n), x(n+1))$, defined by

$$\begin{bmatrix} y(n+1) \\ y(n) \end{bmatrix} = \begin{bmatrix} \sin\frac{\pi}{4} & \cos\frac{\pi}{4} \\ \cos\frac{\pi}{4} & -\sin\frac{\pi}{4} \end{bmatrix} \begin{bmatrix} x(n+1) \\ x(n) \end{bmatrix}, \quad n = 0, \ldots, N-2, \tag{4.20}$$

where $y(n)$ lies on the axis perpendicular to the line of identity. Then, the standard deviations $\sigma_{y,0}$ and $\sigma_{y,1}$ of $y(n)$ and $y(n+1)$, respectively, describe the shape of the cluster. The standard deviations are defined by

$$\sigma_{y,j} = \sqrt{\frac{1}{N-1} \sum_{n=0}^{N-2} (y(n+j) - \bar{m}_y)^2}, \quad j = 0, 1, \tag{4.21}$$

where $\bar{m}_y$ denotes the mean value of $y(n)$. In a broader sense, $\sigma_{y,0}$ may be interpreted as a parameter characterizing the short-term variability of the RR intervals, whereas $\sigma_{y,1}$ characterizes long-term variability [71, 78].

On the other hand, the point distribution in AF differs significantly from that in normal sinus rhythm, implying that the assumption of a cluster with elliptic shape looses its meaning. Still, $\sigma_{y,0}$ has been employed as a detection parameter to quantify short-term variability [50], see also [79], but not $\sigma_{y,1}$ since it reflects a much coarser time scale than does $\sigma_{y,0}$. The transformation in (4.20) implies that successive RR intervals should be differenced,

$$y(n) = \frac{1}{\sqrt{2}}(x(n+1) - x(n)) = \frac{\Delta x(n)}{\sqrt{2}}, \tag{4.22}$$

and, therefore, the mean value of $y(n)$ is close to zero. Hence, the standard deviation $\sigma_{y,0}$ is well-approximated by

$$\sigma_{y,0} \approx \sqrt{\frac{1}{2(N-1)} \sum_{n=1}^{N-1} \Delta x^2(n)}, \tag{4.23}$$

which describes the dispersion of points around the diagonal line in the Poincaré plot. It is evident that $\sigma_{y,0}$, apart from different normalization factors, is identical to P_{RMSSD} in (4.3) and employed in [32] but then without any reference to the Poincaré plot. When distance measures are used for characterizing the plot $(x(n), x(n+1))$, the differenced RR interval series $\Delta x(n)$ is a quantity appearing naturally.

The idea of fitting an ellipse to the Poincaré plot stems from the analysis of long-term ECG data. When adapting this idea to AF detection, the resulting plot must be based on much fewer RR intervals (i.e., only those inside the detection window), leading to that the shape of the Poincaré plot becomes dot-like rather than ellipse-like, see Fig. 4.4. Still, the ellipse-inspired analysis of RR intervals has been considered for AF detection.

Another geometrical detection parameter inspired by the Poincaré plot is based on the Euclidean distance between two successive points $(x(n), x(n+1))$ and $(x(n+1), x(n+2))$, describing the local rate of change in the RR interval series [50]. This parameter, denoted σ_c, is defined as the mean of all Euclidean distances contained in the detection window,

$$\sigma_c = \frac{1}{N-2} \sum_{n=1}^{N-2} \sqrt{\Delta x^2(n) + \Delta x^2(n+1))}, \tag{4.24}$$

$$= \frac{1}{N-2} \sum_{n=1}^{N-2} \sqrt{\sum_{k=0}^{1} \Delta x^2(n+k)}, \tag{4.25}$$

which, similar to $\sigma_{y,0}$, represents a measure of RR interval dispersion. Before use in AF detection, both $\sigma_{y,0}$ and σ_c have been "normalized" by the mean RR interval length $\bar{m}_x$,[6] exemplified by

$$\sigma'_{y,0} = \frac{\sigma_{y,0}}{\bar{m}_x}. \tag{4.26}$$

Thus, similar to the coefficient of sample entropy in (4.16) and the simplified and heart rate modified sample entropy in (4.19), the parameters $\sigma'_{y,0}$ and σ'_c are designed so that an increase in heart rate contributes to improved detection performance.

4.2.3 *Time-Varying Coherence Function*

A linear systems approach to AF detection is provided by exploring the difference in spectral coherence of the RR intervals in two adjacent windows: the spectral coherence remains high as long as normal sinus rhythm is present in both windows, whereas it changes rather abruptly at the time when an AF episode either begins or ends. This approach was proposed in [37], benefitting from previously presented results on how to estimate the time-varying coherence function (TVCF) from the time-varying transfer functions obtained from the samples of two adjacent windows [80].

Assuming that the data in the two windows are viewed as the input and output signals of a linear system, denoted $x(n)$ and $y(n)$, respectively, the time-varying coherence function is defined by

$$C_{xy}(\omega, n) = \frac{|S_{xy}(\omega, n)|^2}{S_x(\omega, n)S_y(\omega, n)}, \tag{4.27}$$

where $S_{xy}(\omega, n)$ is the time-varying cross-spectrum between $x(n)$ and $y(n)$, and $S_x(\omega, n)$ and $S_y(\omega, n)$ are the time-varying spectra of $x(n)$ and $y(n)$, respectively. Conversely, when $y(n)$ is viewed as the input signal and $x(n)$ as the output signal, the time-varying coherence function is defined by

$$C_{yx}(\omega, n) = \frac{|S_{yx}(\omega, n)|^2}{S_x(\omega, n)S_y(\omega, n)}. \tag{4.28}$$

Accounting for the fact that the time-varying coherence function can be computed both forwards and backwards, an overall TVCF can be defined by

$$C^2(\omega, n) = C_{xy}(\omega, n)C_{yx}(\omega, n). \tag{4.29}$$

Introducing the two time-varying transfer functions characterizing the linear system when either $x(n)$ or $y(n)$ is the input signal,

---

[6] In [50], the computation of $\bar{m}_x$ includes all RR intervals in the window except the first and last RR intervals, i.e., $x(0)$ and $x(N-1)$; however, the interpretation of $\bar{m}_x$ is similar to that otherwise used in this chapter.

$$H_{x\to y}(\omega, n) = \frac{S_{xy}(\omega, n)}{S_x(\omega, n)}, \tag{4.30}$$

$$H_{y\to x}(\omega, n) = \frac{S_{yx}(\omega, n)}{S_y(\omega, n)}, \tag{4.31}$$

the overall TVCF can be expressed as [80]

$$C^2(\omega, n) = |H_{x\to y}(\omega, n) H_{y\to x}(\omega, n)|^2. \tag{4.32}$$

The two filters $H_{x\to y}(\omega, n)$ and $H_{y\to x}(\omega, n)$ can be determined using a model-based approach in which the samples of the two windows are assumed to be characterized by an autoregressive moving average (ARMA) model. This approach is preferred over a spectrogram-based approach due to its better frequency resolution, provided that the ARMA model is adequate for the analyzed data. Both the model parameters and the model order are determined using an optimization technique developed especially for the identification of time-varying linear systems [81]. Results have demonstrated that the model order estimate depends on the length of the detection window: longer windows require higher model orders.

Figure 4.6 illustrates one of the essential properties of $C^2(\omega, n)$, namely that the variation across the frequency axis is almost nonexistent in normal sinus rhythm,

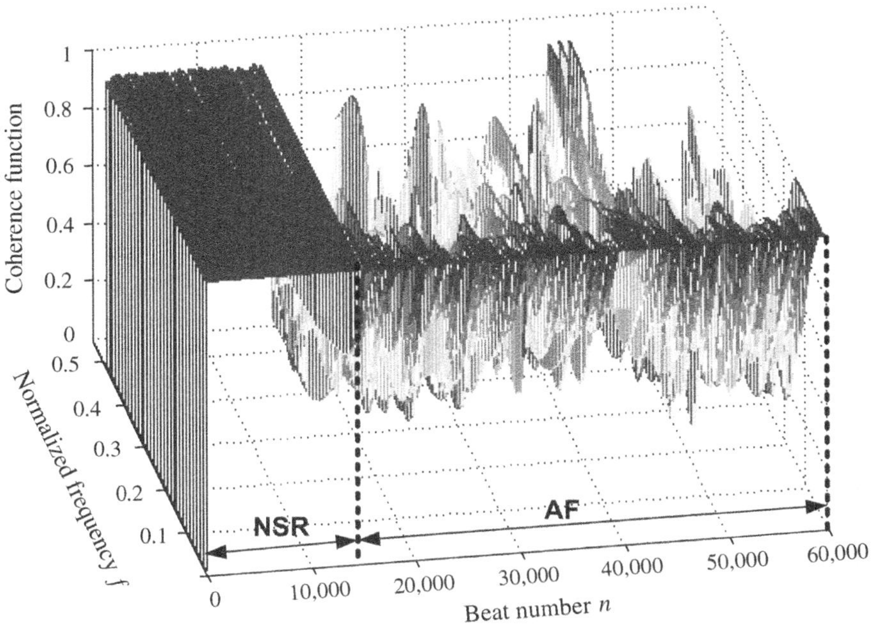

Fig. 4.6 Time-varying coherence function $C^2(\omega, n)$ computed from of an RR interval series containing a transition from normal sinus rhythm (NSR) to AF ($\omega = 2\pi f$). Both detection windows contain 128 beats, and slide with 128 beats at a time. (Reprinted from [37] with permission)

whereas the variation increases at the onset of the AF episode—an increase which becomes more pronounced at higher frequencies. Based on this observation, the variance of $C^2(\omega, n)$ is computed across the frequency axis for each beat n, and used as detection parameter.

4.2.4 Parameter Time Series Exemplified

For an 80-min ambulatory ECG recording with two AF episodes and several runs of ectopic beats, the time series of different detection parameters are displayed in Fig. 4.7. The series are computed using a 128-beat sliding detection window, except for I_{SSampEn} which is computed using an 8-beat window [40]; the window slides one beat at a time.

A number of observations can be made from Fig. 4.7, first and foremost that normal sinus rhythm and AF episodes are easily distinguished in all series. Another observation is that the impact of the runs of ectopic beats, for example, those occurring before the second AF episode, differ quite considerably between the series: while the impact is small for I_{SSampEn}, it is quite substantial for P_{CV} and P_{NMASD} since the ectopic beats are manifested by parameter values which actually exceed those belonging to the AF episodes. Thus, to reduce the number of false alarms, techniques for handling the influence of ectopic beats need to be implemented, see Sect. 4.2.5. Yet another observation to be made from Fig. 4.7 is that I_{ShEn} has more pronounced "background" fluctuations in normal sinus rhythm than the other detection parameters.

4.2.5 Ectopic Beat Handling

An important aspect to address in rhythm-based AF detection is the presence of ectopic beats, often abundant in numbers. The inclusion of a processing block excluding or flagging RR intervals related to VPBs and APBs can, as already pointed out, considerably improve the specificity of a detector. At the same time, ectopic beat handling must not alter the RR intervals which form an AF episode so that the sensitivity is lowered.

In many detectors, no explicit strategy is implemented for handling ectopic beats, but the parameters characterizing rhythm irregularity are fed directly to the classifier, see, e.g., [34, 36, 46–48]. When the ΔRR interval histogram constitutes the basis for detection, rhythms with frequent VPBs are sometimes falsely detected as AF when the Kolmogorov–Smirnov test is involved [31]. The source of the problem is the compensatory pause which accompanies most types of VPB, leading to a negative ΔRR interval immediately followed by a positive. Consequently, the histogram bears resemblance to a histogram determined in AF. It has been noted that the cumulative RR interval histogram determined from rhythms with frequent VPBs exhibits a "prominent shoulder" at around 400–600 ms, while the AF histogram usually does

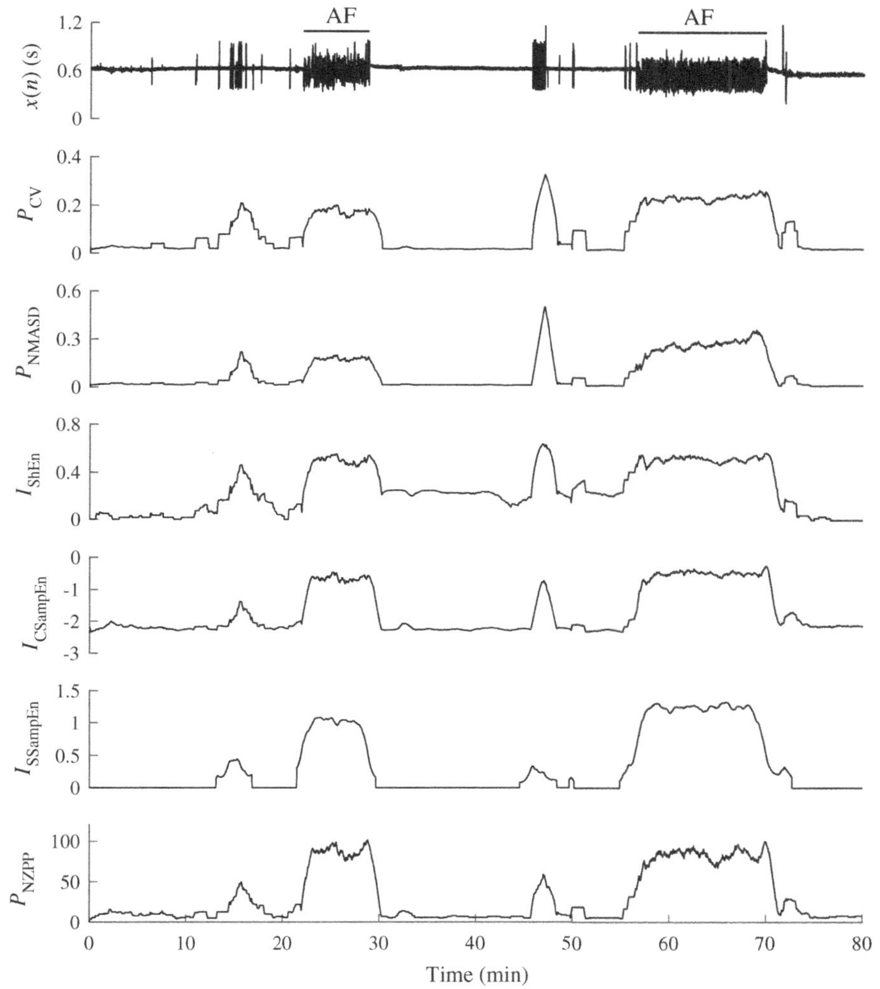

Fig. 4.7 An RR interval series $x(n)$ (top diagram) and related series of different detection parameters: coefficient of variation P_{CV}, normalized mean of absolute successive differences P_{NMASD}, Shannon entropy I_{ShEn} (16 bins, [0.2, 1.7] s, step 0.1 s), coefficient of sample entropy $I_{CSampEn}$ ($m = 1$, $r = 0.03$ s), simplified sample entropy $I_{SSampEn}$ ($r = 0.03$ s), and number of nonzero bins in the Poincaré plot P_{NZPP} (bin size 25 ms)

not [31]. Preliminary results showed that the number of VPB-related false alarms can be reduced by introducing a test on the height and width of a potential shoulder; however, no details have been provided on how to implement a test for identifying a prominent shoulder.

When the Poincaré plot is the starting point for computing a detection parameter, the bin population pattern may be considered for singling out ectopic beats. For example, bigeminy is manifested by clustered points populating just a few bins [33, 51], whereas AF is manifested by points which are much more scattered. When the Poincaré plot is defined by $(x(n), x(n+1))$, changes in heart rate within the detection window smears the clustered points related to ectopic beats, which in turn increases the number of false alarms; this problem is likely to be less pronounced when the plot is defined by $(\Delta x(n), \Delta x(n+1))$.

One of the first rhythm-based detectors to involve handling of ectopic beat was described in [32], see also [37], embracing three different ratio series defined by successive RR intervals. In order to eliminate a VPB, preceded by a short RR interval $x(n)$ and followed by a compensatory pause $x(n+1)$, the following three conditions need to be fulfilled for $x(n)$ and $x(n+1)$ to be excluded from the RR interval series:

$$\frac{x(n)}{x(n-1)} < \gamma_1, \tag{4.33}$$

$$\frac{x(n+1)}{x(n)} > \gamma_{99}, \tag{4.34}$$

$$\frac{x(n+1)}{x(n+2)} > \gamma_{25}. \tag{4.35}$$

The thresholds γ_1, γ_{25}, and γ_{99} denote the 1st, 25th, and 99th percentiles, respectively, of the RR interval ratio histogram of the current detection window. Obviously, these percentiles are increasingly difficult to determine with sufficient reliability as the window becomes shorter. The application of the conditions in (4.33)–(4.35) is illustrated in Fig. 4.8a and b for an RR interval series containing bigeminy and ectopic beats, and then followed by an AF episode. The ectopic beats are eliminated in the thinned output series, whereas the episode of bigeminy is characterized by much flattened RR intervals and reduced irregularity of AF.

Median filtering may be used to eliminate occasional ectopic beats from the RR interval series, while preserving the sharp changes that typically characterize the onset and end of an AF episode. Such filters have been implemented with lengths ranging from 3 [40] to 17 [38], where longer median filters offer better elimination of ectopic beats, but increases the risk of missed brief AF episodes. Therefore, bearing in mind the growing interest in detection of brief episodes, short median filters are to be preferred. Figure 4.8c and d illustrate how the RR interval series is altered when using 3- and 17-point median filters, respectively. The ectopic beats are eliminated in the filtered output, but the episode of bigeminy is largely unaltered and the irregularity of AF is much reduced, especially for the 17-point filter.

In addition to eliminating ectopic beats with median filtering, a set of ad hoc tests, similar to those in (4.33)–(4.35), have been suggested which are also based on the series of ratios of successive RR intervals [35]. The sequence of RR interval ratios is determined for common non-AF arrhythmias, e.g., bi- and trigeminy, and used to build a database with template patterns. The sequence of ratios inside the detection window is correlated to all the template patterns, and the presence of AF is ruled out whenever a sufficient number of correlation matches are found. In this approach,

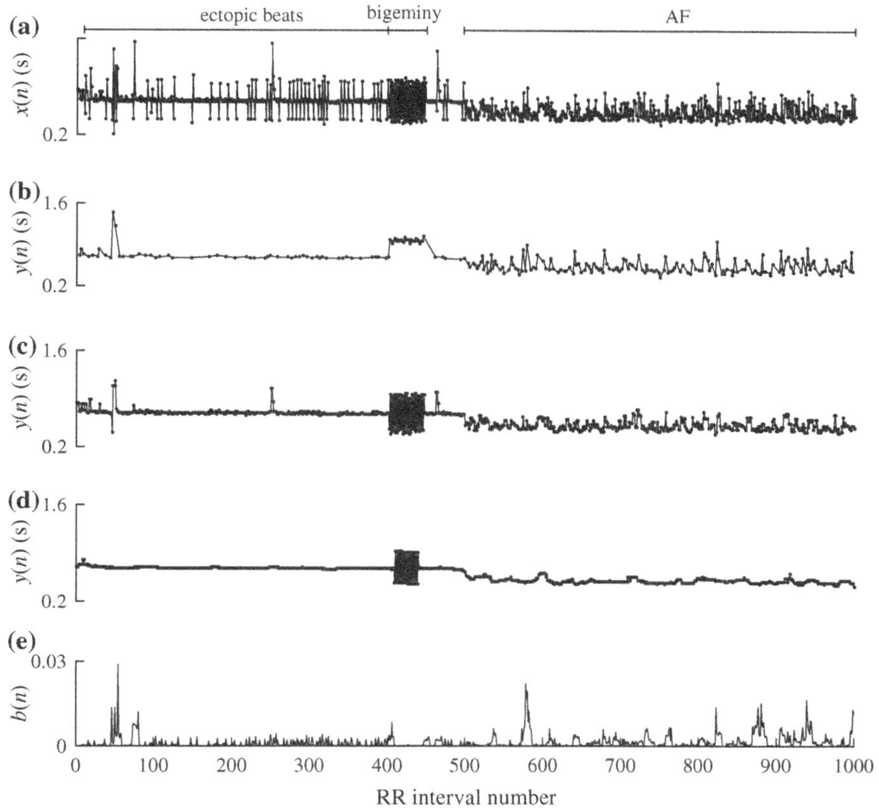

Fig. 4.8 **a** An RR interval series $x(n)$ containing ectopic beats and bigeminy, followed by an AF episode with onset at about interval #500. **b** The output $y(n)$ when applying the three conditions in (4.33)–(4.35) to $x(n)$. **c** The output $y(n)$ from 3-point median filtering and **d** 17-point median filtering. **e** The function $b(n)$ in (4.36), whose only purpose is to flag when bigeminy is present, is computed for $M = 8$; this function does not replace $x(n)$. It should be noted that the output samples in (**b**) are thinned in time compared to $x(n)$, whereas no thinning is introduced in (**c**)–(**e**)

ectopic beat handling is part of the classifier, since no processed RR interval sequence results. Several thresholds need to be set before the tests can be applied—settings whose influence on performance remain to be established.

A simple flag function has been proposed to indicate whether the observed rhythm is likely to be in AF, defined by [40]

$$b(n) = \left(\frac{\sum_{m=0}^{M-1} x_m(n-m)}{\sum_{m=0}^{M-1} x(n-m)} - 1 \right)^2, \quad n = M, \ldots, N-1, \tag{4.36}$$

where n is the end time of the sliding detection window, M is an even-valued integer, and $x_m(n)$ is the output of a three-point median filter. For regular rhythms as well as for bigeminy, the ratio in (4.36) is approximately equal to 1 since $x_m(n)$ and $x(n)$ resemble each other; thus, $b(n)$ is approximately equal to 0. On the other hand, in AF, the variability in $x_m(n)$ is lower than that in $x(n)$ due to the median filtering, and, as a consequence, $b(n)$ increases to indicate AF presence. The squaring operation in (4.36) is introduced to improve the differentiation of AF from non-AF rhythms. In contrast to the criteria in (4.33)–(4.35), resulting in the exclusion of RR intervals, the purpose of $b(n)$ is to serve as a weighting function suitable for use in signal fusion. Figure 4.8e illustrates the behavior of $b(n)$ in the presence of an episode of bigeminy, being flagged by values close to zero.

Given that rhythm-based AF detection is the predominant mode of operation in mHealth monitoring devices and implantable loop recorders, further development of techniques for better handling ectopic beats is warranted.

4.2.6 Classification

The most common approach to designing a classifier is to simply apply one or several threshold tests to the parameters ("features") selected for AF detection. Information on RR interval irregularity is often condensed into one single feature, see, e.g., [31, 33, 34, 38, 40], but as many as nine features, with nine accompanying threshold tests, have also been considered [35]. The threshold values can be determined by optimizing a suitable performance measure, e.g., the area under the receiver operating characteristic (ROC) (Sect. 4.5), with respect to the features of interest using a training data set. The optimized thresholds are then used to evaluate performance on a test data set. Alternatively, the determination of a threshold may be based on some underlying statistical assumptions associated with the feature [31].

When the classifier involves many features, the question arises whether a feature conveys unique information or correlates with the other features. If correlated, which is often the case, the features can be decorrelated using principal component analysis (PCA) so that only the most relevant features are retained, i.e., the dimensionality of the feature vector is reduced. It is well-known that low-dimensional feature vectors generalize better to data not presented during training, thereby leading to more robust detection performance [82]. Another obvious advantage is that fewer features imply less computations. Although feature selection has been considered in AF detection, then involving an improved version of the sequential forward floating selection algorithm [46], this approach has yet to find its way into AF detection on a broader scale.

A simple approach to understanding the relevance of individual features in multi-feature threshold testing is to establish their relative contribution to detection performance, for example, by determining the performance with and without a test involving a certain feature. Such an insight may help to render the detector structure more effective, of particular importance when the detector is aimed at implementation in a low power device. In rhythm-based AF detection, no study has yet reported on the significance of individual tests, whereas one study has presented results on rhythm and morphology based detection, demonstrating that rhythm irregularity plays a more significant role in detection [83].

Another, even simpler, approach to understanding the relevance of a feature is to determine the histograms of the feature for RR intervals observed in either AF or non-AF rhythms, using some suitable database [38, 40, 41, 49]. Then, the extent by which these two histograms overlap serves as a preliminary indication of the feature's discriminatory power. The histograms of different parameters, previously described in this chapter, are presented in Fig. 4.9. Using AFDB, the parameters are computed from the RR intervals contained in a sliding 128-beat window, except I_{SSampEn} which is computed in a sliding 8-beat window. Visual inspection of Fig. 4.9 shows that the least histogram overlap is exhibited by I_{SSampEn}, and therefore this parameter is particularly well-suited for AF detection. Interestingly, the simple-structured feature P_{NZPP}, defined by the number of nonzero bins in the Poincaré plot, is also associated with a small overlap. On the other hand, I_{ShEn} is associated with the largest overlap, thus questioning its suitability for use in AF detection. When the Shannon entropy is computed from a symbolic sequence, determined either from the RR intervals or the instantaneous heart rate, the histogram overlap has been found to decrease, see [38, 41].

In addition to using a traditional classifier defined by a set of threshold tests, pattern classification techniques have been investigated for AF detection, including support vector machines (SVMs) [39, 50, 84] and linear discriminant analysis (LDA) [36]; the former technique has the advantage of offering better flexibility as the decision boundaries can be nonlinear [85]. In these studies, the dimension of the feature vector ranges from 2 to as large as 24. It should be noted that LDA-based classification requires many more computations for training than does simple threshold testing, as the sample mean vector and the covariance matrix for both non-AF and AF data are needed to compute the discriminant function. For SVM, only two design parameters need to be set, both related to the degree with which misclassifications should be penalized [50, 84].

From Table 4.1, it is evident that detection based on a single threshold test offers performance superior to detection based on a classifier incorporating multi-threshold tests or an SVM. For example, the single-test detector in [40] performs better than does the detector using an SVM [39]. At a first glance, this result may stand out as unexpected as an SVM offers so much more freedom with respect to the location of the decision boundaries, and therefore an SVM should perform better. A possible explanation to this result may be that the SVM does not generalize well from training to testing when a small or nonrepresentative training set has been used. A more likely

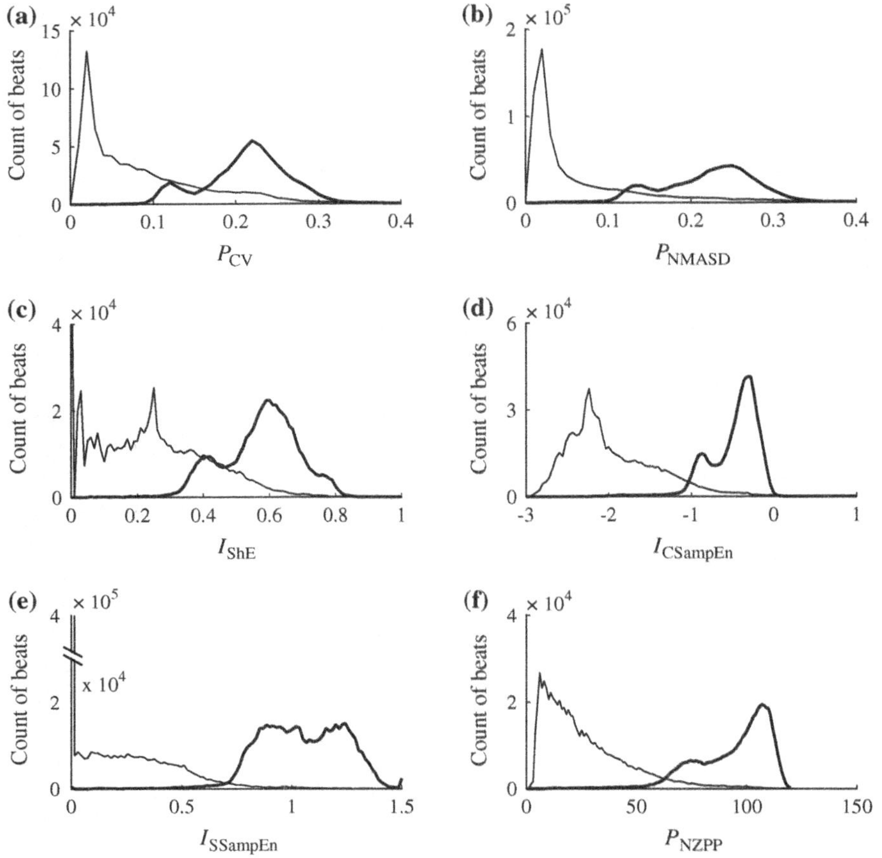

Fig. 4.9 Histograms for six different detection parameters, determined either in AF (thick line) or non-AF (thin line). **a** Coefficient of variation P_{CV}, **b** normalized mean of absolute successive differences P_{NMASD}, **c** Shannon entropy I_{ShEn}, **d** coefficient of sample entropy $I_{CSampEn}$, **e** simplified sample entropy $I_{SSampEn}$, and **f** number of nonzero bins in the Poincaré plot P_{NZPP}. The values used to compute the parameter time series displayed in Fig. 4.7 were also used in this figure

explanation, though unrelated to the SVM, is that less powerful features were used, leading to inadequate handling of non-AF rhythms.

Detectors involving machine learning techniques have yet to demonstrate performance exceeding that of classical threshold-based AF detection. However, this relation may very well change in the future since databases for training are continuously growing—a change which implies time-consuming and meticulous work by expert cardiologists to ensure that the databases are adequately annotated.

None of the above-mentioned approaches to classification offer built-in immunity to non-AF rhythms such as bi- and trigeminy, frequent APBs and VPBs, supraventricular tachycardia, and atrioventricular junctional rhythms, and, therefore, ectopic beat handling prior to classification will have significant repercussions on performance. This aspect is illustrated by considering the performance of the detector in [40] when implemented with and without such handling. In that detector, the fusion of I_{SSampEn}, computed using a sliding 8-beat window, and $b(n)$, indicating the likelihood of AF presence, results in a parameter which is subjected to simple thresholding. Using the MIT–BIH Normal Sinus Rhythm Database (NSRDB), containing several occurrences of bigeminy, cf. Sect. 3.1, the incorporation of $b(n)$ in the detector leads to a dramatic improvement in performance since the specificity increases from 93.2 to 98.6%, whereas the sensitivity remains essentially the same (this is a previously unpublished result).

4.3 Rhythm and Morphology Based AF Detection

Although AF is accompanied by changes in both rhythm and atrial wave morphology, rhythm-based detection continues to be the preferred mode of operation since the RR intervals can be determined much more reliably in noisy signals than information on atrial activity [38, 86]. Since rhythm-based detectors tend to produce false alarms in sinus rhythms with ectopic beats, complete atrioventricular block, as well as in patients with prescribed ventricular rate-controlling medication, it is natural to also analyze whether P waves are absent and/or f waves are present so that the false alarm rate can be reduced. Thus, information on atrial wave morphology needs to be included in the decision process, illustrated by the block diagram of an AF detector in Fig. 4.10a. While the performance of rhythm-based AF detectors is not critically dependent on the lead selected for signal processing, lead selection is crucial when morphologic information is involved since f waves have much lower amplitude in leads positioned farther away from the atria; such lead-dependence is less pronounced for P wave amplitude.

Only a handful of AF detectors have been designed in which information on both rhythm and atrial wave morphology are subject to analysis. The performance reported in the literature must be regarded as rather disappointing since, indeed, none of the detectors achieve performance superior to that of a well-performing rhythm-based detector, see Table 4.3. This result may be explained by the use of detector structures not accounting for the fact that the noise level usually changes over time. As a consequence, measurements characterizing atrial activity are not always reliable, but may actually contribute to worsen the performance rather than to improve it [87]. Hence, an important guiding design principle is to account for the prevailing noise level in the detector structure, implying that information on atrial activity becomes less influential when decisions are made at higher noise levels, and vice versa. Ultimately, when the noise level exceeds a certain threshold, the detector structure should simplify to one based on only the RR interval series, cf. Sect. 4.2. Pursuing the design of a detector accounting for noise calls for the development of a noise level estimator. The noise-dependent mode of operation of an AF detector is described by the block diagram in Fig. 4.10b.

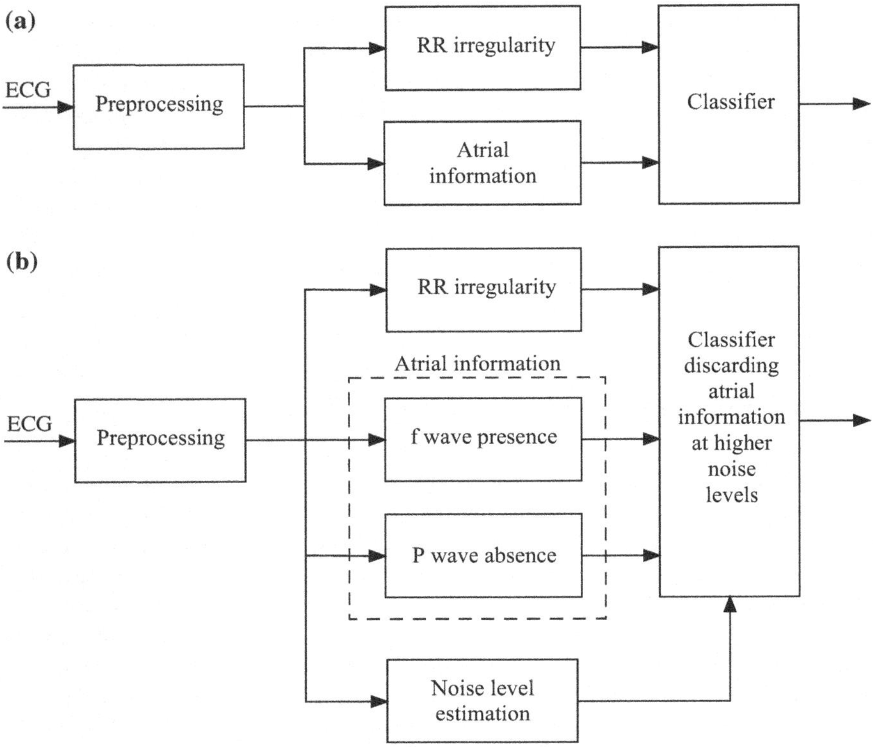

Fig. 4.10 General structure of AF detectors described in the literature. **a** Block diagram of a detector exploring atrial morphology independently of the prevailing noise level. **b** Block diagram of a detector whose classifier is designed to increasingly discard information on atrial wave morphology as the noise level increases

This section provides an overview of the building blocks required for processing information on atrial wave morphology, as well as for estimating the noise level. Some detectors explore information on either P waves or f waves, while others explore both types of waves.

4.3.1 P Wave Detection Information

The problem of P wave detection/delineation has been thoroughly treated in the literature, with emphasis on automated interpretation of diagnostic ECGs where highly accurate measurements of P wave amplitude and duration are of critical importance [91–93]. The prediction of patients prone to AF based on P wave morphology represents another, more immediate application where accurate measurements are

Table 4.3 The performance of five detectors based on both rhythm and morphology, together with the performance figures of rhythm-based detection already presented in Table 4.1. The subset $AFDB_1$ is defined in Table 4.1, $AFDB_2$ is identical to AFDB, except that records 00735 and 03665 are excluded since they do not include ECG signals, only RR interval information, $AFDB_3$ contains only 20 of the 25 records since five records do not have sufficient sinus rhythm data for training, and $AFDB_4$ excludes a huge number of unspecified non-AF segments to balance the sizes of AF and non-AF records. The difficulties associated with comparing detection performance are considered in Sect. 4.6, applying especially to the best-performing detector

Method by	Year	Database	Se (%)	Sp (%)
Dash et al. [32]	2009	$AFDB_1$	94.4	95.1
Lian et al. [33]	2011	AFDB	95.8	96.4
Lake and Moorman [34]	2011	AFDB	91	94
Huang et al. [35]	2011	AFDB	96.1	98.1
Shouldice et al. [36]	2012	AFDB	92	96
Lee et al. [37]	2013	$AFDB_1$	98.2	97.7
Zhou et al. [38]	2014	AFDB	96.9	98.3
Asgari et al. [39]	2015	AFDB	97.0	97.1
Petrėnas et al. [40]	2015	AFDB	97.1	98.3
Zhou et al. [41]	2015	AFDB	97.4	98.4
Babaeizadeh et al. [87]	2009	$AFDB_2$	93	98
Carvalho et al. [83]	2012	$AFDB_2$	93.8	96.1
Ladavich and Ghoraani [88]	2015	$AFDB_3$	98.1	91.7
Ródenas et al. [89]	2015	$AFDB_2$	96.5	94.2
Xia et al. [90]	2018	$AFDB_4$	98.3	98.2

essential [94–96]. In AF detection, however, the demands on accuracy are more relaxed since the absence of P waves can be established without first having to estimate P wave onset and end.

A straightforward approach to determining whether P waves are absent is to use a measure reflecting morphologic similarity between the samples in two consecutive "PR intervals", with the correlation coefficient and the mean square difference as examples of such a measure [87]. In sinus rhythm, P wave morphology is usually stable from one beat to the next, and, therefore, such a measure would indicate a high degree of similarity. In AF, on the other hand, P waves are replaced with f waves which are unsynchronized with the QRS complexes, and, consequently, the degree of similarity between two PR intervals is much lower. Once pairwise comparison has been performed for all beats in the detection window, the average of the resulting similarity measurements can be compared to a threshold to determine whether P waves are absent.

In a related approach, the samples of the PR interval are correlated to the samples of a fixed P wave template [83]. The template is determined by averaging all annotated P waves of a huge annotated database [97, 98]; further considerations on template-based P wave detection can be found in [99]. By analyzing the sequence of correlation coefficients determined from all beats in the detection window, a P wave is detected whenever the correlation coefficient exceeds a fixed threshold. P waves are considered absent when the P wave occurrence ratio, defined as the number of detected P waves to the total number of beats in the window, falls below another fixed threshold.

Rather than quantifying P wave absence directly in the ECG signal, as is usually the case, it can be quantified in a signal resulting from PQRST cancellation of the ECG, thus composed of PQRST-related residuals in normal sinus rhythm and f waves in AF [100]. In this approach, the term "P wave absence" has a different meaning since the input signal no longer contains P waves; however, the term is still useful since an "imaginary" PR interval can be analyzed. It has been shown that PQRST cancellation can be accomplished by means of an echo state network which offers the advantage of handling substantial variation in normal beat morphology as well as the presence of ectopic beats [101]; for a description of the echo state network, see Sect. 5.5.3. In the canceled signal, all possible pairwise combinations of the PR intervals are considered in the detection window, not just the pairs defined by consecutive PR intervals as in [87]. The squared error is computed for pairs of PR intervals, and then averaged over all possible combinations to produce a measure of P wave absence. The PR interval has a fixed location relative to the fiducial point of the QRS complex, with its onset and end preceding the fiducial point by 250 and 50 ms, respectively.

A radically different approach to AF detection is to completely leave out all rhythm information and only explore whether P waves are absent [88, 89].[7] The main motivation for pursuing this approach is that rhythm information may not be discriminative enough to reliably detect AF in patients on rate-controlled medication or with pacemaker, where rhythm irregularity is reduced. It is obvious from Table 4.3 that these two detectors have performance inferior to the best-performing rhythm-based detectors.

As many as nine features have been employed for describing different P wave properties: six features describing P wave amplitude in contiguous 20 ms intervals, and three features describing variance, skewness, and kurtosis of the samples in the PR interval, located, as above, at a fixed distance from the QRS fiducial point [88]. In contrast to the three above-mentioned approaches, which all produce a simple scalar parameter for determining P wave absence, this approach is considerably more complicated as a training phase is required for each patient before AF detection can take place. This phase involves a Gaussian mixture model whose model parameters have to be determined from a half hour long ECG segment containing sinus rhythm; each P wave is represented by the nine-dimensional feature vector. In the testing phase, the Mahalanobi distance between the features of the candidate P wave and the features of the patient-specific P wave model is computed, indicating P wave absence when the distance is sufficiently large.[8]

---

[7] Strictly speaking, this type of detector does not explore both rhythm and morphology. However, since information on P wave absence is still required, the detector is described in this section.

[8] The idea of studying the deviation from "normality", i.e., whether a P wave is absent, is closely related to the concept of novelty detection [102, 103].

The entropy of different scales, *wavelet entropy*, constitutes a set of features explored in AF detection [89]. In order to compute the wavelet entropy, the samples in the TQ interval are first subject to wavelet decomposition [21], resulting in the wavelet coefficients $w_{i,k}$, where i and k denote scale and time, respectively. Then, the relative energy E_i is computed for each scale,

$$E_i = \frac{\sum_{k=0}^{K_i-1} w_{i,k}^2}{\sum_{l=1}^{J}\sum_{k=0}^{K_l-1} w_{l,k}^2}, \quad i = 1, \ldots, J, \tag{4.37}$$

where J denotes the number of scales, and K_l denotes the length of $w_{l,k}$ at scale l. The wavelet entropy is obtained as the Shannon entropy of E_i, cf. (4.7), except that the probabilities $p(x_i)$ are replaced by the relative energies E_i, which, by definition, sum to 1. Statistical analysis of AFDB showed that TQ intervals with P waves were associated with significantly lower wavelet entropies than TQ intervals with f waves. This finding is due to that the relative energy is much more concentrated to one scale for P waves than for f waves.

The variability of the length of the PR interval may serve as an indirect measure of P wave absence [87]. Obviously, this length can only be determined when a P wave is present, requiring that the onset of the P wave and the onset of the QRS complex have first been determined. While PR interval variability is undefined in AF, a surrogate measure may be used in which the onset of an f wave is treated as the onset of a P wave, leading to a PR interval variability which is much larger in AF than in normal sinus rhythm. Considering the imprecise definition of PR interval variability in AF, it is doubtful whether this measure is sufficiently powerful for AF detection.

The above-mentioned techniques for determining P wave absence vary quite substantially in complexity, ranging from simple similarity measures to advanced, statistical modeling of P waves. When a similarity measure is computed between the samples of two PR intervals, e.g., the correlation coefficient or the mean square difference, no particular polarity or morphology of the P wave is favored. This is an important advantage when the objective is to quantify a rather unspecific concept such as "P wave absence." On the other hand, a template-based similarity measure can be expected to perform less well in rhythms with varying P wave morphology, but also for morphologies which are approximately orthogonal (in mathematical terms) to the template, i.e., the correlation coefficient is approximately zero although a P wave is present. Statistical modeling of P wave properties offers more degrees of freedom than the template-based approach, however, such modeling also requires training in each patient on lengthy data which have to be recorded in sinus rhythm; such data is not always is available.

It is obvious that information on P wave absence becomes increasingly unreliable as the noise level increases, eventually reaching a "breakdown" level that differs from one technique to another depending on the robustness of the design. For example, a template-based approach is likely more robust to noise than an approach where P wave onset needs to be determined. In addition, information on P wave absence is more reliable when extracted from more than one lead: by analyzing two leads instead of one, the specificity of a P wave based AF detector has been shown to increase from 91.7 to 94.6%, while the sensitivity remained essentially the same [88].

4.3.2 f Wave Detection Information

The sparse use of f wave information in AF detection is due to the difficulty to reliably characterize low amplitude f waves in the presence of noise, as well as to reliably determine f wave characteristics from the TQ interval. Not only is it challenging to determine the endpoint of the T wave in AF, but the TQ interval becomes increasingly shorter as the heart rate increases. Eventually, the TQ interval may have shrunk to such an extent that the f waves are completely concealed by ventricular activity, thus precluding further analysis. This problem can, however, be addressed by means of f wave extraction—a signal processing operation which is thoroughly reviewed in Chap. 5. While f wave extraction facilitates AF detection, it also increases the complexity of the detector structure so that it may no longer be feasible to implement in a battery-powered device.

Basal time domain information on f wave presence can be obtained by counting the number of f waves in the TQ interval, with f waves considered present whenever the count exceeds one, otherwise absent [104]. The width of a signal fluctuation must exceed a certain threshold to be counted as an f wave; in [104], f wave width is defined as the time elapsed between two level crossings. In order to avoid that noise fluctuations are counted, the amplitude of a fluctuation must exceed an adaptive threshold related to both the amplitude of the TQ interval and the peak amplitude of the T wave. Another means to combat false counts of f waves is to first bandpass filter the observed signal so that baseline wander and noise of muscular origin are reduced. However, even with such filtering, it is well-known that f wave analysis relying on level crossing patterns remains vulnerable to noise since the spectral content of filtered muscle noise overlaps with that of f waves [21]. The consequences of a vanishing TQ interval at higher heart rates, i.e., a count of zero f waves, was not addressed in [104].

Spectral characterization is another approach to determining f wave presence, assuming that f wave extraction is first performed so that all samples in the detection window are suitable for spectral analysis, not just samples in the TQ interval [83, 100]. Since the spectral peak corresponding to the f wave repetition rate (dominant AF frequency, DAF) is typically the largest, parameters describing signal bandwidth have been proposed as a measure of f wave presence. Figure 4.11a illustrates the spectrum of an extracted f wave signal. In this example, the DAF, located at 6 Hz, is the main spectral feature, but important information may also be conveyed by the second and third harmonics, see Sect. 6.3.2. In general, two or more harmonics are more likely to be present in patients with paroxysmal AF than in patients with permanent AF.

The *normalized spectral concentration* is defined by [100], see also [105, 106],

$$F_{\text{SC}} = \int_{\Omega_a} P'_{\hat{d}}(\omega)\, d\omega, \tag{4.38}$$

where $P'_{\hat{d}}(\omega)$ denotes the normalized power spectrum of the extracted f wave signal $\hat{d}(n)$, defined by

$$P'_{\hat{d}}(\omega) = \frac{1}{\sigma^2_{\hat{d}}} P_{\hat{d}}(\omega), \tag{4.39}$$

and $\sigma^2_{\hat{d}}$ the variance of $\hat{d}(n)$. The integration interval Ω_a is centered around the dominant spectral peak located within the interval $[\omega_{a,0}, \omega_{a,1}]$, usually chosen to be [4, 12] Hz. When f waves are present, the spectral concentration is closer to 1, whereas it is closer to 0 when sinus rhythm is present. The power spectrum $P_{\hat{d}}(\omega)$ may be estimated using a nonparametric technique, e.g., Welch's method, or a parametric technique, e.g., Burg's method [107].

Spectral entropy is another parameter used for determining f wave presence [83], defined by

$$F_{\text{SE}} = -\int_{\Omega_a} P'_{\hat{d}}(\omega) \ln(P'_{\hat{d}}(\omega))\, d\omega. \tag{4.40}$$

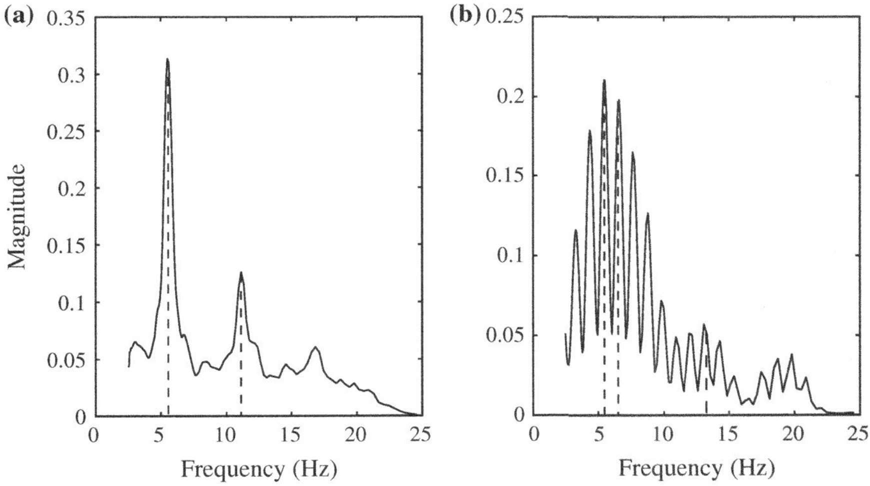

Fig. 4.11 The power spectrum of **a** an extracted f wave signal, and **b** a QRST-cancelled signal observed in sinus rhythm. The two largest spectral peaks are indicated with vertical lines

As the bandwidth of $P'_{\hat{d}}(\omega)$ becomes increasingly narrower in Ω_a, and thus more likely to reflect AF, the spectral entropy becomes increasingly smaller.

Unlike F_{SE}, the *Kullberg–Leibler divergence*, also known as *relative spectral entropy*, accounts for the similarity between $P'_{\hat{d}}(\omega)$ and a template power spectrum $P'_t(\omega)$ [83], defined by

$$F_{\mathrm{KL}} = \int_{\Omega_a} P'_{\hat{d}}(\omega) \ln \left(\frac{P'_{\hat{d}}(\omega)}{P'_t(\omega)} \right) d\omega. \tag{4.41}$$

Ideally, the template power spectrum $P'_t(\omega)$ should be determined so that it is representative of f waves for all patients, e.g., by computing a gross power spectrum from a huge database with high quality f waves. However, not only varies the DAF substantially from patient to patient, but so does f wave morphology. As a result, the practical utility of a template power spectrum is limited, and the information on f wave presence conveyed by F_{KL} can hardly be viewed as representative. In [83], $P'_t(\omega)$ was determined from AFDB and used, in combination with F_{SE}, to decide whether f waves are present. The dominant peak of $P'_t(\omega)$ was found to be located at about 2 Hz, which is far below the expected range of the DAF.

Though not developed specifically for determining f wave presence in AF detection, a set of simple threshold tests have been proposed for judging whether the structure of $P_{\hat{d}}(\omega)$ relates to AF [108]. The tests involve the following ad hoc spectral parameters:

- The SNR, where "signal" is defined as the mean of the magnitudes of the first and second harmonics, and "noise" as the magnitude halfway between the two harmonics.
- The deviation of the second largest peak in $P_{\hat{d}}(\omega)$ from the expected position of the second harmonic, aiming at excluding signal segments with a "ringing" spectrum, e.g., due to P waves occurring at slow rates.
- The ratio between the magnitudes of the second largest and the largest peak in $P_{\hat{d}}(\omega)$, detecting when the second harmonic is too large.
- The squared error between the spectrum of the sliding window and an exponentially averaged spectrum based on past signal segments not containing muscle noise or residuals due to poor f wave extraction.

The spectrum in Fig. 4.11a fulfills the above four tests to be considered an AF spectrum, whereas the spectrum in Fig. 4.11b does not; the test outcome is correct in both cases.

The additional value of including information on atrial wave morphology in AF detection is illustrated in Fig. 4.12, where ECGs with either several APBs or respiratory sinus arrhythmia are analyzed. Using the fuzzy logic detector in [100] which processes information on P wave absence and f wave presence, none of the two non-AF rhythms are detected as AF, whereas both are falsely detected as AF when the coefficient of sample entropy I_{CSampEn} of the RR intervals is used as detection parameter [34]. The decision functions of the two detectors are displayed in Fig. 4.12.

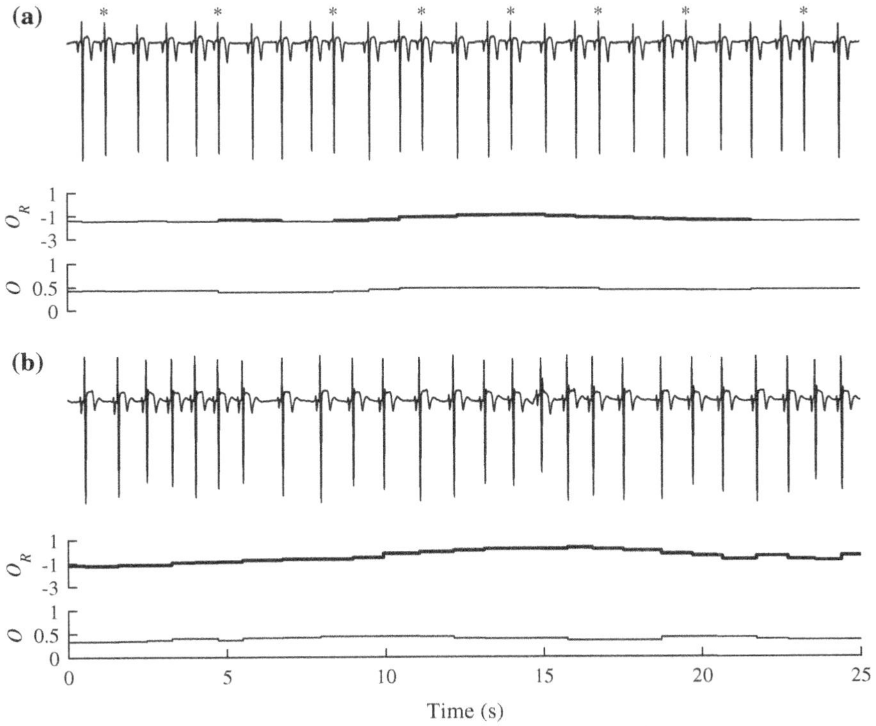

Fig. 4.12 Non-AF arrhythmias causing false alarms in rhythm-based detection, but not in rhythm and morphology based detection: **a** Frequent atrial premature beats (marked with "$*$"), and **b** respiratory sinus arrhythmia. Atrial fibrillation is detected (thicker line) whenever the decision function, denoted O_R for rhythm-based detection [34] and O for rhythm and morphology based detection [100], exceeds the detection threshold

4.3.3 Noise Level Estimation

Although an AF detector must operate at highly varying noise levels, remarkably little attention has been paid to the problem of how to adjusting detector operation relative to such variation. Rather, the observed ECG signal is processed in the same way, irrespective of the prevailing noise level [83, 87, 88]. One explanation to this structural omission may be related to the challenge of how to integrate noise information into the classifier so that information on atrial wave morphology becomes increasingly discarded as the noise level increases, see Fig. 4.10. Another, more fundamental explanation may be related to the development of the noise level estimator itself, which should be designed so that the estimate actually reflects the noise level, but not the cardiac activity.

One of the very few AF detectors operating in a noise-dependent mode was proposed in [100]. In that detector, the extracted f wave signal $\hat{d}(n)$, produced by an echo state network, serves as the starting point for estimating the noise level. The estimator is defined by the root mean square value $R_{\hat{d}}$ of $\hat{d}(n)$, weighted by a ratio of spectral entropies:

$$\hat{N}_{\text{WRMS}} = R_{\hat{d}} \cdot \frac{\displaystyle\int_{\Omega_n} P_{\hat{d}}(\omega) \log_2 P_{\hat{d}}(\omega)\, d\omega}{\displaystyle\int_{\Omega_a} P_{\hat{d}}(\omega) \log_2 P_{\hat{d}}(\omega)\, d\omega}. \tag{4.42}$$

The numerator is computed in a spectral band dominated by noise, defined by $\Omega_n \in [\omega_{n,0}, \omega_{n,1}]$, and the denominator in a spectral band dominated by f waves, cf. (4.38). The definitions of spectral entropy in (4.40) and (4.42) differ with respect to the logarithm—a difference with little importance from a practical viewpoint. The estimator $\hat{N}_{\text{WRMS}}$ produces smaller values when $P_{\hat{d}}(\omega)$ reflects the presence of f waves, but larger values when muscle noise and motion artifacts are present. Figure 4.13 illustrates the estimation of noise level, demonstrating that the estimate tracks the changes in noise level during the last 15 s, while it remains uninfluenced by the f waves of the first AF episode.

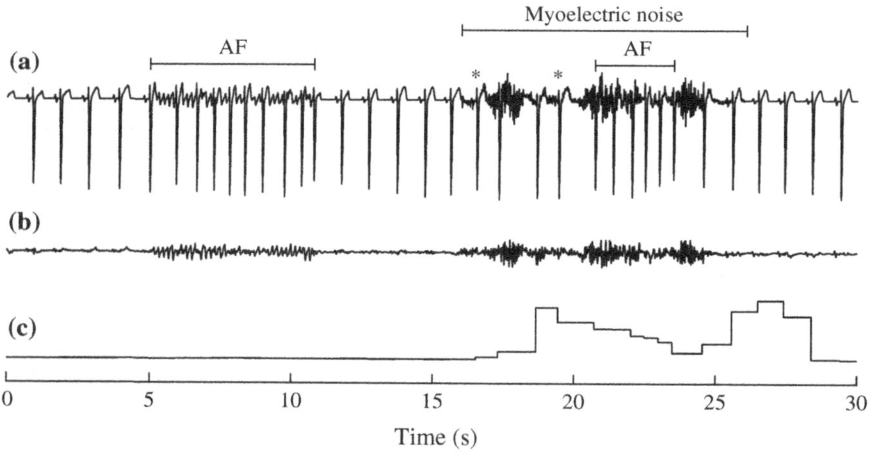

Fig. 4.13 Noise level estimation based on (4.42). **a** The first 15 s of the signal are noise-free, then followed by a 10 s burst of myoelectric noise. The second AF episode is preceded by two atrial premature beats (marked with "∗"). **b** f wave signal extracted using an echo state network. **c** The noise level estimate $\hat{N}$, defined in (4.42), is delayed due to that it is computed in a sliding 5-beat window

The wavelet entropy of the samples in the TQ interval can, in addition to quantifying P wave absence (Sect. 4.3.1), be used as a noise level estimator. While the energy of P waves is mostly confined to one scale, the noise energy is more evenly distributed across the different scales, implying that noise is associated with higher wavelet entropy than P waves. It should be emphasized that the wavelet entropy measures signal organization, and, therefore, contrary to the estimator in (4.42), not proportional to noise level.

If the purpose of the noise level estimator is instead to provide information on whether the RR interval sequence can be reliably analyzed for AF detection, other approaches to noise level estimation may be considered [109–115]. For example, the noise level can be associated with the differences in output from two different QRS detectors, where one is tuned to be more sensitive to noise than the other; large differences in QRS detection then represents an indirect measure of a high noise level [111]. Thus, this type of signal quality index does not have to be integrated into the classifier of the AF detector, but can be treated as independent information indicating whether the samples in the detection window should processed [114]. Given that signal quality assessment is essential for f wave characterization, it is further considered in Sect. 6.5.

4.3.4 Ectopic Beat Handling

Detectors which process information on both rhythm and morphology offer indirect handling of ectopic beats, either through the analysis of P wave absence [87, 88] or the analysis of P wave absence in combination with f wave presence [100]. None of these detectors implement any of the techniques for ectopic beat handling previously described in Sect. 4.2.5 for rhythm-based AF detection. When detection is confined to analysis of P wave absence, the number of false detections due to frequent APBs can be considerably reduced since an APB is preceded by a P wave, on condition that the detector can cope with P wave morphologies that differ from the dominant morphology in normal sinus rhythm [100]. A complication arises, however, when APB prematurity is so pronounced that the P wave is hidden in the preceding T wave, thereby increasing the risk of falsely detecting frequent APBs as AF. In addition, frequent VPBs increase the risk of false detections since VPBs are not preceded by a P wave. Despite these complications, detectors using information on both P wave absence and f wave presence are likely to perform better in ectopic rhythms than would a rhythm-based detector.

If AF detection is implemented in a system for automated ECG analysis, whether for resting or continuous long-term recordings, classification of beat morphology is a built-in functionality which may be utilized for excluding segments with VPBs before AF detection is performed. Such exclusion can also be based on beat classification performed jointly with AF detection [83]. Alternatively, the output of the built-in beat morphology classifier can be used to augment the feature vector created for AF detection.[9]

4.3.5 Classification

The considerations concerning classification in rhythm-based AF detection earlier discussed in Sect. 4.2.6 are equally valid for detection based on both rhythm and atrial wave morphology. With morphologic information included in the feature vector, the noise level should also be included so that the reliability of the parameters describing P wave absence and f wave presence can be assessed by the classifier. However, such an approach has not yet permeated the design of detectors, but classifiers are rather trained on data with considerable variation in noise level, with the objective to produce a fixed classifier suitable for use on data with both low and high noise levels.

One of the very first rhythm and morphology based detectors was described in [87], where the decisions were based on a feature vector composed of one rhythm parameter (the transition probability matrix of a stationary first-order Markov process [42]) and two P wave related parameters (P wave similarity and PR interval variability), see Sect. 4.3.1. A regression decision tree technique was considered for classification, implemented as a series of simple threshold tests, without involving any assumptions on the statistical distribution of the features.

In order to classify more accurately the nine P wave amplitude features described in Sect. 4.3.1, a multivariate mixture model was introduced in [88]. In this model, the features are characterized by a PDF defined as a sum of Gaussians, where each Gaussian is defined by its mean vector and covariance matrix. The model parameters, as well as the number of Gaussians in the sum, are determined by the expectation–maximization algorithm, requiring that a patient-specific training phase is first performed [85]. Once the statistical model has been identified, the likelihood of P wave absence is evaluated for each beat in the detection window. Based on the combined likelihood for all beats in the window, a decision is taken whether an AF episode is present.

The first detector architecture to offer joint processing of features describing rhythm irregularity, P wave absence, as well as f wave presence, was proposed in [83, 121]. A feedforward artificial neural network (ANN) was used as classifier, trained on a subset of records from AFDB.

A comparison of the performance figures listed in Table 4.3 is unfortunately not straightforward since both sensitivity and specificity differ from detector to detector. Nonetheless, the performance figures clearly indicate that detectors based on both

[9]While classification of beat morphology is not reviewed here, it deserves to be mentioned that this classification problem has received, and continues to receive, considerable attention in the literature, see, e.g., [116–120].

rhythm and morphology do not offer performance superior to that of rhythm-based detectors. In fact, the much earlier presented rhythm-based detector in [31] offers better performance than does the detector in [83], where account is made of both P wave and f wave information. This, rather disappointing result may be explained by the use of decision boundaries not adjusted in relation to the prevailing noise level. Interestingly, the authors of [83, 87–89] all point out noise as an important source to performance degradation of their respective detectors, although none of the detectors were designed to account for noise.

In fact, few of the above-mentioned detectors have a structure which lends itself to the handling of noise information. For example, it is unclear how an ANN-based classifier trained on signals with low noise levels generalizes to signals with higher levels. This observation is likely to apply also to classifiers based on a regression decision tree or a Gaussian mixture model.[10]

The first AF detector to account for information on noise level was proposed in [100], having a structure which agrees with that displayed in Fig. 4.10b. The information fed to the classifier consists of four different parameters, describing rhythm irregularity, P wave absence, f wave presence, and noise level as defined by (4.42). The classification is based on a Mamdani-type fuzzy logic in which the four input parameter values are mapped by a membership function to indicate the degree of belonging to a certain fuzzy set. For the parameters describing rhythm irregularity, P wave absence, f wave presence, the fuzzy sets relate to sinus rhythm and AF, whereas the fuzzy set relates to low level and high level for the noise parameter. The fuzzified parameter values are then combined using a set of fuzzy if–then rules, producing an output between 0 and 1 reflecting the likelihood that the detection window contains AF. With simplicity as the guiding star, the fuzzy rules are defined such that more weight is assigned to rhythm irregularity, and less weight to P wave absence and f wave presence, when the noise level is high, and vice versa when the noise level is low [100]. An AF episode is detected whenever the output exceeds a fixed threshold, which, for the example presented in Fig. 4.14 as well as for the overall detector evaluation, was simply set to 0.5.

An important advantage with the fuzzy logic classifier is that no training phase is required. On the other hand, the membership functions and fuzzy rules are defined by a large number of parameters which need to be set to reflect basic knowledge on AF. It should be noted that detector in [100] has not been subject to performance evaluation on AFDB since the method for f wave extraction requires a reference lead with negligible atrial waves which is not available in all recordings of that database.

Another approach to noise-dependent classification is to simply exclude beats whose noise level exceeds a certain fixed threshold [89]. The noise threshold is chosen so that the agreement with manual annotation of noisy beats is optimized. In noisy ECG segments, detector operation is suspended as information on P wave

---

[10]Neither is noise level taken into account in classification of beat morphology, even though it is well-known that certain beats are difficult to cluster due to excessive noise. This problem was indirectly addressed in [120], where an elegant technique based on switching Kalman filters was proposed for detecting “strange” beat morphologies falling outside the well-established clusters.

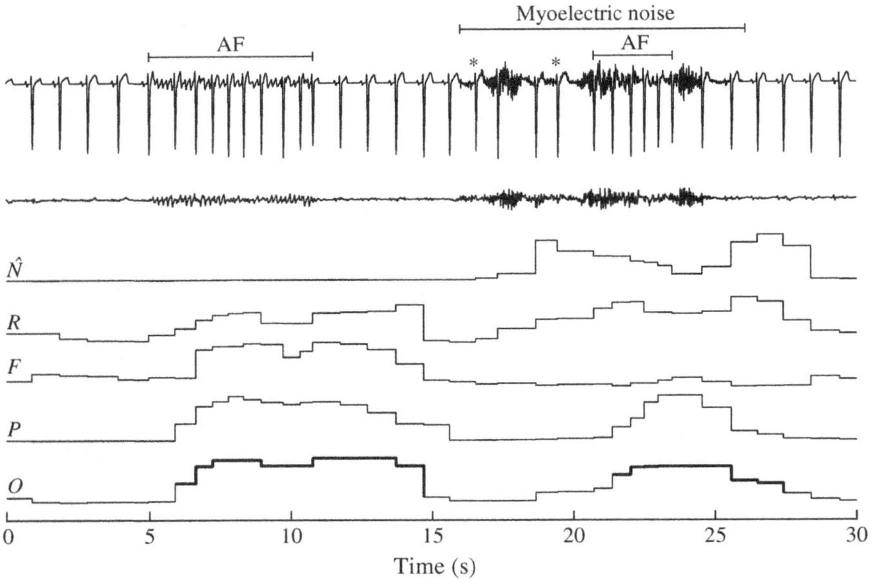

Fig. 4.14 Rhythm and morphology based AF detection using a fuzzy logic classifier. The example in Fig. 4.13 is here extended to also include trends on rhythm irregularity (R), f wave presence (F), and P wave absence (P)

absence cannot be determined. This property stands in contrast to the detector in [100] which continues to operate at higher noise levels, but then "resorting" to information on rhythm irregularity.

It should be noted that the most recent rhythm and morphology based detector listed in Table 4.3 offers slightly better performance than do any of the other detectors. This detector is based on a deep convolution neural network whose input is either the short-term Fourier transform (STFT) or the stationary wavelet transform of consecutive 5 s segments of the ECG signal, i.e., the input signal contains both atrial and ventricular activity [90]. Thus, the design of the detector is not driven by physiology—none of the three properties mentioned in the beginning of this chapter are taken into consideration—but emphasis is given to general ECG properties as well as nonphysiological aspects such as whether color or greyscale should be used to represent the STFT. While this approach to AF detection has potential, the performance figures must be called into question for reasons related to the use of a subset of AFDB in combination with tenfold cross-validation, further discussed in Sect. 4.6.

4.4 Implementation Aspects

When AF detection is to be implemented in a battery-powered, portable device, aspects such as computationally efficient algorithms and minimized memory usage are essential to ensure so that the device can operate continuously over an extended period of time. These requirements become even more crucial when AF detection is to be implemented in an implantable device, for example, a loop recorder. However, details on detector implementation are sparse in the literature, and those which have been published apply to rhythm-based detection where the input data, i.e., the RR series, has a very low rate, thus requiring few computations. On the other hand, for detectors exploring both rhythm and morphology, the input data rate is dramatically higher since the analysis of atrial wave morphology requires that the original ECG samples are available.

Thus, the amount of computations differs vastly between AF detectors, ranging from the simple rhythm-based detector using bin counts of the RR-based Poincaré plot to make decisions [33] to the detector using an echo state network for f wave extraction and fuzzy logic for decision-making [100]. The former detector can be implemented without multiplications, whereas the latter detector requires a huge amount of floating point multiplications as well as much memory to implement the different processing steps. Detailed information on the required amount of computations and memory is lacking for most detectors, with the exception of the rhythm-based detector exploring the combination of symbolic dynamics and the Shannon entropy as detection principle [41]. The computational complexity is analyzed by determining the number of arithmetic operations, shifts, and conditional expressions required per RR interval. Another, much more sweeping approach is to determine the time required by the central processing unit (CPU) and the amount of memory consumed during AF detection [122]. However, figures on CPU time and memory consumption are heavily system-dependent, and, therefore, it is difficult to make a fair comparison to the figures reported in other studies.

Hardware implementation of an invasive AF detector not only must consider requirements on computational complexity, but also energy dissipation when operating in idle and active mode. Idle energy is dominated by the leakage drawn by the memory retaining data, and active energy is minimized by reducing computational complexity. For a rhythm-based AF detector, with its low input data rate, minimization of computational complexity may, in fact, turn out to be less of a concern than minimization of required memory.

The rhythm-based detector in [32], using the number of turning points N_{TP}, the root mean square of successive differences P_{RMSSD}, and the Shannon entropy I_{ShEn} as parameters for characterizing the RR interval series, has been implemented in hardware, resulting in a fabricated application-specific integrated circuit (ASIC) optimized for ultra-low voltage operation [123]. The main reason for choosing the detector in [32] for implementation was that no storage of data was required for online training. It was demonstrated that the three parameters can be efficiently implemented thanks to that resource sharing of arithmetic units reduces the requirements of memory capacity, and that time multiplexing efficiently implements the arithmetic operations required to evaluate the conditions in (4.33)–(4.35) to remove VPBs. A potential AF episode is detected when all three threshold tests are fulfilled, each test involving one parameter. Rather than computing all three parameters first, only the parameter with the lowest cost from an energy consumption perspective is computed and tested. If the test is not fulfilled, the computation of the other parameters is unnecessary, and so on; N_{TP} was found to be the parameter with the lowest cost. The results suggested that the energy required to operate the detector for several years is well within what is provided by the battery of an implantable device [123].

4.5 Performance Measures

The predominant approach to quantifying detection performance is to compare the labels of the detected beats to those of the annotated beats contained in the database—the labels being either AF or non-AF. Such a comparison results in the following four counts,

$$\begin{aligned}
N_{\mathrm{TP}} &= \#\text{beats in AF correctly detected as AF (true positive)},\\
N_{\mathrm{TN}} &= \#\text{beats in non-AF correctly detected as non-AF (true negative)},\\
N_{\mathrm{FP}} &= \#\text{beats in non-AF falsely detected as AF (false positive)},\\
N_{\mathrm{FN}} &= \#\text{beats in AF falsely detected as non-AF (false negative)},
\end{aligned}$$

which are required for computing the two most commonly used performance measures,

$$\text{Sensitivity} = \frac{N_{\mathrm{TP}}}{N_{\mathrm{TP}} + N_{\mathrm{FN}}}, \tag{4.43}$$

$$\text{Specificity} = \frac{N_{\mathrm{TN}}}{N_{\mathrm{FP}} + N_{\mathrm{TN}}}. \tag{4.44}$$

Performance is often studied by displaying sensitivity versus (1−specificity) for different values of a detection threshold, resulting in the ROC [34, 37]. From this curve, the threshold value achieving the desired trade-off between sensitivity and specificity can be chosen. The ROC is sometimes condensed into an overall, scalar measure defined as the area under the curve (AUC), where an area of 1 represents perfect performance and an area of 0.5 random performance. The AUC is considered a robust performance measure because all possible detection thresholds are involved. In AF detection, certain parameter values have been determined by maximizing the AUC [39, 41, 51, 88].

In addition, the following measures have been employed to describe detection performance:

$$\text{Positive predictive value} = \frac{N_{\text{TP}}}{N_{\text{TP}} + N_{\text{FP}}}, \tag{4.45}$$

$$\text{Detection accuracy} = \frac{N_{\text{TP}} + N_{\text{TN}}}{N_{\text{TP}} + N_{\text{FN}} + N_{\text{FP}} + N_{\text{TN}}}. \tag{4.46}$$

It should be noted that detection accuracy should only be used when the two classes AF and non-AF have approximately the same size. Otherwise, Matthews correlation coefficient may be a better choice to evaluate the performance of binary classifiers such as the ones used in AF detection [124, 125].

Sensitivity and specificity based on the counts from a beat-to-beat comparison obviously convey important information on detection performance; however, these two measures also suffer from the disadvantage of not reflecting the episodic nature of paroxysmal AF. This is illustrated by the following scenario where an ECG recording is assumed to contain two AF episodes, one hour-long and another just 10-beat-long. The detector correctly identifies the long episode, but misses the brief one—a likely scenario given that the window length of most AF detectors precludes the detection of a 10-beat episode. The change in sensitivity due to a missed, brief episode is negligible, and illustrates that performance measures based on a beat-to-beat comparison tend to gloss over when brief episodes are missed. Accordingly, valuable clinical information may be lost. A similar glossing takes place in situations when numerous brief episodes are falsely detected, although the corresponding ROC still indicates almost perfect performance; this drawback is illustrated by the example in Fig. 4.15.

A kindred solution would be to replace the beat-to-beat comparison with an episode-to-episode comparison. Such a replacement will, however, raise a number of questions which need to be resolved: What is the meaning of "true negative" in episode-based detection? To what extent must the detected episode overlap with the annotated episode to be treated as a correct detection? Should a minimum duration be imposed on a detected episode to avoid that single beats, falsely labeled as AF beats, are counted as AF episodes?

Inspired by the work in [126] on performance measures appropriate for evaluating the detection of transient ischemia in long-term ECG recordings, these questions have been discussed in the context of AF detection [36]. Since an episode of non-AF beats has little meaning, the number of true negatives N_{TN} is undefined, and, therefore, only sensitivity and positive predictive value can be computed, requiring that the following, redefined counts are determined:

$$N_{\text{TP}} = \#\text{AF episodes correctly detected as AF episodes (true positive)},$$
$$N_{\text{FP}} = \#\text{non-AF episodes falsely detected as AF episodes (false positive)},$$
$$N_{\text{FN}} = \#\text{AF episodes falsely detected as non-AF episodes (false negative)}.$$

An episode is judged as correctly detected if it overlaps the annotated episode with at least 50%, otherwise the episode is labeled non-AF [33]. While the minimum duration of a detected episode not necessarily has to be stated, it is indirectly determined by

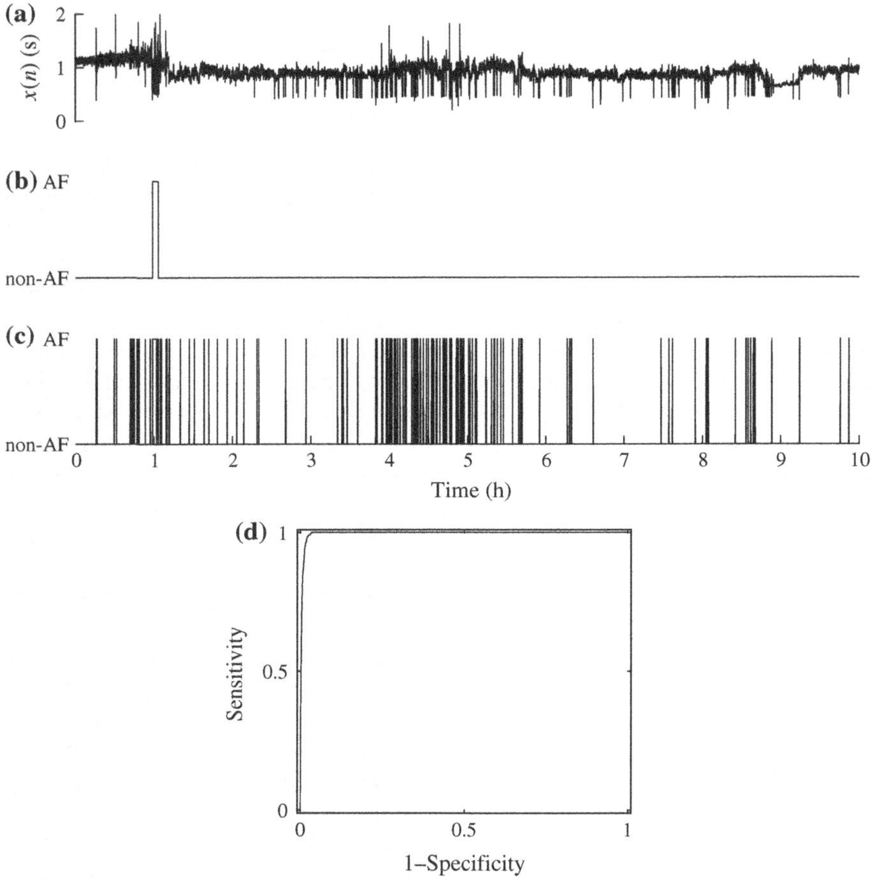

Fig. 4.15 **a** An RR interval series $x(n)$ and **b** AF episode annotation. **c** Output from a detector based on the coefficient of sample entropy (computed in a 12-beat window), and **d** related ROC. The detector correctly identifies the single AF episode, but also produces numerous false detections due to the presence of ectopic beats. Still, the corresponding ROC indicates that almost perfect detection performance is achieved

the choice of window length. For a 100-beat window, the beat-based sensitivity of 0.92, reported in [36] and listed in Table 4.1, dropped to 0.71 when episode-based sensitivity was considered instead. This drop in sensitivity illustrates that the use of a 100-beat window precludes the detection of brief episodes.

Episode-based performance measures have not yet gained a foothold in the literature on AF detection, although such measures provide information which is complementary to beat-based measures. The popularity of beat-based measures may be due to their ease of computation, but also to the many ECG applications where beat-based performance measures have become well-established. However, neither beat-based nor episode-based measures provide information on the detectability of episodes with varying lengths.

The delay between the annotated onset of the episode and the onset produced by the detector represents another type of performance measure which has received attention in the literature [32, 35, 39, 89]. From an algorithmic viewpoint, the time delay introduced by the detector needs to be established to make a comparison with episode onset/end annotations meaningful. From a clinical viewpoint, however, a short time delay is of subordinate importance to the above-mentioned performance measures, since very few ECG applications call for immediate action after the initiation of an episode.

4.6 Detection Performance

4.6.1 ECG Databases

Detection performance is commonly evaluated on one or several publicly available, annotated databases of long-term ECG recordings, where AFDB holds the position as the most popular database. While the availability of public databases certainly facilitates the comparison of performance, conclusions drawn from the performance figures presented in Table 4.3, or the tables presented in e.g., [86, 89, 127], should be made with caution for a number of reasons. Since both specificity and sensitivity differ from one detector to another, performance is not easily compared. Better, though not perfect, is to first compute the ROC for each detector, and then determine the sensitivity at a fixed specificity, or vice versa, which leads to a more relevant comparison.

Another complicating factor is that detection performance is not always established from the analysis of the entire AFDB, but from a subset of varying size. In some studies, records 4936 and 5091 were omitted, since the annotations were deemed to be incorrect ($AFDB_1$) [32, 37]. While rhythm-based detectors can analyze all 25 records of the AFDB, only 23 records can be analyzed by detectors based on rhythm and morphology since two records lack the original ECG signals ($AFDB_2$). Yet another complicating factor is that certain detectors require a minimum length of normal sinus rhythm to fulfill detector training, in one case leading to the exclusion of as many as 5 out of the 25 records ($AFDB_3$) [88]. Moreover, in detector training, it is highly desirable to analyze data sets containing AF and non-AF segments which are balanced in size. A straightforward approach to handling the fact that AFDB contains about 80% more non-AF segments than AF segments is therefore to discard the excess amount of non-AF segments ($AFDB_4$) [90]; unfortunately, the non-AF data set cannot be reproduced in other studies since the segments were randomly excluded. However, such a drastic exclusion of data precludes any meaningful comparison of detection performance—a fact which should be kept in mind when assessing the results in Table 4.3.

From a comparative perspective, the picture becomes even more complicated when performance is evaluated on tiny subsets of RR interval series [67] or beats [127], excerpted from the records in AFDB. Such data excerption not only tends to exaggerate performance figures due to inclusion of better-than-average data quality, but the reproduction of results is not possible due to the lack of detail on what data were actually excerpted.

Other public databases have been analyzed to provide a more complete description of detection performance, notably NSRDB, MIT–BIH Arrhythmia Database (MITDB), and Long-Term AF Database (LTAFDB) [98], see Sect. 3.1. Since NSRDB contains no significant arrhythmias, it can only provide information on specificity, e.g., [32, 33, 35, 37, 41, 51]. The MITDB contains several types of arrhythmia, including AF and atrial flutter, and may be used to evaluate both specificity and sensitivity [32, 33, 37, 41]; however, as pointed out in Sect. 3.1, MITDB contains relatively few AF episodes, and, consequently, performance figures describing episode detection are not representative. The LTAFDB, containing many more and much longer ECG recordings than AFDB, is well-suited for performance evaluation, though not very often used [41].

Some studies involve proprietary ECG databases, acquired to strengthen the results obtained on public databases [37], or used for classifier training [100]. Another reason for acquiring a database is that public databases do not always account for the signal characteristics pertinent to the application of interest.

4.6.2 Training and Evaluation

Widely different approaches have been considered for classifier training and performance evaluation—an observation illustrated by the way different data sets are handled by the detectors listed in Table 4.3. In some studies, either a proprietary database or LTAFDB were used for training, accompanied by a performance evaluation on AFDB [38, 40, 41, 87]. Such an approach is preferred since it avoids that the same patients are used for both training and evaluation. In other studies, no information is provided on the data set used for training [33, 37], whereas AFDB or some other databases is used for evaluation.

With respect to training, AFDB has been used to determine optimal detection thresholds [32, 34], or to select an optimal set of features for classification [36], accompanied by performance evaluation on other databases. Although the results from evaluation are the important ones in these studies, the positively biased results obtained from training on AFDB were also reported. Later on, these results have been included in comparisons of detector performance [38, 88, 89, 128], although the figures are not fully representative. This observation applies even more to the results reported in [35], where AFDB was used for both detector development and evaluation.

In an effort to reduce positive bias, AFDB can be partitioned into different subsets, one for training and another for performance evaluation. The subsets have been formed either by random selection of non-AF/AF segments, division into disjoint subsets of equal size for use in stratified twofold cross-validation [39], or tenfold cross-validation [90]. A small subset of AFDB was used in [83] for training, whereas the entire AFDB was used for evaluation. One of the subsets was used for training in [39] and the other for evaluation, followed by reverse use of the two subsets; the results from the two evaluations were then averaged to yield the overall performance. It is highly questionable whether the performance figures of cross-validation on AFDB can be compared to those obtained for a detector which have been trained on a separate database, especially when considering that AFDB only contains 25 patients [129].

The above-mentioned approaches to training and evaluation are population-based, however, patient-based training may be pursued as well [88]. For each patient in AFDB, detector training was based on the initial part of the ECG record, whereas evaluation was based on the remaining part. However, before training, all beats with "irregularities" were excluded from the training data set using manual review, introducing positive bias in the results. In addition, the practical use of the detector is limited since good-quality signals are not always available for training, nor is manual review prior to AF analysis feasible in clinical routine.

Based on the above considerations, it is evident that a comparison of detection performance is seriously challenged by the presence of positive bias. Independent data sets for training and evaluation should ideally be analyzed, however, not uncommonly, the same patient is part of both data sets. Therefore, as already pointed out, caution should be exercised when comparing detection performance, e.g., with respect to sensitivity and specificity as in Table 4.3.

It deserves to be noted that AF detectors using adaptive filtering for f wave extraction, such as the one in [100], cannot be trained and evaluated on AFDB since none of the two leads is appropriate for use as a reference lead, i.e., none of the leads contains negligible atrial activity. This problem may be addressed using a proprietary multi-lead database for training, and simulated multi-lead signals for performance evaluation [100].

4.6.3 Simulated ECG Signals

Performance evaluation is typically based on real ECG signals annotated with respect to the onset and end of AF episodes, whereas simulated ECG signals are rarely used. This stands in contrast to the evaluation of f wave extraction performance, where simulated signals are frequently used—the main reason being that manual annotations are irrelevant in f wave extraction. Nonetheless, simulated ECG signals have a place in AF detection since certain properties of clinical or technical significance, e.g., atrial ectopy, episode duration, and noise level, can be easily controlled in such signals, whereas public databases may not allow adequate investigation of these properties.

Detection accuracy has been investigated on simulated signals with different noise levels, both with and without the presence of APBs [100]. The results put spotlight on the importance of proper handling of APBs, and show that detection based on both rhythm and morphology provides much higher accuracy than does rhythm-based detection in the presence of APBs, especially at lower noise levels where P wave absence and f wave presence can be reliably estimated. A similar relationship exists between detection accuracy and episode duration, i.e., detection based on both rhythm and morphology provides much higher accuracy in finding brief episodes of varying duration than does rhythm-based detection (5, 10, 20, and 30-beat duration were investigated).

Simulated ECG signals can also serve as a means to establish the SNR below which AF detector operation no longer is recommended. In one of the few studies to address this issue, simulated muscle noise was added to real ECGs, contained in LTAFDB, at different SNRs [114]. The noisy ECG signals were then used to evaluate the influence of noise on QRS detection, as well as on rhythm-based AF detection. The results suggested an essentially linear reduction in AF detection accuracy with respect to SNR when expressed in terms of decibels. The evaluation of performance in noise is even more important for AF detectors analyzing both rhythm and morphology.

4.6.4 Brief AF Episodes

Despite the clinical interest in occult PAF and related risk of future stroke, little attention has been paid to the detection of brief AF episodes. Although AFDB contains a few brief episodes, it is completely dominated by long episodes, cf. Fig. 3.1b, so that missed brief episodes have little influence on beat-based performance measures. Interestingly, some studies report the number of missed brief episodes: 30 out of the 254 episodes in $AFDB_1$ were missed by the detector in [32], all missed episodes having a duration less than 75 beats. In another study [35], 32 out of the 299 episodes in AFDB were missed, the main reason again being missed brief episodes (durations from 4 to 62 beats).

Indirect evaluation of performance with respect to brief episodes can be accomplished by analyzing the influence of different lengths of the detection window on performance. The window length imposes a minimum duration on AF episode detectability. While the exact relationship between window length and episode duration depends on the detection principle used, an episode with a duration of about half the window length or shorter will, in general, be missed, illustrated in Fig. 4.16. The choice of window length is a trade-off: a shorter window facilitates the detection of brief AF episodes, whereas a longer window implies more reliable parameter estimates (assuming that the window contains the same rhythm), but also a larger amount of computations.

Over the years, the trend has been to design detectors with increasingly shorter windows, primarily motivated by the wish to reduce the time to decision [32, 89] and the amount of computations [84]. The recommended window length in rhythm-

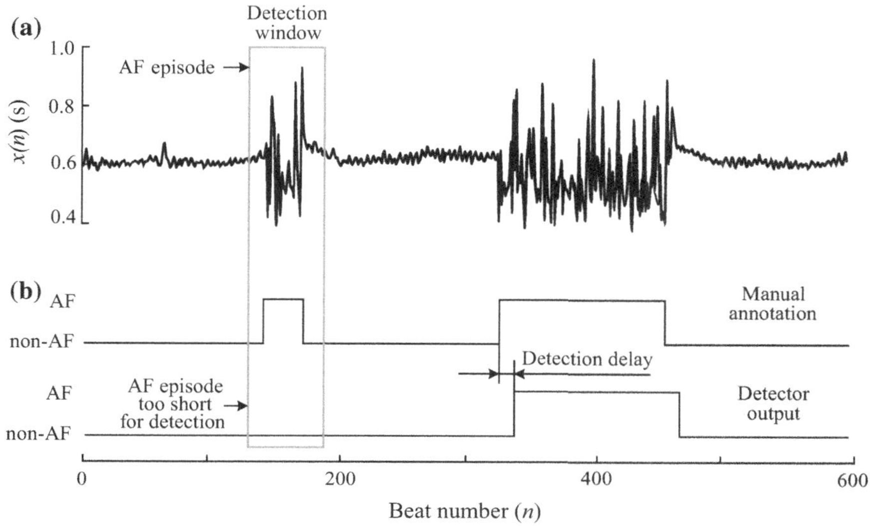

Fig. 4.16 **a** RR series from record 4043 of AFDB containing a brief AF episode (20 beats) and a longer AF episode. The sliding detection window, displayed as a box, is too wide to allow detection of the first episode. **b** Annotation of the RR interval series and detector output. The delay in detecting the second AF episode is indicated. (Reprinted from [32] with permission)

based detection has decreased from 180 s in 1992 [130] to just 8 beats in 2015 [40], whereas, for rhythm and morphology based detectors, even shorter window lengths has been considered, i.e., 5 beats [100].

The degradation in performance when using a short window is well-illustrated by the detector based on the time-varying coherence function [37], briefly described in Sect. 4.2.3. Using a 128-beat window, sensitivity of 98.2% and specificity of 97.7% were obtained on $AFDB_1$, see Table 4.1. Using instead a 32-beat window, the sensitivity and specificity dropped to 96.7% and 96.1%, respectively. Despite the degradation in performance, the authors concluded that a shorter window is still of interest, since it will likely provide a more accurate description of AF burden. For the simple-structured detector exploring the distribution of the Poincaré point population [33], the use of a 128-beat window resulted in a sensitivity of 95.9% and a specificity of 95.4%, dropping to 94.4% and 92.6%, respectively, for a 32-beat window.

Alternatively, direct evaluation of performance can be accomplished by means of simulated ECG signals in paroxysmal AF, where episode duration is controlled by a set of model parameters [40]. The direct approach to evaluation is illustrated in Fig. 4.17, where detection accuracy is presented as a function of median episode duration, denoted T_E, for two different AF detectors. The simulated signals are produced by the model described in Sect. 3.3, and constructed from either synthetic or real components. Figure 4.17 underlines not only the expected result that shorter episodes imply decreased detection accuracy, but it also demonstrates that a detector

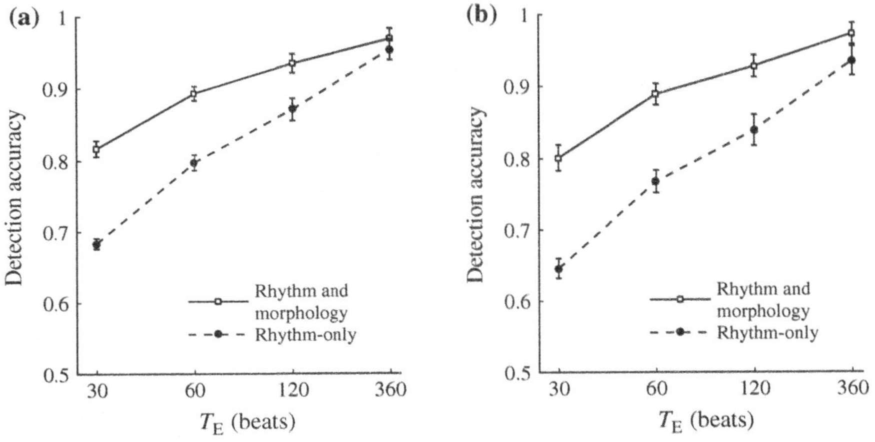

Fig. 4.17 Detection accuracy as a function of median episode duration T_E when the ECG signals are generated using **a** synthetic and **b** real components. The noise level is set to 20 μV RMS. The rhythm-based detector is described in [40], and the detector based on both rhythm and morphology in [100]

based on rhythm and morphology performs better than a detector based on rhythm-only; the difference in performance increases as T_E becomes increasingly shorter. Comparing Fig. 4.17a and b, it is obvious that detection accuracy is essentially independent of whether synthetic or real components are used to produce the simulated ECG. However, as T_E becomes increasingly shorter, the difference in performance between the detector based on rhythm and morphology and the detector based on rhythm-only becomes increasingly larger for real components than for synthetic components. This drop in performance is likely explained by the pathological rhythms present in the database from which the RR interval series were extracted.

4.7 Additional Detector Information

Certain ECG signal properties have been explored for the purpose of predicting either the onset or the end of an AF episode. Similar to heart rate, the properties are not of immediate importance to detector design, but may be integrated in the detector, for example, using a threshold whose level is adjusted in relation to the proneness with which a transition occurs from sinus rhythm to AF, or vice versa. Whether such integration improves detection performance remains to be demonstrated. Considering that more than 90% of all AF episodes are triggered by APBs [131–135], successful prediction of AF onset can be accomplished with a simple test on whether the rate of APBs, not followed by a regular RR interval, increases. This test is combined with other tests on runs of atrial bigeminy/trigeminy and the duration of short runs of paroxysmal atrial tachycardia [136].

Another approach to predicting the onset of paroxysmal AF is to analyze changes in heart rate variability (HRV) which may precede an AF episode. Indeed, in many patients, AF onset is immediately preceded by a significant reduction of the ratio between the low and the high frequency HRV components [137–139], a pattern which is not detectable after spontaneous recovery of sinus rhythm [140]. Alternatively, changes in HRV may be characterized by entropy, with results suggesting that AF onset is preceded by reduced complexity of the RR intervals [141], see also [142]. In yet another approach, AF onset could be predicted by combining spectral, bispectral, and nonlinear features, using a machine learning technique for classification of the preceding HRV pattern [143]. For the above-mentioned studies on APB- and HRV-based prediction, a 30-min segment immediately preceding AF onset is usually considered for evaluating prediction performance.

Different P wave properties have been explored for predicting AF. For example, changes in P wave morphology due to abnormal interatrial conduction are observed in patients bound to develop AF [144, 145], prolongation of the maximum P wave duration may predict recurrent AF [146–148], as well as shortening of the minimum P wave duration [147, 149]. Moreover, changes in the dynamics of P wave morphology may predict AF onset [96, 148]. However, changes in P wave properties occur over a much longer time frame than changes associated with APB- and HRV-based prediction: the former type of changes occurs over weeks to months, whereas the latter over minutes. Hence, information on P wave related changes are less useful in AF detection.

The prediction of AF termination takes its starting point in the analysis of f wave properties, and typically requires that the ventricular activity has been cancelled before prediction can take place. Among the properties explored, the DAF has been found to exhibit gradual slowing just before termination of paroxysmal or persistent AF [150, 151]. Results from studying the wavelet entropy of f waves, employed as a measure reflecting unpredictability in time as well as frequency, suggest that f waves are characterized by decreasing entropy as the termination is approaching [152].

Information on physical activity will most likely play a role in AF detection in the quest to reduce the number of falsely detected episodes, especially since accelerometers are nowadays standard implementation in ECG devices. Although it remains to be demonstrated that AF detectors analyzing both bioelectrical and physical information offer better performance, a preliminary study shows that the number of falsely classified arrhythmias due to noise and artifacts can be considerably reduced when accelerometer information is taken into account [153]. The potential of accelerometer information in AF detection is further supported by results showing that AF episodes can be detected from accelerometers attached to the chest [154], or from an electromechanical vibration sensor attached to a bed mattress [155], without involving the analysis of the ECG.

References

1. A. Haeberlin, L. Roten, M. Schilling, F. Scarcia, T. Niederhauser, R. Vogel, J. Fuhrer, H. Tanner, Software-based detection of atrial fibrillation in long-term ECGs. Heart Rhythm **11**, 933–938 (2014)
2. K.M. Stein, J. Walden, N. Lippman, B.B. Lerman, Ventricular response in atrial fibrillation: random or deterministic? Am. J. Physiol. **277**, H452–458 (1999)
3. J. Hayano, F. Yamasaki, S. Sakata, A. Okada, S. Mukai, T. Fujinami, Spectral characteristics of ventricular response to atrial fibrillation. Am. J. Physiol. **273**, H2811–2816 (1997)
4. V. Fuster, L.E. Rydén, D.S. Cannom, H.J. Crijns, A.B. Curtis et al., ACC/AHA/ESC 2006 guidelines for the management of patients with atrial fibrillation: a report of the American College of Cardiology/American Heart Association Task Force on practice guidelines and the European Society of Cardiology Committee for Practice Guidelines developed in collaboration with the European Heart Rhythm Association and the Heart Rhythm Society Heart Association Task Force on practice guidelines and the European Society of Cardiology Committee for Practice Guidelines developed in collaboration with the European Heart Rhythm Association and the Heart Rhythm Society. Europace **8**, 651–745 (2006)
5. C.T. January, L.S. Wann, J.S. Alpert, H. Calkins, J.E. Cigarroa et al., 2014 AHA/ACC/HRS guideline for the management of patients with atrial fibrillation: a report of the American College of Cardiology/American Heart Association Task Force on Practice Guidelines and the Heart Rhythm Society. Circulation **130**, 2071–2104 (2014)
6. P. Kirchhof, S. Benussi, D. Kotecha, A. Ahlsson, D. Atar, B. Casadei, M. Castella, H.C. Diener, H. Heidbuchel, J. Hendriks, G. Hindricks, A.S. Manolis, J. Oldgren, B.A. Popescu, U. Schotten, B. Van Putte, P. Vardas, S. Agewall, J. Camm, G. Baron Esquivias, W. Budts, S. Carerj, F. Casselman, A. Coca, R. De Caterina, S. Deftereos, D. Dobrev, J.M. Ferro, G. Filippatos, D. Fitzsimons, B. Gorenek, M. Guenoun, S.H. Hohnloser, P. Kolh, G.Y. Lip, A. Manolis, J. McMurray, P. Ponikowski, R. Rosenhek, F. Ruschitzka, I. Savelieva, S. Sharma, P. Suwalski, J.L. Tamargo, C.J. Taylor, I.C. Van Gelder, A.A. Voors, S. Windecker, J.L. Zamorano, K. Zeppenfeld, 2016 ESC guidelines for the management of atrial fibrillation developed in collaboration with EACTS. Eur. Heart J. **37**, 2893–2962 (2016)
7. R.C.S. Seet, P.A. Friedman, A.A. Rabinstein, Prolonged rhythm monitoring for the detection of occult paroxysmal atrial fibrillation in ischemic stroke of unknown cause. Circulation **124**, 477–486 (2011)
8. J.W. Keach, S.M. Bradley, M.P. Turakhia, T.M. Maddox, Early detection of occult atrial fibrillation and stroke prevention. Heart **101**, 1097–1102 (2015)
9. J.G. Andrade, T. Field, P. Khairy, Detection of occult atrial fibrillation in patients with embolic stroke of uncertain source: a work in progress. Front. Physiol. **1**, 1–9 (2015)
10. D.J. Miller, K. Shah, S. Modi, A. Mahajan, S. Zahoor, M. Affan, The evolution and application of cardiac monitoring for occult atrial fibrillation in cryptogenic stroke and TIA. Curr. Treat. Options Neurol. **18**, 17 (2016)
11. J.O. Cerasuolo, L.E. Cipriano, L.A. Sposato, The complexity of atrial fibrillation newly diagnosed after ischemic stroke and transient ischemic attack: advances and uncertainties. Curr. Opin. Neurol. **30**, 28–37 (2017)
12. A.H. Tayal, M. Tian, K.M. Kelly, S.C. Jones, D.G. Wright, D. Singh, J. Jarouse, J. Brillman, S. Murali, R. Gupta, Atrial fibrillation detected by mobile cardiac outpatient telemetry in cryptogenic TIA or stroke. Neurology **71**, 1696–1701 (2008)
13. A.A. Rabinstein, J.E. Fugate, J. Mandrekar, J.D. Burns, R.C. Seet, S.A. Dupont, T.J. Kauffman, S.J. Asirvatham, P.A. Friedman, Paroxysmal atrial fibrillation in cryptogenic stroke: a case control study. J. Stroke Cerebrovascular Dis. **22**, 1405–1411 (2013)
14. A.H. Abdul-Rahim, K.R. Lees, Paroxysmal atrial fibrillation after ischemic stroke: how should we hunt for it? Expert Rev. Cardiovasc. Ther. **11**, 485–494 (2013)
15. C.G. Favilla, E. Ingala, J. Jara, E. Fessler, B. Cucchiara, S.R. Messé, M.T. Mullen, A. Prasad, J. Siegler, M.D. Hutchinson, S.E. Kasner, Predictors of finding occult atrial fibrillation after cryptogenic stroke. Stroke **46**, 1210–1215 (2015)

16. E.I. Charitos, U. Stierle, P.D. Ziegler, M. Baldewig, D.R. Robinson, H. Sievers, T. Hanke, A comprehensive evaluation of rhythm monitoring strategies for the detection of atrial fibrillation recurrence: insights from 647 continuously monitored patients and implications for monitoring after therapeutic interventions. Circulation **126**, 806–814 (2012)
17. T. Etgen, M. Hochreiter, M. Mundel, T. Freudenberger, Insertable cardiac event recorder in detection of atrial fibrillation after cryptogenic stroke: an audit report. Stroke **44**, 2007–2009 (2013)
18. J. Reiffel, A. Verma, J.L. Halperin, B. Gersh, S. Tombul, J. Carrithers, L. Sherfesee, P. Kowey, Rationale and design of REVEAL AF: a prospective study of previously undiagnosed atrial fibrillation as documented by an insertable cardiac monitor in high-risk patients. Am. Heart J. **167**, 22–27 (2014)
19. A.C. Flint, N.M. Banki, X. Ren, V.A. Rao, A.S. Go, Detection of paroxysmal atrial fibrillation by 30-day event monitoring in cryptogenic ischemic stroke: The stroke and monitoring for PAF in real time (SMART) registry. Stroke **43**, 2788–2790 (2012)
20. S.B. Silverman, Paroxysmal atrial fibrillation: Novel strategies for monitoring and implications for treatment in stroke. Curr. Treat. Options Cardio. Med. **18**, 1–13 (2016)
21. L. Sörnmo, P. Laguna, *Bioelectrical Signal Processing in Cardiac and Neurological Applications* (Elsevier (Academic Press), Amsterdam, 2005)
22. N. Lowres, L. Neubeck, J. Redfern, S.B. Freedman, Screening to identify unknown atrial fibrillation. A systematic review. Thromb. Haemost. **110**, 213–222 (2013)
23. B. Vaes, S. Stalpaert, K. Tavernier, B. Thaels, D. Lapeire, W. Mullens, J. Degryse, The diagnostic accuracy of the MyDiagnostick to detect atrial fibrillation in primary care. BMC Fam. Pract. **15**, 113 (2014)
24. F. Kaasenbrood, M.H.F.H. Rutten, L.J. Gerhards, A.W. Hoes, R.G. Tieleman, Yield of screening for atrial fibrillation in primary care with a hand-held, single-lead electrocardiogram device during influenza vaccination. Europace **18**, 1514–1520 (2016)
25. L. Desteghe, Z. Raymaekers, M. Lutin, J. Vijgen, D. Dilling-Boer, P. Koopman, J. Schurmans, P. Vanduynhoven, P. Dendale, H. Heidbuchel, Performance of handheld electrocardiogram devices to detect atrial fibrillation in a cardiology and geriatric ward setting. Europace **19**, 29–39 (2017)
26. E. Svennberg, J. Engdahl, F. Al-Khalili, L. Friberg, V. Frykman, M. Rosenqvist, Mass screening for untreated atrial fibrillation: the STROKESTOP study. Circulation **131**, 2176–2184 (2015)
27. E. Svennberg, M. Stridh, J. Engdahl, F. Al-Khalili, L. Friberg, V. Frykman, M. Rosenquist, Safe automatic one-lead electrocardiogram analysis in screening for atrial fibrillation. Europace **19**, 1449–1453 (2016)
28. S.R. Steinhubl, R.R. Mehta, G.S. Ebner, M.M. Ballesteros, J. Waalen, G. Steinberg, P. Van Crocker, Jr., E. Felicione, C. T. Carter, S. Edmonds, J. P. Honcz, G. D. Miralles, D. Talantov, T. C. Sarich, E. J. Topol, Rationale and design of a home-based trial using wearable sensors to detect asymptomatic atrial fibrillation in a targeted population: the mHealth screening to prevent strokes (mSToPS) trial. Am. Heart J. **175**, 77–85 (2016)
29. M.P. Turakhia, D.W. Kaiser, Transforming the care of atrial fibrillation with mobile health. J. Interv. Card. Electrophysiol. **47**, 45–50 (2016)
30. G. D. Clifford, C. Liu, B. Moody, L.-W. H. Lehman, I. Silva, Q. Li, A. Johnson, and R. G. Mark, "AF classification from a short single lead ECG recording: the PhysioNet Computing in Cardiology Challenge 2017, in *Proceedings of Computing in Cardiology*, vol. 44 (2017)
31. K. Tateno, L. Glass, Automatic detection of atrial fibrillation using the coefficient of variation and density histograms of RR and deltaRR intervals. Med. Biol. Eng. Comput. **39**, 664–671 (2001)
32. S. Dash, K.H. Chon, S. Lu, E.A. Raeder, Automatic real time detection of atrial fibrillation. Ann. Biomed. Eng. **37**, 1701–1709 (2009)
33. J. Lian, L. Wang, D. Muessig, A simple method to detect atrial fibrillation using RR intervals. Am. J. Cardiol. **107**, 1494–1497 (2011)

34. D. E. Lake, J.R. Moorman, Accurate estimation of entropy in very short physiological time series: the problem of atrial fibrillation detection in implanted ventricular devices. Am. J. Physiol. (Heart Circ. Physiol.) **300**: H319–H325 (2011)
35. C. Huang, S. Ye, H. Chen, D. Li, F. He, Y. Tu, A novel method for detection of the transition between atrial fibrillation and sinus rhythm. IEEE Trans. Biomed. Eng. **58**, 1113–1119 (2011)
36. R.B. Shouldice, C. Heneghan, P. de Chazal, Automatic detection of paroxysmal atrial fibrillation, in *Atrial fibrillation – basic research and clinical applications*. (J. Choi, ed.), chap. 7, pp. 125–146, InTech (2012)
37. J. Lee, Y. Nam, D.D. McManus, K.H. Chon, Time-varying coherence function for atrial fibrillation detection. IEEE Trans. Biomed. Eng. **60**, 2783–2793 (2013)
38. X. Zhou, H. Ding, B. Ung, E. Pickwell-MacPherson, Y. Zhang, Automatic online detection of atrial fibrillation based on symbolic dynamics and shannon entropy. Biomed. Eng. Online **13**, 18 (2014)
39. S. Asgari, A. Mehrni, M. Moussavi, Automatic detection of atrial fibrillation using stationary wavelet transform and support vector machine. Comput. Biol. Med. **60**, 132–142 (2015)
40. A. Petrėnas, V. Marozas, L. Sörnmo, Low-complexity detection of atrial fibrillation in continuous long-term monitoring. Comput. Biol. Med. **65**, 184–191 (2015)
41. X. Zhou, H. Ding, W. Wu, Y. Zhang, A real-time atrial fibrillation detection algorithm based on the instantaneous state of heart rate. PLoS ONE **10**, e0136544 (2015)
42. G.B. Moody, R.G. Mark, A new method for detecting atrial fibrillation using R-R intervals. in *Proceedings of Computers in Cardiology* vol. 10, pp. 227–230 (1983)
43. S. Cerutti, L.T. Mainardi, A. Porta, A.M. Bianchi, Analysis of the dynamics of RR interval series for the detection of atrial fibrillation episodes, in *Proceedings of Computers in Cardiology*, vol. 24, pp. 7–80 (1997)
44. S. Shkurovich, A.V. Sahakian, S. Swiryn, Detection of atrial activity from high-voltage leads of implantable ventricular defibrillators using a cancellation technique. IEEE Trans. Biomed. Eng. **45**, 229–234 (1998)
45. D. Duverney, J.M. Gaspoz, V. Pichot, F. Roche, R. Brion, A. Antoniadis, J.C. Barthelemy, High accuracy of automatic detection of atrial fibrillation using wavelet transform of heart rate intervals. Pacing Clin. Electrophysiol. **25**, 457–462 (2002)
46. F. Yaghouby, A. Ayatollahi, R. Bahramali, M. Yaghouby, A.H. Alavi, Towards automatic detection of atrial fibrillation: a hybrid computational approach. Comput. Biol. Med. **40**, 919–930 (2010)
47. C.-T. Lin, K.-C. Chang, C.-L. Lin, C.-C. Chiang, S.-W. Lu, S.-S. Chang, B.-S. Lin, H.-Y. Liang, R.-J. Chen, Y.-T. Lee, L.-W. Ko, An intelligent telecardiology system using a wearable and wireless ECG to detect atrial fibrillation. IEEE Trans. Info. Tech. Biomed. **14**, 726–733 (2010)
48. P. Langley, M. Dewhurst, L.D. Marco, P. Adams, F. Dewhurst, J. Mwita, R. Walker, A. Murray, Accuracy of algorithms for detection of atrial fibrillation from short duration beat interval recordings. Med. Eng. Phys. **34**, 1441–1447 (2012)
49. J. Lee, B. Reyes, D. McManus, O. Mathias, K. Chon, Atrial fibrillation detection using an iPhone 4S. IEEE Trans. Biomed. Eng. **60**, 203–206 (2013)
50. J. Park, S. Lee, M. Jeon, Atrial fibrillation detection by heart rate variability in Poincaré plot. Biomed. Eng. Online **8**, 1–12 (2009)
51. S. Sarkar, D. Ritscher, R. Mehra, A detector for a chronic implantable atrial tachyarrhythmia monitor. IEEE Trans. Biomed. Eng. **55**, 1219–1224 (2008)
52. M.S. Kendall, A. Stuart, J.K. Ord, *The Advanced Theory of Statistics*, vol. 3, 4th edn. (High Wycombe: Charles Griffin, 1983)
53. S.B. Olsson, N. Cai, M. Dohnal, K.K. Talwar, Noninvasive support for and characterization of multiple intranodal pathways in patients with mitral valve disease and atrial fibrillation. Eur. Heart J. **7**, 320–333 (1986)
54. N. Cai, M. Dohnal, S.B. Olsson, Methodological aspects of the use of heart rate stratified RR interval histograms in the analysis of atrioventricular conduction during atrial fibrillation. Cardiovasc. Res. **21**, 455–462 (1987)

55. J. Dickinson Gibbons and S. Chakraborti, *Nonparametric Statistical Inference*, 5th edn. (Chapman and Hall/CRC, 2010)
56. J. Tebbenjohanns, B. Schumacher, T. Korte, M. Niehaus, D. Pfeiffer, Bimodal RR interval distribution in chronic atrial fibrillation: impact of dual atrioventricular nodal physiology on long-term rate control after catheter ablation of the posterior atrionodal input. J. Cardiovasc. Electrophysiol. **11**, 497–503 (2000)
57. S. Rokas, S. Gaitanidou, S. Chatzidou, C. Pamboucas, D. Achtipis, S. Stamatelopoulos, Atrioventricular node modification in patients with chronic atrial fibrillation: role of morphology of RR interval variation. Circulation **103**, 2942–2948 (2001)
58. V.D.A. Corino, F. Sandberg, L.T. Mainardi, L. Sörnmo, An atrioventricular node model for analysis of the ventricular response during atrial fibrillation. IEEE Trans. Biomed. Eng. **58**, 3386–3395 (2011)
59. C.E. Shannon, A mathematical theory of communication. Bell Sys. Tech. J. **27**, 379–423 (1948)
60. S.J. Richman, J.R. Moorman, Physiological time-series analysis using approximate entropy and sample entropy. Am. J. Physiol. **278**, H2039–H2049 (2000)
61. D.E. Lake, J.S. Richman, M.P. Griffin, J.R. Moorman, Sample entropy analysis of neonatal heart rate variability. Am. J. Physiol. Regul. Integr. Comp. Physiol. **283**, R789–R797 (2002)
62. D.E. Lake, Renyi entropy measures of heart rate Gaussianity. IEEE Trans. Biomed. Eng. **53**, 21–27 (2006)
63. M. S. Pincus, A.L. Goldberger, Physiological time-series analysis: what does regularity quantify? Am. J. Physiol. 266 (Heart Circ. Physiol.) **35**: H1643–H1656 (1994)
64. W. Chen, Z. Wang, H. Xie, W. Yu, Characterization of surface EMG signal based on fuzzy entropy. IEEE Trans. Neural Sys. Rehab. Eng. **15**, 266–272 (2007)
65. A. Avolio, Heart rate variability and stroke: strange attractors with loss of complexity. J. Hypertension **31**, 1529–1531 (2013)
66. M. Julián, R. Alcaraz, J.J. Rieta, Comparative assessment of nonlinear metrics to quantify organization-related events in surface electrocardiograms of atrial fibrillation. Comput. Biol. Med. **48**, 66–76 (2014)
67. L. Hong-wei, S. Ying, L. Min, L. Pi-ding, Z. Zheng, A probability density function method for detecting atrial fibrillation using R-R intervals. Med. Eng. Phys. **31**, 116–123 (2009)
68. P. Grassberger, I. Procaccia, Characterization of strange attractors. Phys. Rev. Lett. **50**, 346–349 (1983)
69. T. Anan, K. Sunagawa, H. Araki, M. Nakamura, Arrhythmia analysis by successive RR plotting. J. Electrocardiol. **23**, 243–248 (1990)
70. P.W. Kamen, H. Krum, A.M. Tonkin, Poincaré plot of heart rate variability allows quantitative display of parasympathetic nervous activity in humans. Clin. Sci. (Lond.) **91**, 201–208 (1996)
71. M. Brennan, M. Palaniswami, P. Kamen, Poincaré plot interpretation using a physiological model of HRV based on a network of oscillators. Am. J. Physiol. Heart Circ. Physiol. **283**, H1873–H1886 (2002)
72. M. Malik, Standard measurements of heart rate variability, in *Dynamic electrocardiography* ed. by M. Malik, A.J. Camm, chap. 2, (Wiley–Blackwell, New York, 2004), pp. 13–21
73. K. Monahan, Y. Song, K. Loparo, R. Mehra, F.E. Harrell Jr., S. Redline, Automated detection of atrial fibrillation from the electrocardiogram channel of polysomnograms. Sleep Breath **20**, 515–522 (2015)
74. C.K. Karmakar, A.H. Khandoker, J. Gubbi, M. Palaniswami, Complex correlation measure: a novel descriptor for Poincaré plot. BioMed. Eng. Online **8**, 37–48 (2009)
75. L. Zhang, T. Guo, B. Xi, Y. Fan, K. Wang, J. Bi, Y. Wang, Automatic recognition of cardiac arrhythmias based on the geometric patterns of Poincaré plots. Physiol. Meas. **36**, 283–301 (2015)
76. R. Mehra, J. Gillberg, P. Ziegler, S. Sarkar, Algorithms for atrial tachyarrhythmia detection for long-term monitoring with implantable devices, in *Understanding atrial fibrillation: the signal processing contribution* ed. by L.T. Mainardi, L.Sörnmo, S. Cerutti, chap. 8 (Morgan & Claypool, San Francisco, 2008), pp. 175–214

77. H. Käsmacher, S. Wiese, M. Lahl, Monitoring the complexity of ventricular response in atrial fibrillation. Discrete Dynamics Nature Soc. **4**, 63–75 (2000)
78. M. Brennan, M. Palaniswami, P. Kamen, Do existing measures of Poincaré plot geometry reflect nonlinear features of heart rate variability? IEEE Trans. Biomed. Eng. **48**, 1342–1347 (2001)
79. R.A. Thuraisingham, An electrocardiogram marker to detect paroxysmal atrial fibrillation. J. Electrocardiol. **40**, 344–347 (2007)
80. H. Zhao, S. Lu, R. Zou, K. Ju, K.H. Chon, Estimation of time-varying coherence function using time-varying transfer functions. Ann. Biomed. Eng. **33**, 1582–1594 (2005)
81. R. Zou, H. Wang, K.H. Chon, A robust time-varying identification algorithm using basis functions. Ann. Biomed. Eng **31**, 840–853 (2003)
82. F. van der Heijden, R.P.W. Duin, D. de Ridder, D.M.J. Tax, *Classification, Parameter Estimation and State Estimation–An Engineering Approach using Matlab* (Wiley, New York, 2005)
83. P. Carvalho, J. Henriques, R. Couceiro, M. Harris, M. Antunes, J. Habetha, Model-based atrial fibrillation detection, in *ECG signal processing, classification and interpretation* ed. by A. Gacek, W. Pedrycz (Springer London, 2012), pp. 99–133
84. R. Colloca, A.E.W. Johnson, L. Mainardi, G.D. Clifford, A support vector machine approach for reliable detection of atrial fibrillation events, in *Proceedings of Computing in Cardiology*, vol. 40, pp. 1047–1050 (2013)
85. R.O. Duda, P.E. Hart, D.G. Stork, *Pattern Classification*, 2nd edn. (Wiley–Interscience, New York, 2001)
86. N. Larburu, T. Lopetegi, I. Romero, Comparative study of algorithms for atrial fibrillation detection, in *Proceedings of Computing in Cardiology*, vol. 38, pp. 265–268 (2011)
87. S. Babaeizadeh, R.E. Gregg, E.D. Helfenbein, J.M. Lindauer, S.H. Zhou, Improvements in atrial fibrillation detection for real-time monitoring. J. Electrocardiol. **42**, 522–526 (2009)
88. S. Ladavich, B. Ghoraani, Rate-independent detection of atrial fibrillation by statistical modeling of atrial activity. Biomed. Signal Process. Control **18**, 274–281 (2015)
89. J. Ródenas, M. García, R. Alcaraz, J.J. Rieta, Wavelet entropy automatically detects episodes of atrial fibrillation from single-lead electrocardiograms. Entropy **17**, 6179–6199 (2015)
90. Y. Xia, N. Wulan, K. Wang, H. Zhang, Detecting atrial fibrillation by deep convolutional neural networks. Comput. Biol. Med. **93**, 84–92 (2018)
91. P. Laguna, R. Jané, P. Caminal, Automatic detection of wave boundaries in multilead ECG signals: validation with the CSE database. Comput. Biomed. Res. **27**, 45–60 (1994)
92. J.P. Martínez, R. Almeida, S. Olmos, A.P. Rocha, P. Laguna, A wavelet-based ECG delineator: evaluation on standard databases. IEEE Trans. Biomed. Eng. **51**, 570–581 (2004)
93. J. Dumont, A. Hernández, G. Carrault, Improving ECG beats delineation with an evolutionary optimization process. IEEE Trans. Biomed. Eng. **57**, 607–615 (2010)
94. L. Clavier, J.-M. Boucher, R. Lepage, J.-J. Blanc, J.-C. Cornily, Automatic P-wave analysis of patients prone to atrial fibrillation. Med. Biol. Eng. Comput. **40**, 63–71 (2002)
95. F. Censi, G. Calcagnini, C. Ricci, R.P. Ricci, M. Santini, A. Grammatico, P. Bartolini, P-wave morphology assessment by a Gaussian functions-based model in atrial fibrillation patients. IEEE Trans. Biomed. Eng. **54**, 663–671 (2007)
96. A. Martínez, D. Abásolo, R. Alcaraz, J.J. Rieta, Alteration of the P-wave non-linear dynamics near the onset of paroxysmal atrial fibrillation. Med. Eng. Phys. **37**, 692–697 (2015)
97. P. Laguna, R. G. Mark, A. L. Goldberger, and G. B. Moody, "A database for evaluation of algorithms for measurement of QT and other waveform intervals in the ECG, in *Proceedings of Computers in Cardiology*, Vol. 23, pp. 673–676 (1997)
98. A.L. Goldberger, L.A. Amaral, L. Glass, J.M. Hausdorff, P.C. Ivanov, R.G. Mark, J.E. Mietus, G.B. Moody, C.K. Peng, H.E. Stanley, PhysioBank, PhysioToolkit, and PhysioNet: components of a new research resource for complex physiologic signals. Circulation **101**, E215–220 (2000)
99. I. Dotsinsky, Atrial wave detection algorithm for discovery of some rhythm abnormalities. Physiol. Meas **28**, 595–610 (2007)

100. A. Petrėnas, L. Sörnmo, A. Lukoševičius, V. Marozas, Detection of occult paroxysmal atrial fibrillation. Med. Biol. Eng. Comput. **53**, 287–297 (2015)
101. A. Petrėnas, V. Marozas, L. Sörnmo, A. Lukoševičius, An echo state neural network for QRST cancellation during atrial fibrillation. IEEE Trans. Biomed. Eng. **59**, 2950–2957 (2012)
102. S. Roberts, L. Tarassenko, A probabilistic resource allocating network for novelty detection. Neural Comput. **6**, 270–284 (1994)
103. M. Markou, S. Singh, Novelty detection: a review. Signal Process. **83**, 2481–2497 (2003)
104. X. Du, N. Rao, M. Qian, D. Liu, J. Li, W. Feng, L. Yin, X. Chen, A novel method for real-time atrial fibrillation detection in electrocardiograms using multiple parameters. Ann. Noninvasive Electrocardiol. **19**, 217–225 (2014)
105. F. Castells, J.J. Rieta, J. Millet, V. Zarzoso, Spatiotemporal blind source separation approach to atrial activity estimation in atrial tachyarrhythmias. IEEE Trans. Biomed. Eng. **52**, 258–267 (2005)
106. R. Llinares, J. Igual, J. Miró-Borrás, A fixed point algorithm for extracting the atrial activity in the frequency domain. Comput. Biol. Med. **40**, 943–949 (2010)
107. S.M. Kay, *Modern Spectral Estimation Theory and Application* (Prentice-Hall, New Jersey, 1999)
108. M. Stridh, A. Bollmann, S.B. Olsson, L. Sörnmo, Time-frequency analysis of atrial tachyarrhythmias: detection and feature extraction. IEEE Eng. Med. Biol. Mag. **25**, 31–39 (2006)
109. Q. Li, R.G. Mark, G.D. Clifford, Robust heart rate estimation from multiple asynchronous noisy sources using signal quality indices and a Kalman filter. Physiol. Meas. **29**, 15–32 (2008)
110. G.D. Clifford, D. Clifton, Wireless technology in disease management and medicine. Ann. Rev. Med. **63**, 479–92 (2012)
111. J. Behar, J. Oster, Q. Li, G.D. Clifford, ECG signal quality during arrhythmia and its application to false alarm reduction. IEEE Trans. Biomed. Eng. **60**, 1660–1666 (2013)
112. A.E.W. Johnson, J. Behar, F. Andreotti, G.D. Clifford, J. Oster, Multimodal heart beat detection using signal quality indices. Physiol. Meas. **36**, 1665–1677 (2015)
113. C. Orphanidou, T. Bonnici, P. Charlton, D. Clifton, D. Vallance, L. Tarassenko, Signal-quality indices for the electrocardiogram and photoplethysmogram: derivation and applications to wireless monitoring. IEEE J. Biomed. Health Inform. **19**, 832–838 (2015)
114. J. Oster, G.D. Clifford, Impact of the presence of noise on RR interval-based atrial fibrillation detection. J. Electrocardiol. **48**, 947–951 (2015)
115. N. Gambarotta, F. Aletti, G. Baselli, M. Ferrario, A review of methods for the signal quality assessment to improve reliability of heart rate and blood pressures derived parameters. Med. Biol. Eng. Comput. **54**, 1025–1035 (2016)
116. S.H. Rappaport, L. Gillick, G.B. Moody, R.G. Mark, QRS morphology classification: quantitative evaluation of different strategies, in *Proceedings of Computers in Cardiology*, Vol. 9, pp. 33–38 (1982)
117. M. Lagerholm, C. Peterson, G. Braccini, L. Edenbrandt, L. Sörnmo, Clustering ECG complexes using Hermite functions and self-organizing maps. IEEE Trans. Biomed. Eng. **47**, 838–848 (2000)
118. P. de Chazal, M. O'Dwyer, R.B. Reilly, Automatic classification of heartbeats using ECG morphology and heartbeat interval features. IEEE Trans. Biomed. Eng. **51**, 1196–1206 (2004)
119. M. Llamedo, J.P. Martínez, Heartbeat classification using feature selection driven by database generalization criteria. IEEE Trans. Biomed. Eng. **58**, 616–625 (2011)
120. J. Oster, J. Behar, O. Sayadi, S. Nemati, A.E.W. Johnson, G.D. Clifford, Semisupervised ECG ventricular beat classification with novelty detection based on switching Kalman filters. IEEE Trans. Biomed. Eng. **62**, 2125–2134 (2015)
121. R. Couceiro, P. Carvalho, J. Henriques, M. Antunes, M. Harris, J. Habertha, Detection of atrial fibrillation using model-based ECG analysis, in *Proceedings of International Conference on Pattern Recognition (ICPR)*, vol. 19, pp. 1–5 (2008)
122. T. Jeon, B. Kim, M. Jeon, B.-G. Lee, Implementation of a portable device for real-time ECG signal analysis. Biomed. Eng. Online **13**, 1–13 (2014)

123. O. Andersson, K.H. Chon, L. Sörnmo, J. Neves Rodrigues, A 290 mV sub-V_T ASIC for real-time atrial fibrillation detection. IEEE Trans. Biomed. Circ. Sys. **9**: 377–386 (2015)
124. B. Matthews, Comparison of the predicted and observed secondary structure of T4 phage lysozyme. Biochimica et Biophysica Acta **405**, 442–451 (1975)
125. P. Baldi, S. Brunak, Y. Chauvin, C.A. Andersen, H. Nielsen, Assessing the accuracy of prediction algorithms for classification: an overview. Bioinformatics **16**, 412–424 (2000)
126. F. Jager, G.B. Moody, A. Taddei, R. Mark, Performance measures for algorithms to detect transient ischemic ST segment changes, in *Proceedings of Computers in Cardiology*, vol. 18, pp. 369–372 (1991)
127. R.J. Martis, U.R. Acharya, H. Prasad, C.K. Chua, C.M. Lim, Automated detection of atrial fibrillation using Bayesian paradigm. Knowledge-Based Syst. **54**, 269–275 (2013)
128. M. García, J. Ródenas, R. Alcaraz, J.J. Rieta, Application of the relative wavelet energy to heart rate independent detection of atrial fibrillation. Comput. Meth. Progr. Biomed. **131**, 157–168 (2016)
129. A. Isaksson, M. Wallman, H. Göransson, M.G. Gustafsson, Cross-validation and bootstrapping are unreliable in small sample classification. Pattern Recogn. Letter **29**, 1960–1965 (2008)
130. J. Slocum, A.V. Sahakian, S. Swiryn, Diagnosis of atrial fibrillation from surface electrocardiograms based on computer-detected atrial activity. J. Electrocardiol. **25**, 1–8 (1992)
131. M. Haïssaguerre, P. Jaïs, D.C. Shah, A. Takahashi, M. Hocini, G. Quiniou, S. Garrigue, A. Le Mouroux, P. Le Métayer, J. Clémenty, Spontaneous initiation of atrial fibrillation by ectopic beats originating in the pulmonary veins. N. Engl. J. Med. **339**, 659–666 (1998)
132. S.-A. Chen, M.-H. Hsieh, C.-T. Tai, C.-F. Tsai, V.S. Prakash, W.-C. Yu, T.-L. Hsu, Y.-A. Ding, M.-S. Chang, Initiation of atrial fibrillation by ectopic beats originating from the pulmonary veins: electrophysiological characteristics, pharmacological responses, and effects of radiofrequency ablation. Circulation **100**, 1879–1886 (1999)
133. C. Kolb, S. Nürnberger, G. Ndrepepa, B. Zrenner, A. Schömig, C. Schmitt, Modes of initiation of paroxysmal atrial fibrillation from analysis of spontaneously occurring episodes using a 12-lead Holter monitoring system. Am. J. Cardiol. **88**, 853–857 (2001)
134. D. Wallmann, D. Tüller, K. Wustmann, P. Meier, J. Isenegger, M. Arnold, H.P. Mattle, E. Delacrétaz, Frequent atrial premature beats predict paroxysmal atrial fibrillation in stroke patients. An opportunity for a new diagnostic strategy. Stroke **38**, 2292–2294 (2007)
135. K.T. Nguyen, E. Vittinghoff, T.A. Dewland, M.C. Mandyam, P.K. Stein, E.Z. Soliman, S.R. Heckbert, G.M. Marcus, Electrocardiographic predictors of incident atrial fibrillation. Am. J. Cardiol. **118**, 714–719 (2016)
136. T. Thong, J. McNames, M. Aboy, B. Goldstein, Prediction of paroxysmal atrial fibrillation by analysis of atrial premature complexes. IEEE Trans. Biomed. Eng. **4**, 561–569 (2004)
137. J.L. Huang, W.C. Wen, W.L. Lee, M.S. Chang, S.A. Chen, Changes of autonomic tone before the onset of paroxysmal atrial fibrillation. Int. J. Cardiol. **66**, 275–283 (1998)
138. M. Bettoni, M. Zimmermann, Autonomic tone variations before the onset of paroxysmal atrial fibrillation. Circulation **105**, 2753–2759 (2002)
139. C. Gallo, P.P. Bocchino, M. Magnano, L. Gaido, D. Zema, A. Battaglia, M. Anselmino, F. Gaita, Autonomic tone activity before the onset of atrial fibrillation. J. Cardiovasc. Electrophysiol. **28**, 304–314 (2017)
140. F. Lombardi, D. Tarricone, F. Tundo, F. Colombo, S. Belletti, C. Fiorentini, Autonomic nervous system and paroxysmal atrial fibrillation: a study based on the analysis of RR interval changes before, during and after paroxysmal atrial fibrillation. Eur. Heart J. **25**, 1242–1248 (2004)
141. D.-G. Shin, C.-S. Yoo, S.-H. Yi, J.-H. Bae, Y.-J. Kim, J.-S. Park, G.-R. Hong, Prediction of paroxysmal atrial fibrillation using nonlinear analysis of the R-R interval dynamics before the spontaneous onset of atrial fibrillation. Circ. J. **70**, 94–99 (2006)
142. Y. Chesnokov, Complexity and spectral analysis of the heart rate variability dynamics for distant prediction of paroxysmal atrial fibrillation with artificial intelligence methods. Artif. Intell. Med. **43**, 151–165 (2008)

143. M. Mohebbi, H. Ghassemian, Prediction of paroxysmal atrial fibrillation based on non-linear analysis and spectrum and bispectrum features of the heart rate variability signal. Comput. Meth. Progr. Biomed. **5**, 40–49 (2012)
144. F. Holmqvist, P.G. Platonov, J. Carlson, W. Zareba, A.J. Moss, Altered interatrial conduction detected in MADIT II patients bound to develop atrial fibrillation. Ann. Noninvasive Electrocardiol. **14**:268–275 (2009)
145. F. Holmqvist, P.G. Platonov, S. McNitt, S. Polonsky, J. Carlson, W. Zareba, A.J. Moss, Abnormal P-wave morphology is a predictor of atrial fibrillation development and cardiac death in MADIT II patients. Ann. Noninvasive Electrocardiol. **15**: 63–72 (2010)
146. H. Gonna, M.M. Gallagher, X.H. Guo, Y.G. Yap, K. Hnatkova, A.J. Camm, P-wave abnormality predicts recurrence of atrial fibrillation after electrical cardioversion: a prospective study. Ann. Noninvasive Electrocardiol. **19**, 57–62 (2014)
147. J.B. Nielsen, J.T. Kühl, A. Pietersen, C. Graff, B. Lind, J.J. Struijk, M.S. Olesen, M.F. Sinner, T.N. Bachmann, S. Haunsø, B.G. Nordestgaard, P.T. Ellinor, J.H. Svendsen, K.F. Kofoed, L. Køber, A.G. Holst, P-wave duration and the risk of atrial fibrillation: results from the copenhagen ECG study. Heart Rhythm **12**, 1887–1895 (2015)
148. G. Conte, A. Luca, S. Yazdani, M.L. Caputo, F. Regol, T. Moccetti, L. Kappenberger, J.-M. Vesin, A. Auricchio, Usefulness of P-wave duration and morphologic variability to identify patients prone to paroxysmal atrial fibrillation. Am. J. Cardiol. **119**, 275–279 (2017)
149. I.C.Y. Chang, E. Austin, B. Krishnan, D.G. Benditt, C.N. Quay, L.H. Ling, L.Y. Chen, Shorter minimum P-wave duration is associated with paroxysmal lone atrial fibrillation. J. Electrocardiol. **47**, 106–112 (2014)
150. F. Nilsson, M. Stridh, A. Bollmann, L. Sörnmo, Predicting spontaneous termination of atrial fibrillation using the surface ECG. Med. Eng. Phys. **26**, 802–808 (2006)
151. P. G. Platonov, V.D.A. Corino, M. Seifert, F. Holmqvist, L. Sörnmo, Atrial fibrillatory rate in the clinical context: natural course and prediction of intervention outcome. Europace **16**: iv110–iv119 (2014)
152. R. Alcaraz, J.J. Rieta, Application of wavelet entropy to predict atrial fibrillation progression from the surface ECG. Comput. Math. Meth. Med. **13**, 1–9 (2012)
153. T. Tanantong, E. Nantajeewarawat, S. Thiemjarus, False alarm reduction in BSN-based cardiac monitoring using signal quality and activity type information. Sensors **15**, 3952–3974 (2015)
154. T. Koivisto, M. Pänkäälä, T. Hurnanen, T. Vasankari, T. Kiviniemi, A. Saraste, J. Airaksinen, Automatic detection of atrial fibrillation using MEMS accelerometer. in *Proceedings of Computing in Cardiology*, vol. 42, pp. 829–832 (2015)
155. C. Brueser, J. Diesel, M.D.H. Zink, S. Winter, P. Schauerte, S. Leonhardt, Automatic detection of atrial fibrillation in cardiac vibration signals. IEEE J. Biomed. Health Inform. **17**, 162–171 (2013)

Chapter 5
Extraction of f Waves

Leif Sörnmo, Andrius Petrėnas, Pablo Laguna and Vaidotas Marozas

5.1 Introduction

For many years, the level of detail in clinical f wave interpretation was confined to whether f wave amplitude was "coarse" or "fine", whereas atrial fibrillatory rate (AFR) was not considered although it must be viewed as a fundamental characteristic of AF. Two clinical studies, both published in 1998, contributed to changing this oversight by independently proposing spectral analysis of an f wave signal as a means to infer information on atrial refractoriness [1] and to predict the outcome of drug-based cardioversion [2]. The f wave signal was produced by computing an average QRST complex, obtained from an ensemble of beats with similar morphology, and subtracting the resulting average from each beat of the original ECG. This operation was motivated by the fact that the QRST complexes, most of the time, overshadow the f waves due to their much larger amplitude. Following the two above-mentioned studies, much attention has been paid to addressing the problem of how to extract

L. Sörnmo (✉)
Department of Biomedical Engineering and Center for Integrative Electrocardiology, Lund University, Lund, Sweden
e-mail: leif.sornmo@bme.lth.se

A. Petrėnas · V. Marozas
Biomedical Engineering Institute, Kaunas University of Technology, Kaunas, Lithuania
e-mail: andrius.petrenas@ktu.lt

V. Marozas
e-mail: vaidotas.marozas@ktu.lt

P. Laguna
Biomedical Signal Interpretation and Computational Simulation (BSICoS), Aragón Institute of Engineering Research (I3A), Centro de Investigacíon Biomédica en Red de Bioingeniería, Biomateriales y Nanomedicina (CIBER-BBN), Zaragoza University, Zaragoza, Spain
e-mail: laguna@unizar.es

L. Sörnmo (ed.), *Atrial Fibrillation from an Engineering Perspective*, Series in BioEngineering, https://doi.org/10.1007/978-3-319-68515-1_5

an f wave signal free of ventricular activity. As a result, a rich selection of methods have seen the light of the day which has helped to accelerate the development of more sophisticated techniques for f wave characterization, described in Chap. 6.

In this chapter, the notion *f wave extraction* is preferred as the purpose of the described methods is to extract f waves in the presence of ventricular activity, whereas the notion "atrial activity extraction" also includes P waves and therefore is too broad; indeed, the problem of P wave extraction has its own particular prerequisites. In the same way, the notions "QRST cancellation" and "ventricular activity cancellation" are avoided as much as possible since they can apply to P wave extraction as well. It is obvious from the literature that some methods put emphasis on the QRST cancellation process, whereas other methods put emphasis on the f wave extraction process.

The design of a method for f wave extraction is strongly influenced by the specific operating conditions. Most methods require that AF has been first detected, see Chap. 4, meaning that the starting point for extraction is an input signal which is a priori known to contain f waves, although the f waves may wax and wane and sometimes completely disappear. If an ECG in sinus rhythm is subject to extraction, the resulting signal may or may not contain the original P waves, depending on the method used. Although atrial flutter bears certain resemblance to AF, it is unlikely that the extracted signal will contain flutter waves ("F waves") since several extraction methods rely heavily on the assumption that the atrial and ventricular rates are decoupled. Atrial flutter is manifested by a reentrant wave circulating in the atria and arriving at the AV node in a much more regular fashion than the multiple excitation wavelets in AF. As a result, the ventricular rate is usually a fraction of the atrial flutter repetition rate, while this is not the case for the AFR. Very few methods for f wave extraction have been specifically evaluated with respect to performance in atrial flutter.

The available number of leads is an important design factor. While f wave extraction was initially performed in individual leads, it soon became obvious that joint processing of multiple leads offers certain advantages. In particular, the property that the atrial and ventricular activities originate from different electrical sources can be explored with signal separation techniques, provided that at least two, but preferably many more, leads are available. Respiratory-induced modulation of the QRS amplitude, caused by alterations in the electrical axis of the heart, introduces unwanted residuals in the extracted signal of a single lead. By analyzing two or more leads, the influence of such amplitude modulation can be considerably reduced, and, accordingly, a more accurate f wave signal extracted.

The presence of noise and artifacts, e.g., due to muscular activity or electrode motion, imposes an important limitation on performance. Surprisingly, not much attention has been paid to exploring the relationship between signal-to-noise ratio (SNR) and the performance of an extraction method. Typically, performance has been investigated at rather high SNRs, without any attempt to establish the "breakdown" SNR below which robust operation of the method is no longer possible.

Most extraction methods are designed to process ECG recordings whose total duration exceeds one minute, so as to provide sufficient time for learning the properties of the unwanted interference, notably the ventricular activity. However, such methods may very well be inappropriate for the processing of a standard 10-s resting ECG, since such a short time period is insufficient for producing a reliable estimate of the ventricular activity. Hence, total signal duration represents yet another design factor which needs to be considered.

Ventricular premature beats (VPBs) represent a major challenge to any extraction method, the reason being that such beats have morphologies which usually deviate considerably from beats originating from the atria. As a result, the presence of VPBs may introduce large-amplitude, QRS-related residuals in the extracted signal, with serious repercussions on subsequent f wave characterization. For occasional VPBs, it may be acceptable to simply discard the intervals related to such beats. However, a much more elaborate design strategy is required when VPBs are abundant, especially for multiform VPBs.

For methods processing the ECG signal on a beat-by-beat basis, it is essential to ensure that the extracted f wave signal does not exhibit jumps at the boundaries between successive beats. This problem is particularly pronounced when the ventricular activity is estimated from short RR intervals [3]. On the other hand, jumps are less likely to occur when the ECG signal is processed on a sequential, sample-by-sample basis.

It is, of course, desirable that an extraction method can satisfactorily handle both single- and multiple-lead recordings, the presence of VPBs, short-duration recordings, as well as any of the other factors mentioned above. However, the methods described in the literature have various limitations which call for awareness of the conditions under which a method is planned to operate. Indeed, different extraction methods may be needed in certain ECG applications. Another important consideration is the amount of computations which can differ dramatically from one method to another, ranging from just a few additions to several hundreds of multiplications to process one sample of the ECG, with implications on the applicability of a method.

This chapter provides a comprehensive overview of methods for f wave extraction, divided into the following categories:

- average beat subtraction and variants (Sect. 5.2),
- interpolation (Sect. 5.3)
- extended Kalman filtering (Sect. 5.4)
- adaptive filtering (Sect. 5.5),
- principal component analysis (Sect. 5.6),
- singular spectral analysis (Sect. 5.7),
- autoregressive modeling and prediction error analysis (Sect. 5.8), and
- independent component analysis (Sect. 5.9).

For some of these categories, QRS detection, beat morphology classification, and time alignment of beats with similar morphology need to be performed before f wave extraction. A review of methods for handling these well-investigated preprocessing steps can be found in [4, 5].

Depending on heart rate and f wave repetition rate, a number of f waves can usually be observed in the TQ interval without any interfering ventricular activity. Hence, some extraction methods either analyze the samples of the TQ intervals for the purpose of producing interpolated samples in the QT interval, or simply confine the entire analysis to the TQ intervals, for example, the approach originally considered for estimating the power spectrum of f waves [6]. However, most extraction methods account for the fact that AF is an ongoing process, and, therefore, make use of all samples in the ECG, including those in the QT interval. The latter approach should not only lead to more reliable f wave characterization, but also circumvent the problems associated with delineating the TQ interval and a vanishing TQ interval at higher heart rates.

In contrast to AF detection, where annotated ECG databases have been available since long to facilitate performance evaluation, it is considerably more challenging to evaluate methods for f wave extraction since the accuracy of the extracted signal is not easily related to manual annotations. As a consequence, indirect performance measures have been investigated, for example, by comparing the amplitude of the signal extracted within the QRS interval to the amplitude in the surrounding TQ intervals. Performance evaluation can also be addressed by employing simulated ECG signals in AF, for example, produced by the simulation model described in Sect. 3.4. Such signals make it possible to compute the sample-by-sample error between the extracted and the simulated f wave signal, which in turn can be condensed into a suitably defined error measure. Sections 5.10 and 5.11 describe different performance measures, developed either for real or simulated ECG signals, and different approaches to performance evaluation, respectively.

The early review of f wave extraction methods published in [7] is significantly expanded below to cover state-of-the-art research, as well as to provide a much more detailed description of the methods. Another early review describes extraction methods developed for electrograms [8].

5.2 Average Beat Subtraction and Variants

Average beat subtraction (ABS) is the most well-known method for extraction of f waves in individual leads, having become one of the fixtures in the biomedical signal processing toolbox. The method was originally developed for detection of atrioventricular dissociated ventricular tachycardia in which the P waves are dissociated from the QRS complexes [9–11], but later employed for the processing of ECG signals in AF [12]. Thanks to its implementation simplicity, ABS continues to be used in clinically oriented studies on AF, both for the analysis of surface ECGs and electrograms, see, e.g., [1, 2, 13–20]. Since ABS suffers from several major limitations, different variants have been developed with the goal to offer better performance, described in Sects. 5.2.1–5.2.5. However, entirely different signal processing principles have also been explored.

In the ABS method, as well as variants, the well-known signal-plus-noise model is the starting point for finding an estimate of the QRST complex $s(n)$ to be subtracted from the ECG signal; the resulting estimate, denoted $\hat{s}(n)$, is sometimes referred to as a QRST template. In this model, each beat $x_i(n)$ of the observed signal is assumed to be composed of $s(n)$ and noise $z_i(n)$,

$$x_i(n) = s(n) + z_i(n), \quad i = 1, \ldots, M, \quad n = 0, \ldots, N-1, \tag{5.1}$$

where M is the number of beats in the ensemble, and N is the number of samples in each beat. From a conceptual viewpoint, it is advantageous to decompose the noise $z_i(n)$ into an f wave signal $d_i(n)$, being the desired quantity for extraction, and noise $v_i(n)$ of extracardiac origin,

$$z_i(n) = d_i(n) + v_i(n), \tag{5.2}$$

where both terms are usually modeled as random processes. The noise $z_i(n)$ is modeled as a zero-mean stationary process with variance σ_z^2, assumed to be uncorrelated from beat to beat,

$$E[z_i(n)z_j(n)] = \sigma_z^2\delta(i-j), \quad i, j = 1, \ldots, M, \tag{5.3}$$

where

$$\delta(i) = \begin{cases} 1, & i = 0, \\ 0, & i \neq 0. \end{cases} \tag{5.4}$$

The structure of the resulting estimator depends on the statistical assumptions made on $s(n)$ and $z_i(n)$. In general, an increasingly more detailed statistical characterization of the different components implies that more statistical parameters need to be determined from the ECG signal, which in turn may jeopardize performance in certain situations.

In its general form, the linear estimator of $s(n)$ is given by

$$\hat{s}(n) = \sum_{m=1}^{M} w_m(n)x_m(n), \tag{5.5}$$

where the weights $w_m(n)$ differ from beat to beat as well as from sample to sample. To ensure that the estimator in (5.5) is unbiased, $w_m(n)$ must fulfill the following constraint:

$$\sum_{m=1}^{M} w_m(n) = 1. \tag{5.6}$$

For the i-th beat, an estimate of the f wave signal is obtained by subtracting the QRST template $\hat{s}(n)$ from $x_i(n)$,

$$\hat{d}_i(n) = x_i(n) - \hat{s}(n), \tag{5.7}$$

where the noise component $v_i(n)$ in (5.2) has been neglected.

The *ensemble average*, being central to the ABS method, is computed by simply setting all weights in (5.5) to the same value,

$$w_i(n) = \frac{1}{M}, \quad i = 1, \ldots, M, \quad n = 0, \ldots, N-1. \tag{5.8}$$

The usefulness of this approach rests on the assumption that the ventricular activity, modeled by $s(n)$, is decoupled from the atrial activity, modeled by $z_i(n)$. Moreover, $s(n)$ is viewed as a deterministic, but unknown, signal. The main steps involved with ABS are illustrated in Fig. 5.1.

As the ensemble size M grows, the attenuation of the f waves in $\hat{s}(n)$ becomes increasingly better. Using simulated f waves, it has been shown that the error between the true and the estimated f wave signals no longer improves significantly when more than 30–40 beats are averaged [21], see also [22]. When VPBs are present, each type of morphology needs to be associated with its specific ensemble average to make beat

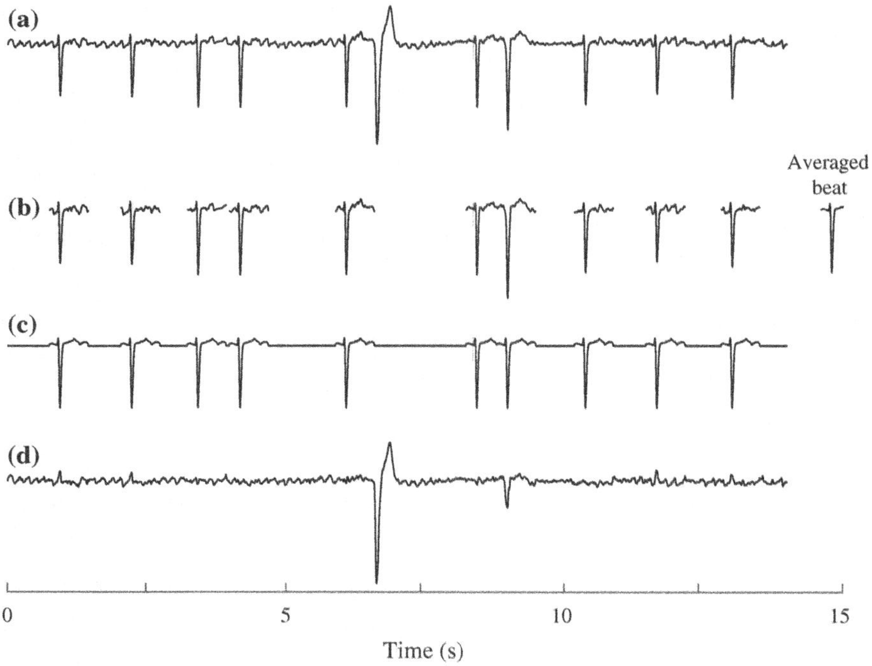

Fig. 5.1 Steps involved with average beat subtraction. **a** Original ECG, and **b** QRST complexes with similar morphology used for computing an averaged QRST complex. **c** Ventricular signal, constructed from the averaged QRST complexes and subtracted from the ECG in (**a**) to produce **d** the extracted f wave signal. Since only beats with similar (dominant) morphology are averaged, the VPB with deviating morphology, occurring after about 6 s, remains unprocessed in the f wave signal. The other ectopic beat has a more dominant morphology which, when subtracted, causes a large-amplitude, QRS-related residual

subtraction meaningful. Not uncommonly, VPBs are sparse in numbers which cause the ventricular activity to be insufficiently reduced in the extracted f wave signal, or not reduced at all if only a single VPB occurs, cf. Fig. 5.1. The computation of morphology-specific beat averages must be preceded by clustering of the different beat morphologies found in the ECG signal.

While the length of a beat varies dramatically in AF, the signal model in (5.1) does not reflect this variation since a large enough, fixed value of N is assumed. As a result, the number of averaged samples depends on the location within the cardiac cycle, i.e., more samples are available for averaging within the QRS interval than after the T wave, implying that the variance of $\hat{s}(n)$ is larger at the boundaries of the cardiac cycle than within the QRS interval. Thus, the extracted f wave signal can be expected to have better accuracy within the central parts of the cardiac cycle than at its boundaries.

Another popular estimator is the *exponential averager* which, thanks to its beat-recursive structure, can track slow changes in QRST morphology [23]. The exponential averager is defined by

$$\hat{s}_i(n) = \hat{s}_{i-1}(n) + \alpha(x_i(n) - \hat{s}_{i-1}(n)), \quad i = 1, \ldots, M, \tag{5.9}$$

where the weight factor α $(0 < \alpha < 1)$ determines the speed of tracking. The recursion may be initialized by setting $\hat{s}_0(n) = 0$. It may be interesting to note that after processing of all M beats in the ensemble, i.e., $\hat{s}(n) \equiv \hat{s}_M(n)$, the weights associated with the exponential averager are given by

$$w_i(n) = \alpha(1-\alpha)^{M-i}, \quad i = 1, \ldots, M, \quad n = 0, \ldots, N-1. \tag{5.10}$$

5.2.1 *Noise-Dependent Weights*

The above fixed-variance assumption of $z_i(n)$ can be relaxed so that the variance is allowed to change from beat to beat, i.e., σ_z^2 is replaced by $\sigma_{z,i}^2$. This relaxation may be motivated by the variation in noise level often encountered in ambulatory recordings as well as the waxing and waning nature of the f waves. The weights $w_i(n)$ of the linear estimator can be determined by minimizing the mean square error (MSE) between $s(n)$ and $\hat{s}(n)$,

$$E\left[(s(n) - \hat{s}(n))^2\right] = E\left[\left(s(n) - \sum_{i=1}^{M} w_i(n)x_i(n)\right)^2\right]. \tag{5.11}$$

Accounting for the constraint in (5.6) which states that the weights must sum to one, it can be shown that the optimal weights are inversely proportional to the noise variance of each beat [4, Chap. 4]:

$$w_i(n) = \frac{\dfrac{1}{\sigma^2_{z,i}}}{\displaystyle\sum_{m=1}^{M} \frac{1}{\sigma^2_{z,m}}}, \quad i = 1, \ldots, M, \quad n = 0, \ldots, N-1. \tag{5.12}$$

This approach to estimating $s(n)$, known as *weighted averaging*, requires prior knowledge of the noise variance $\sigma^2_{z,i}$. An estimate of $\sigma^2_{z,i}$ may be determined, for example, from the samples of the TQ intervals enclosing the i-th beat, on condition that these two intervals are sufficiently long. Obviously, this approach assumes that the noise in these two intervals are representative for the QRST interval. It is well-known that weighted averaging is particularly efficient when large differences in noise variance exist within the ensemble of beats. When the noise has fixed variance, i.e., $\sigma^2_{z,i} \equiv \sigma^2_z$, the weights $w_i(n)$ in (5.12) simplify to the fixed weights of ensemble averaging given in (5.8) .

An adaptive version of weighted averaging has been developed in which both the QRST complex $s(n)$ and the weights $w_i(n)$ are estimated on a sample-by-sample basis [24]. Recursive minimization of the MSE criterion is performed, involving a constraint which ensures that the estimated signal amplitude remains unaltered by the averaging algorithm. This approach results in weights which are allowed to change from sample to sample within each beat, thus differing from the weights in (5.12) which are fixed within each beat. The improvement in SNR of $\hat{s}(n)$ was found to be particularly pronounced when small ensemble sizes were analyzed—a result with significance when slow variations in QRST morphology are of particular interest to track, which would have been smoothed out by ensemble averaging.

5.2.2 Signal- and Noise-Dependent Weighted Averaging

Another variant of the signal-plus-noise model in (5.1) is based on the assumption that the QRST complex $s(n)$ changes from beat to beat,

$$s(n) \to s_i(n),$$

implying that the weights in (5.5) are generalized to $w_{i,m}(n)$ when computing the estimate of the i-th beat $\hat{s}_i(n)$,

$$\hat{s}_i(n) = \sum_{m=1}^{M} w_{i,m}(n)x_m(n), \quad i = 1, \ldots, M. \tag{5.13}$$

Moreover, $s_i(n)$ is assumed to be random in nature, rather than deterministic as earlier assumed, and characterized by a time-varying variance which allows $s_i(n)$ to have larger variability in certain intervals. However, the variance pattern is assumed to be the same in all beats [21],[1]

$$E\left[(s_i(n) - E[s_i(n)])^2\right] = \sigma_s^2(n), \quad n = 0, \ldots, N-1. \tag{5.14}$$

In this model, $z_i(n)$ only accounts for f waves, i.e., $z_i(n) \equiv d_i(n)$, modeled as a zero-mean, stochastic process with identical variance σ_d^2 of all beats. As before, $z_i(n)$ is assumed to be uncorrelated from beat to beat, and the ventricular and atrial activities are assumed to be mutually uncorrelated,

$$E[s_i(n)d_i(n)] = 0, \quad i = 1, \ldots, M. \tag{5.15}$$

Since this method involves statistical assumptions on both $s_i(n)$ and $z_i(n)$, it is here referred to as *signal- and noise-dependent weighted averaging*.

To proceed, the MSE between the f wave signal $d_i(n)$ and its corresponding estimate $\hat{d}_i(n)$ [21],

$$E\left[e_i^2(n)\right] = E\left[(d_i(n) - \hat{d}_i(n))^2\right], \tag{5.16}$$

is explored. Since

$$\begin{aligned} \hat{d}_i(n) &= x_i(n) - \sum_{m=1}^{M} w_{i,m}(n)x_m(n) \\ &= s_i(n) + d_i(n) - \sum_{m=1}^{M} w_{i,m}(n)x_m(n), \end{aligned} \tag{5.17}$$

the MSE in (5.16) can alternatively be expressed as

$$E\left[e_i^2(n; w_{i,m}(n))\right] = E\left[\left(\sum_{m=1}^{M} w_{i,m}(n)s_m(n) + \sum_{m=1}^{M} w_{i,m}(n)d_m(n) - s_i(n)\right)^2\right], \tag{5.18}$$

where the dependence of $e_i(n)$ on $w_{i,m}(n)$ has been made explicit. The difference in the definitions of error criteria deserves a comment: the MSE in (5.16) explicitly involves the desired f wave signal $\hat{d}_i(n)$, whereas the MSE in (5.11) involves the QRST complex $\hat{s}(n)$ required in (5.7) to produce $\hat{d}_i(n)$. However, when $z_i(n)$ is confined to f waves only, i.e., $z_i(n) \equiv d_i(n)$, the two definitions lead to the same result. It should be noted that

[1]This signal model was also considered for determining a QRST template using a maximum likelihood (ML) approach. Since the performance of the ML-based method was found to be inferior to that of the method based on the MSE criterion described in this section, the interested reader is referred to [21] for further details.

$$x_i(n) = s_i(n) + d_i(n) = \hat{s}_i(n) + \hat{d}_i(n). \tag{5.19}$$

Using (5.15) and (5.14), the MSE becomes

$$E\left[e_i^2(n; w_{i,m}(n))\right] = \sigma_s^2(n)\left(1 - 2w_{i,i}(n) + \sum_{m=1}^{M} w_{i,m}^2(n)\right) + \sigma_d^2 \sum_{m=1}^{M} w_{i,m}^2(n). \tag{5.20}$$

Since the weights of the linear estimator must sum to one, cf. (5.6), the minimization of $E\left[e_i^2(n; w_{i,m}(n))\right]$ is accomplished by introducing a function (the Lagrange function) which involves this sum as a constraint,

$$\mathscr{L}(w_i(n)) = E\left[e_i^2(n; w_{i,m}(n))\right] + \lambda\left(1 - \sum_{m=1}^{M} w_{i,m}(n)\right), \tag{5.21}$$

where λ is the Lagrange multiplier. The optimal weights are given by [21]

$$w_{i,m}(n) = \frac{1}{M}\frac{\sigma_d^2}{\sigma_d^2 + \sigma_s^2(n)} + \frac{\sigma_s^2(n)}{\sigma_d^2 + \sigma_s^2(n)}\delta(i - m), \quad i = 1, \ldots, M. \tag{5.22}$$

Using this approach to f wave extraction, all beats in the ensemble are weighted in the same way, except the current beat, i.e., $i = m$, which is assigned more weight due to the inclusion of the positive-valued quantity $\sigma_s^2(n)/(\sigma_d^2 + \sigma_s^2(n))$. While $w_{i,m}(n)$ varies over time across each beat, the same weight function is applied to all beats, except for the current beat, since the variance σ_d^2 is assumed to be fixed. As the number of beats M increases, $w_{i,m}(n)$ becomes increasingly more dominated by $\sigma_s^2(n)$ through the second term in (5.22). When $\sigma_s^2(n) = 0$, reflecting a perfectly homogenous beat ensemble, the weights in (5.22) become identical to those of ensemble averaging, i.e., $w_{i,m}(n) = 1/M$. On the other hand, when $\sigma_s^2(n) \gg \sigma_d^2$, all weights become approximately zero, except the weight of the current beat which is assigned a value close to one. When large beat-to-beat variation in morphology is encountered, i.e., when $\sigma_s^2(n)$ is large, this method assigns more weight to the current beat in the weighted average than to the other beats of the ensemble.

Before applying the weights in (5.22), the statistical parameters $\sigma_s^2(n)$ and σ_d^2 need to be determined from the observed signal. An estimate of σ_d^2 may be obtained by computing the sample variance of several, concatenated TQ intervals [21, 25], cf. also Sect. 5.2.1. Such an approach becomes less useful at higher heart rates when the TQ intervals are short and may vanish. However, this particular problem is largely ignored when the PTB database is used for evaluation of extraction performance [26], with synthetic f waves added to non-AF ECGs [21], since high heart rates are uncommon in this database. Hence, the results on performance should be interpreted with caution.

An estimate of $\sigma_s^2(n)$ can be obtained by first computing the ensemble variance of the beats $x_1(n), \ldots, x_M(n)$, $n = 0, \ldots, N-1$, and then subtracting the variance σ_d^2 which describes the f waves. To ensure that $\hat{\sigma}_s^2(n)$ is always positive-valued, the following estimator can be used [21]:

$$\hat{\sigma}_s^2(n) = \max\left(\frac{1}{M-1}\sum_{m=1}^{M}(x_m(n) - \hat{s}(n))^2 - \hat{\sigma}_d^2, 0\right), \tag{5.23}$$

where $\hat{s}(n)$ denotes the ensemble average. While the computation of $\hat{\sigma}_s^2(n)$ is straightforward, it has been shown that $\hat{\sigma}_s^2(n)$ is more sensitive to temporal misalignment of the beats in the ensemble than is $\hat{s}(n)$, leading to an overestimation of $\hat{\sigma}_s^2(n)$ [27]. In that study, a sampling rate of 3 kHz was recommended to ensure that sampling-related misalignment does not influence $\hat{\sigma}_s^2(n)$, i.e., a much higher sampling rate than what is used in most modern ECG devices (1 kHz). However, it is straightforward to perform digital interpolation to obtain the recommended sampling rate.

Since low-amplitude T waves are difficult to detect, and consequently to align, $\sigma_s^2(n)$ may not be reliably estimated. This problem can be handled by using the time-dependent weights in (5.22) when processing the QRS interval, whereas the fixed weights of ABS are used when processing the T wave interval [21]—a solution requiring two templates to accomplish f wave extraction. Separate processing of these two intervals has been considered before, but then in the context of spatiotemporal alignment, see Sect. 5.2.4.

As already noted, the performance of signal- and noise-dependent weighted averaging was evaluated using simulated f waves, generated by the harmonic model described in Sect. 3.4, added to non-AF ECGs of the PTB database. The results showed that the normalized MSE between the extracted and the model f wave signals is about 25% lower than that of ABS. The main explanation to this improvement is related to the inclusion of the ensemble variance $\hat{\sigma}_s^2(n)$ in the computation of $w_{i,m}(n)$, which should increase as the morphologic beat-to-beat variability increases. No information was provided on whether classification of beat morphology is required before forming the beat ensemble, and, therefore, it remains to be quantified to what extent beats with deviating morphology influence f wave extraction in terms of QRS-related residuals. However, when such beats are encountered, smaller QRS-related residuals are likely to result for $w_{i,m}(n)$ than for the weights used in ABS or weighted ABS.

Figure 5.2 illustrates the performance of signal- and noise-dependent weighted averaging when an ECG signal with considerable variation in f wave amplitude is processed. It is interesting to note that the extracted f wave signal does not contain as much QRS-related residuals as does the ABS-produced f wave signal.

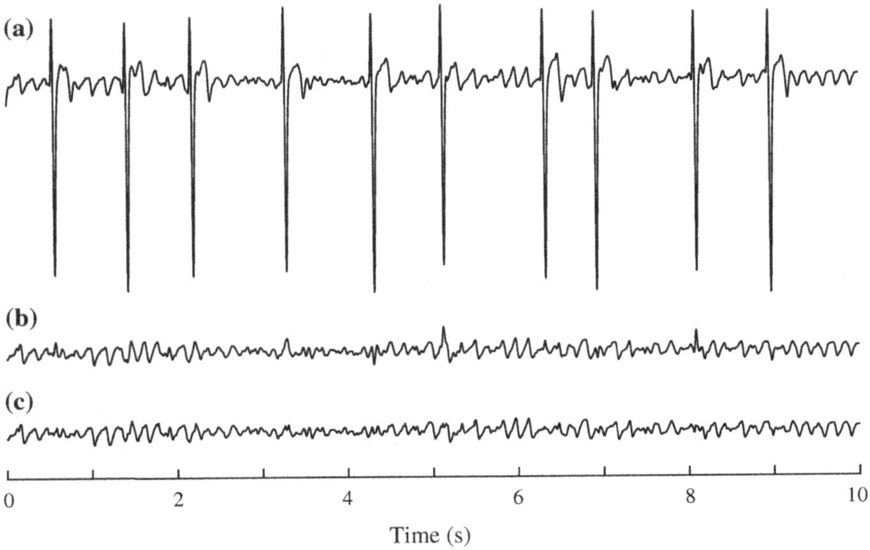

Fig. 5.2 Signal- and noise-dependent weighted averaging. **a** Original ECG and extracted f wave signal obtained by **b** average beat subtraction, and **c** signal- and noise-dependent weighted averaging

5.2.3 Spatiotemporal QRST Cancellation

Average beat subtraction relies on the assumption that the ensemble average of time-aligned beats can represent individual beats adequately. Due to variations in the orientation of the heart's electrical axis, minor changes in QRST morphology are often observed, thus questioning the validity of this assumption. The variations are primarily induced by respiratory activity: the electrical axis of the QRS complex varies as much as 10° during inspiration in the transversal plane, and, consequently, influences the precordial leads quite considerably [28]. In general, V_2 is more sensitive to changes in position and orientation of the heart than the other five precordial leads [29]. Given that ABS processes signals on a lead-by-lead basis, the respiratory-induced variations in the electrical axis will sometimes cause large-amplitude QRS-related residuals. Since the variations occur on a beat-to-beat basis, recursive beat-to-beat update methods such as exponential averaging, cf. (5.9), do not offer a satisfactory solution to this problem. However, the use of signal- and noise-dependent weighted averaging, see Sect. 5.2.2, is able to handle minor changes in QRST morphology, because a larger weight is assigned to the current beat than to the other beats in the ensemble when computing the beat template $\hat{s}_i(n)$. It should be noted, though, that the potentially better handling is not the result of a model describing how respiration influences different leads, but of the assumptions made on the signal and noise properties, leading to the proposed weighting scheme.

Spatiotemporal cancellation (STC) [30], based on a statistical method originally developed for vectorcardiographic loop alignment [31], aims at reducing large-amplitude QRS-related residuals by combining the average beat of the processed lead with the average beats of the adjacent lead(s). Consequently, at least two leads are needed to benefit from the spatial analysis performed by this method, preferably more leads. While it has been claimed that STC is unsuitable for use in two- or three-lead Holter recordings [21], several studies have demonstrated that STC can very well be applied to such recordings, see, e.g., [32, 33]. For single-lead recordings, the STC becomes almost identical to ABS, as explained below.

Spatiotemporal cancellation takes its starting point in the L-lead QRST template $\hat{\mathbf{S}}_t$, represented by the $(N + 2\Delta) \times L$ matrix

$$\hat{\mathbf{S}}_t = \left[\hat{\mathbf{s}}_{t,1}\ \hat{\mathbf{s}}_{t,2} \cdots \hat{\mathbf{s}}_{t,L}\right], \tag{5.24}$$

where the column vector $\hat{\mathbf{s}}_{t,l}$ contains $N + 2\Delta$ samples of the l-th lead. The average beat $\hat{\mathbf{S}}_t$ contains 2Δ additional samples to facilitate time alignment of the observed, multi-lead beat $\mathbf{X}$, represented by an $N \times L$ matrix, and $\hat{\mathbf{S}}_t$. A reduction of QRS-related residuals is achieved by modifying $\hat{\mathbf{S}}_t$ with the $N \times (N + 2\Delta)$ shift matrix $\mathbf{J}_\tau$ which accounts for time misalignment, defined by

$$\mathbf{J}_\tau = \left[\mathbf{0}_{N\times(\Delta+\tau)}\ \mathbf{I}_{N\times N}\ \mathbf{0}_{N\times(\Delta-\tau)}\right], \tag{5.25}$$

and the $L \times L$ matrix $\mathbf{P}$ which accounts for spatial misalignment,

$$\hat{\mathbf{S}} = \mathbf{J}_\tau \hat{\mathbf{S}}_t \mathbf{P}. \tag{5.26}$$

The matrices $\mathbf{0}$ and $\mathbf{I}$ denote the zero matrix and the identity matrix, respectively, and the integer time shift τ can handle a misalignment not exceeding $\pm\Delta$. As shown below, $\mathbf{J}_\tau$ and $\mathbf{P}$ are estimated from the observed beat $\mathbf{X}$ after which the resulting estimate of the QRST complex $\hat{\mathbf{S}}$ is subtracted from $\mathbf{X}$ to produce the desired f wave signal.

The matrix $\mathbf{P}$ introduces the following two desirable properties in the cancellation process:

1. shifting of information between leads to compensate for variations in the electrical axis, and
2. scaling to compensate for variations in tissue conductivity and heart position which may affect the amplitude in different leads.

Although $\mathbf{P}$ can be assigned different structures, its definition as the product of an $L \times L$ diagonal amplitude scaling matrix $\mathbf{A}$ and an $L \times L$ rotation matrix $\mathbf{Q}$,

$$\mathbf{P} = \mathbf{A}\mathbf{Q}, \tag{5.27}$$

was found to be associated with the best performance [30]. The matrix $\mathbf{Q}$ is orthogonal and, therefore, $\mathbf{Q}^T = \mathbf{Q}^{-1}$.[2]

The QRST template $\hat{\mathbf{S}}_t$ was obtained by ensemble averaging in [30], however, any other type of ensemble estimator can be used. Thus, spatiotemporal QRST cancellation is a technique for processing one multi-lead beat at a time, on condition that a QRST template is available. This explains why the indexing of beat number, previously used when describing ABS and variants, is dropped in this subsection.

To proceed, each observed beat $\mathbf{X}$ is modeled as the sum of a QRST complex $\mathbf{S}$, f waves $\mathbf{D}$, and noise $\mathbf{V}$,

$$\mathbf{X} = \mathbf{S} + \mathbf{D} + \mathbf{V}, \tag{5.28}$$

being the multi-lead counterpart to the signal model in (5.1) and (5.2). The desired f wave signal $\hat{\mathbf{D}}$ is obtained by first estimating $\mathbf{A}$, $\mathbf{Q}$, and τ from $\mathbf{X}$, and then subtracting $\hat{\mathbf{S}}$ from $\mathbf{X}$,

$$\begin{aligned}\hat{\mathbf{D}} &= \mathbf{X} - \hat{\mathbf{S}} \\ &= \mathbf{X} - \mathbf{J}_{\hat{\tau}}\hat{\mathbf{S}}_t\hat{\mathbf{A}}\hat{\mathbf{Q}},\end{aligned} \tag{5.29}$$

where, just as in (5.7), the noise component $\mathbf{V}$ has been neglected. However, not only will $\mathbf{V}$ limit how well the modified QRST complex $\mathbf{J}_{\hat{\tau}}\hat{\mathbf{S}}_t\hat{\mathbf{A}}\hat{\mathbf{Q}}$ fits $\mathbf{X}$, but so will the f waves $\mathbf{D}$. To mitigate this problem, it is helpful to introduce an intermediate ("quick-and-dirty") estimate $\tilde{\mathbf{D}}$ which, before the estimation of $\mathbf{A}$, $\mathbf{Q}$, and τ, is subtracted from $\mathbf{X}$,

$$\begin{aligned}\mathbf{Y} &= \mathbf{X} - \tilde{\mathbf{D}}, \\ &= \mathbf{S} + (\mathbf{D} - \tilde{\mathbf{D}}) + \mathbf{V}.\end{aligned} \tag{5.30}$$

A straightforward approach to determining $\tilde{\mathbf{D}}$ is to form, for each lead, an f wave signal in the QRST interval through interpolation based on the f waves of the two enclosing TQ-intervals; such interpolation is briefly described at the end of this subsection.

The parameters $\mathbf{A}$, $\mathbf{Q}$, and τ are estimated by minimizing the quadratic error ε^2 between $\mathbf{Y}$ and $\mathbf{J}_\tau\hat{\mathbf{S}}_t\mathbf{A}\mathbf{Q}$, i.e., the right hand side of (5.29), but with $\mathbf{X}$ replaced by the quick-and-dirty-corrected $\mathbf{Y}$,

$$\varepsilon^2 = \|\mathbf{Y} - \mathbf{J}_\tau\hat{\mathbf{S}}_t\mathbf{A}\mathbf{Q}\|_F^2, \tag{5.31}$$

---

[2]For the single-lead case, i.e., $L = 1$, STC simplifies to ABS, except for that misalignment in time can still be handled by $\mathbf{J}_\tau$ and amplitude mismatch to $\hat{\mathbf{S}}_t$ by the scaling factor a_1; the rotation matrix reduces to a scalar equal to one. The single-lead version of STC is closely related to the extraction method based on singular value decomposition, as described in Sect. 5.6.1, which also involves amplitude scaling.

where the Frobenius norm of $\mathbf{X}$ is defined as the trace of $\mathbf{X}\mathbf{X}^T$, i.e., $\|\mathbf{X}\|_F^2 = \text{tr}(\mathbf{X}\mathbf{X}^T)$. In expanded form, the quadratic error is

$$\varepsilon^2 = \text{tr}\left(\mathbf{Y}\mathbf{Y}^T\right) + \text{tr}\left(\mathbf{J}_\tau\hat{\mathbf{S}}_t\mathbf{A}\mathbf{A}^T\hat{\mathbf{S}}_t^T\mathbf{J}_\tau^T\right) - 2\text{tr}\left(\mathbf{A}^T\hat{\mathbf{S}}_t^T\mathbf{J}_\tau^T\mathbf{Y}\mathbf{Q}^T\right), \tag{5.32}$$

where the cross-term results from the following properties of the trace:

$$\text{tr}\left(\mathbf{J}_\tau\hat{\mathbf{S}}_t\mathbf{A}\mathbf{Q}\mathbf{Y}^T\right) = \text{tr}\left(\mathbf{A}^T\hat{\mathbf{S}}_t^T\mathbf{J}_\tau^T\mathbf{Y}\mathbf{Q}^T\right), \tag{5.33}$$

$$\text{tr}\left(\mathbf{Y}(\mathbf{J}_\tau\hat{\mathbf{S}}_t\mathbf{A}\mathbf{Q})^T\right) = \text{tr}\left(\mathbf{A}^T\hat{\mathbf{S}}_t^T\mathbf{J}_\tau^T\mathbf{Y}\mathbf{Q}^T\right). \tag{5.34}$$

Unfortunately, minimization with respect to $\mathbf{Q}$ and $\mathbf{A}$ cannot be done independently, and a closed-form solution is difficult to find. Instead, an alternating, iterative approach can be employed in which ε^2 is minimized with respect to $\mathbf{Q}$ by maximizing the last term in (5.32) under the assumption that $\mathbf{A}$ is known. The maximization is accomplished using singular value decomposition (SVD) by which a general matrix $\mathbf{T}$ is decomposed into two orthonormal matrices $\mathbf{U}$ and $\mathbf{V}$ and a diagonal matrix $\boldsymbol{\Sigma}$ containing singular values,

$$\mathbf{T} = \mathbf{U}\boldsymbol{\Sigma}\mathbf{V}^T. \tag{5.35}$$

By setting $\mathbf{T} = \mathbf{A}^T\hat{\mathbf{S}}_t^T\mathbf{J}_\tau^T\mathbf{Y}$, the cross-term in (5.32) can be expressed as $2\text{tr}(\mathbf{T}\mathbf{Q}^T)$, which is maximized when [34]

$$\hat{\mathbf{Q}} = \mathbf{U}\mathbf{V}^T. \tag{5.36}$$

Next, with an estimate of $\mathbf{Q}$ available, the diagonal entries a_l of $\mathbf{A}$ can be estimated using [30]

$$\hat{a}_l = \left(\left[\mathbf{J}_\tau\hat{\mathbf{S}}_t\right]_l^T\left[\mathbf{J}_\tau\hat{\mathbf{S}}_t\right]_l\right)^{-1}\left(\left[\mathbf{J}_\tau\hat{\mathbf{S}}_t\right]_l^T\left[\mathbf{Y}\hat{\mathbf{Q}}^{-1}\right]_l\right), \quad l = 1, \ldots, L, \tag{5.37}$$

where $[\cdot]_l$ denotes the l-th column of the matrix. An improved estimate of $\mathbf{Q}$ is obtained from (5.36) using $\hat{\mathbf{A}}$, and so on.

Typically, a solution close to $\mathbf{Q} = \mathbf{A} = \mathbf{I}$ is desirable, and, therefore, the iterative procedure is initialized with $\mathbf{A}_0 = \mathbf{I}$, i.e., the assumption implicit to ABS which states that all beats have the same amplitude. The rotation matrix at iteration step k, i.e., $\hat{\mathbf{Q}}_k$, is then calculated from $\mathbf{A}_{k-1}$. Since

$$\|\mathbf{Y} - \mathbf{J}_\tau\hat{\mathbf{S}}_t\mathbf{A}_{k-1}\hat{\mathbf{Q}}_k\|_F^2 \;\leq\; \|\mathbf{Y} - \mathbf{J}_\tau\hat{\mathbf{S}}_t\mathbf{A}_{k-1}\hat{\mathbf{Q}}_{k-1}\|_F^2, \tag{5.38}$$

the error will be less or equal to that of the previous iteration step. When $\mathbf{Q}_k$ is known, $\mathbf{A}_k$ can be calculated and, accordingly,

$$\|\mathbf{Y} - \mathbf{J}_\tau \hat{\mathbf{S}}_t \mathbf{A}_k \hat{\mathbf{Q}}_k\|_F^2 \;\leq\; \|\mathbf{Y} - \mathbf{J}_\tau \hat{\mathbf{S}}_t \mathbf{A}_{k-1} \hat{\mathbf{Q}}_k\|_F^2. \tag{5.39}$$

This procedure is repeated until the difference in the error between two successive iterations becomes sufficiently small. The algorithm will converge since minimization with respect to **A** and **Q** in each step, described by (5.38) and (5.39), will successively lower the error ε^2. In practice, convergence is achieved after five or six iterations, with most of the improvement achieved already after the first iteration.

Finally, minimization with respect to τ is implemented as a grid search in the interval $[-\Delta, \Delta]$. Consequently, $\hat{\mathbf{A}}$ and $\hat{\mathbf{Q}}$ have to be computed for all possible values of τ before the optimal estimates become available. Provided that the accuracy of the fiducial point used for aligning the beats is reasonably good, a value of Δ equal to 5 ms is sufficient.

As already pointed out, before estimating **A** and **Q** it is desirable to determine intermediate, TQ-based f wave signals $\tilde{\mathbf{d}}_1, \ldots, \tilde{\mathbf{d}}_L$ which help reduce the influence of f waves in the estimation process. A simple approach is to replicate the f waves of the TQ interval preceding the QRST complex in the QRST interval, where each replicated sample is multiplied with a weight linearly decreasing from 1 at the interval onset to 0 at the interval end. This procedure is repeated, but in a time-reversed fashion, by replicating the f waves of the TQ interval following the QRST complex in the QRST interval, where each replicated sample is multiplied with a weight linearly decreasing from 1 at the interval end to 0 at the interval onset. The interpolated signal is obtained by summing the two replicated f wave signals in the QRST interval. The f wave cycle length to be replicated is estimated from the autocorrelation function of the two adjacent TQ intervals. If only one of the enclosing intervals is long enough for extrapolation, only that interval is used for reconstruction, whereas the TQ-based f wave signal is set to zero when both intervals are too short. It should be noted that the interpolation methods described in Sect. 5.3, proposed for reducing the influence of large-amplitude QRS-related residuals, may alternatively be considered.

The use of a TQ-based, interpolated signal improves the accuracy of $\hat{\mathbf{A}}$ and $\hat{\mathbf{Q}}$, while not introducing significant f wave distortion [30]. For low-amplitude f waves, the cycle length becomes less well-defined and therefore more difficult to estimate when computing a TQ-based signal. Fortunately, low-amplitude f waves are also less influential on the estimation process. On the other hand, large-amplitude f waves tend to be more "organized" and characterized by a well-defined cycle length, thus lending themselves better to TQ-based signal interpolation.

Although rotation lacks a geometrical interpretation in nonorthogonal leads, it can still reduce QRS-related residuals due to variation in the orientation of electrical axis of the heart. The reduction is illustrated in Fig. 5.3, where the commonly used precordial leads V_1, V_2, and V_3 are processed with ABS and STC.

The parameters **A**, **Q**, and τ are estimated from the samples of the QRS interval since the samples outside this interval have a much lower SNR. However, the spatiotemporally modified template $\hat{\mathbf{S}}$ in (5.26) is still subtracted from the entire cardiac cycle, not just from the QRS interval. This approach offers better cancellation in the QRS interval than ABS, while the cancellation outside the QRS interval may be worse. Hence, the use of two-template STC—one template assigned to the

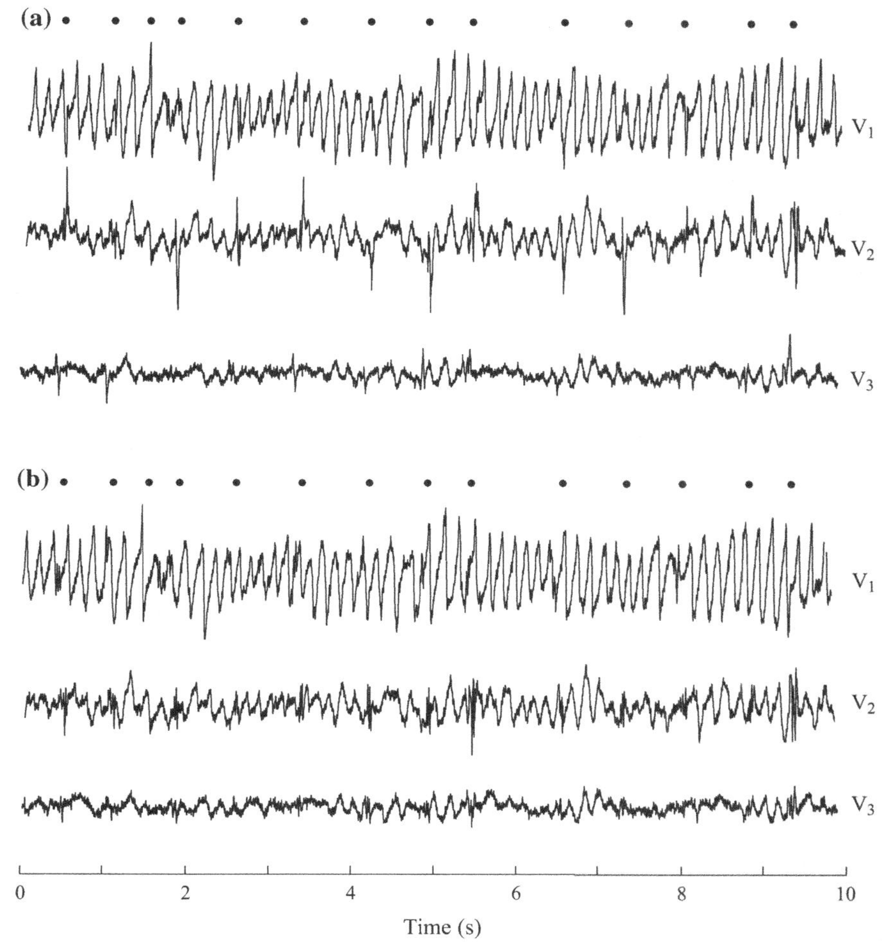

Fig. 5.3 Extraction of f wave signals in leads V_1, V_2, and V_3, using **a** average beat subtraction and **b** spatiotemporal cancellation which reduces the QRS-related residuals. The dots indicate the occurrence time of the QRS complexes

QRS interval and another to the JQ interval, i.e., beginning at the end of the QRS complex ("J point") and ending at the onset of the Q wave of the subsequent beat—may offer a solution to this problem.

5.2.4 Separate Cancellation of the QRS and JQ Intervals

The use of separate templates for cancellation of the ventricular activity in the QRS and JQ intervals was proposed in [35], implemented in [36], and further developed and evaluated in [37]. The two-template approach is motivated by the fact that T wave duration changes with heart rate, while QRS duration remains essentially unchanged. Considering the presence of respiratory-induced variation in the electrical axis of the heart, another important motivation for pursuing this approach is that different amplitude scaling and rotation matrices can be applied in the two intervals which may lead to better overall cancellation of the ventricular activity.[3]

For the method described in [37], the template beat $\hat{\mathbf{S}}_t$ is partitioned into two submatrices,

$$\hat{\mathbf{S}}_t = \left[\hat{\mathbf{S}}_{t,1}\ \hat{\mathbf{S}}_{t,2}\right], \tag{5.40}$$

where $\hat{\mathbf{S}}_{t,1}$ and $\hat{\mathbf{S}}_{t,2}$ contain the ensemble averaged samples of the QRS interval and the JQ interval, respectively. Thus, the T wave interval, originally analyzed in [35], is replaced by the longer JQ interval. These two intervals are processed differently with respect to how the two templates are fitted to the observed signal: Spatiotemporal optimization is performed in the QRS interval, i.e., scaling, rotation, and time alignment, whereas only time alignment is performed in the JQ interval, meaning that the cancellation is identical to ABS in the JQ interval [37]. Because of the often low SNR in the JQ interval, estimation of $\mathbf{A}$ and $\mathbf{Q}$ is not attempted.

Separate cancellation in the QRS and JQ intervals brings into focus the question whether this approach outperforms cancellation based on a single template applied to the entire heartbeat. Using simulated ECG signals, produced by a biophysical model of the atria with ventricular activity added [40], the performance was evaluated in terms of the normalized MSE (NMSE) between model f wave signals and estimated f wave signals, using the energy of the model signal as normalization factor, cf. Sect. 5.10.2. One- and two-template STC were found to produce about the same NMSE in the QRS interval, whereas two-template STC produced a factor of two lower NMSE in the JQ interval. This result is expected since the estimates of $\mathbf{A}$ and $\mathbf{Q}$ determined from the QRS interval cannot be optimal for processing of the JQ interval due to the temporal variation of the electrical axis. However, the better performance of two-template STC in the JQ interval should be viewed in light of the fact that the NMSE was about 10 times smaller in the JQ interval than in the QRS interval, and thus of much less significance when evaluating performance of the f wave extraction as a whole [37]. Clearly, template subtraction rarely causes large-amplitude residuals in the JQ interval, since the T wave is characterized by much lower frequencies than the QRS complex. Given that the extracted f wave signal is commonly prefiltered to remove low-frequency noise, using a bandpass filter with the lower cut-off frequency set to 3–4 Hz, the presence of residuals in the JQ interval is further reduced.

Two-template cancellation requires that the QRS and JQ intervals are first delineated, with the J point as the main challenge. Since fixed locations of the two intervals

[3]Time-dependent scaling and rotation have been considered when evaluating a method for vectorcardiographic loop alignment [38], see also [39]. A mathematical model was proposed in which the time-dependent, angular variation associated with the rotation matrix is assumed to be proportional to the amount of air in the lungs during a respiratory cycle—a property modeled by the product of two sigmoidal functions reflecting inhalation and exhalation, respectively, cf. Sect. 3.4.4.

relative to a fiducial point are in conflict with the large patient-to-patient variability in QRS duration, as well as the large beat-to-beat variability in T wave duration, it may be tempting to employ one of the well-performing delineation algorithms described in the literature, e.g., [41–43]. However, few, if any, algorithms have been evaluated on ECGs in AF, and, therefore, excellent delineation performance in P wave rhythms does not imply excellent delineation performance in AF. In fact, some authors have claimed that accurate delineation of the QRS complex in the presence of f waves is impossible [44], whereas others have arrived to the opposite conclusion and developed a technique for delineating the T wave endpoint in AF, where f waves are suppressed before delineation [45]; no approach was, however, devised for delineation of QRS onset.

Inaccurate QRS delineation causes a jump at the boundary between the QRS complex and the T wave—a problem which, to some extent, can be mitigated by linear, time-invariant lowpass filtering, for example, using a fifth-order Butterworth filter with cut-off frequency at 50 Hz in combination with forward/backward filtering [37].

5.2.5 Residual-Constrained QRS Template

Another approach to addressing the problem of large-amplitude QRS residuals is to modify the single-lead QRS template $\hat{\mathbf{s}}_t$ under certain constraints, so that it better fits the observed beat $\mathbf{x}$. As a result, the amplitude of QRS-related residuals will be reduced [23]. The modified QRS template $\hat{\mathbf{s}}$ results from a linear transformation of $\hat{\mathbf{s}}_t$ defined by the weight matrix $\mathbf{W}$,

$$\hat{\mathbf{s}} = \mathbf{W}\hat{\mathbf{s}}_t, \tag{5.41}$$

where the samples of $\hat{\mathbf{s}}$ and $\hat{\mathbf{s}}_t$ are defined over the interval I_{QRS}. In [23], $\hat{\mathbf{s}}_t$ was determined by exponential averaging of time-aligned beats, but any other ensemble estimator can be used for determining $\hat{\mathbf{s}}_t$. Comparing the linear transformations in (5.26) and (5.41), it is obvious that the former transformation is much more restrictive since the leads in $\hat{\mathbf{S}}_t$ are scaled by their corresponding amplitude factors in $\mathbf{A}$ and rotated by $\mathbf{Q}$, whereas the latter transformation offers much more freedom since each sample of the single-lead template $\hat{\mathbf{s}}_t$ is modified with its own individual weights, without performing any type of multi-lead processing.

The weight matrix $\mathbf{W}$ is determined by maximizing a fitness function $J_F(\mathbf{W})$, defined so that similarity between the f waves in the TQ interval, denoted $\mathbf{x}_0(\approx \mathbf{d}_0)$, and the f waves estimated in I_{QRS}, obtained as $\mathbf{x} - \mathbf{W}\hat{\mathbf{s}}_t$, is rewarded. Wave similarity is assessed by two different ratios, namely the ratio of the signal power computed from the two intervals,

$$g_1(\mathbf{W}) = \frac{\dfrac{1}{N_{QRS}}\|\mathbf{x} - \mathbf{W}\hat{\mathbf{s}}_t\|^2}{\dfrac{1}{N_0}\|\mathbf{x}_0\|^2}, \tag{5.42}$$

and the ratio of the power of the first-differenced version of the signals computed from the two intervals,

$$g_2(\mathbf{W}) = \frac{\dfrac{1}{N_{QRS}} \|\mathbf{x}' - \mathbf{W}\hat{\mathbf{s}}_t'\|^2}{\dfrac{1}{N_0} \|\mathbf{x}_0'\|^2}, \tag{5.43}$$

where N_{QRS} and N_0 denote the lengths of $\mathbf{x}$ and $\mathbf{x}_0$, respectively, and $\|\mathbf{x}\|^2 = \mathbf{x}^T\mathbf{x}$ denotes the Euclidean norm. The first difference of the samples of $\mathbf{x}$, denoted $\mathbf{x}'$, is included in the fitness function so that a high-frequency behavior of QRS-related residuals is penalized. In addition, the fitness function includes a third term which "limits the deformation" of $\hat{\mathbf{s}}$ relative to $\hat{\mathbf{s}}_t$, defined by the normalized crosscorrelation between $\hat{\mathbf{s}}$ $(= \mathbf{W}\hat{\mathbf{s}}_t)$ and $\hat{\mathbf{s}}_t$,

$$g_3(\mathbf{W}) = \frac{\gamma}{\pi} \arccos\left(\frac{\hat{\mathbf{s}}_t^T \mathbf{W}\hat{\mathbf{s}}_t}{\sqrt{\|\hat{\mathbf{s}}_t\|^2 \cdot \|\mathbf{W}\hat{\mathbf{s}}_t\|^2}} \right). \tag{5.44}$$

The arccosine function transforms the range of the normalized crosscorrelation from $[-1, 1]$ to $[0, \pi]$, and the parameter γ determines how dissimilar $\hat{\mathbf{s}}_t$ and $\hat{\mathbf{s}}$ are allowed to be. The three measures $g_1(\mathbf{W})$, $g_2(\mathbf{W})$, and $g_3(\mathbf{W})$ are all designed to reflect similarity in terms of overall signal properties.

The fitness function $J_F(\mathbf{W})$ is defined by

$$J_F(\mathbf{W}) = \alpha_1 J(g_1(\mathbf{W})) + \alpha_2 J(g_2(\mathbf{W})) + \alpha_3 (J(g_3(\mathbf{W})) - 1), \tag{5.45}$$

where the importance of each similarity measure is controlled by the weights α_i. The function $J(x)$ is defined by

$$J(x) = \begin{cases} 1, & x \leq 1, \\ (1-x)^3, & x > 1, \end{cases} \tag{5.46}$$

which becomes increasingly negative as the power ratios in (5.42) and (5.43) become increasingly larger than 1. Consequently, large-amplitude QRS residuals become increasingly more penalized. Moreover, $g_3(\mathbf{W}) = 0$ for perfectly similar waveforms, i.e., $\hat{\mathbf{s}}_t \equiv \hat{\mathbf{s}}$. To penalize dissimilar waveforms, it is necessary to set $\gamma > 2$ for this particular choice of $J(x)$ [23].

The fitness function in (5.45) can be successfully maximized with respect to $\mathbf{W}$ using multi-swarm particle swarm optimization. This technique is well-suited for problems where multiple local maxima are likely to be encountered; see [46, 47] for a general description of this technique, and [23] for a description of its use in signal extraction. In contrast to the methods described below for interpolation, the template $\hat{\mathbf{s}}$ which results from the residual constrained approach does not preserve the spectral properties of the f wave signal in the TQ intervals. Rather, all elements of **W** are determined independently of each other during optimization.

The residual constrained approach was originally developed for cancellation of the ventricular component in bipolar, atrial electrograms. However, there are good reasons to believe that this approach can be used for processing of signals acquired on the body surface. The residual-constrained technique is illustrated in Fig. 5.4, showing that the amplitude of the estimated f waves in the QRS intervals, as expected, is similar to that of the enclosing intervals. It is noted that the signal estimated in the QRS intervals is sometimes composed of higher frequencies than the signal in the enclosing intervals, explained by the independent determination of the elements in **W**.

The optimization is computationally costly since thousands of elements in **W** need to be determined for each beat. Therefore, the method lends itself less well to the processing of continuous, long-term, ambulatory recordings. Assuming a sampling rate of the ECG of 1 kHz, a typical QRS interval contains 100 samples, implying that 10,000 elements needs to be determined in **W** for each QRS complex.

5.3 Interpolation and Large-Amplitude QRS Residuals

Spatiotemporal QRST cancellation, whether involving one or two templates, offers a considerable reduction in the amplitude of QRS-related residuals, leading to more

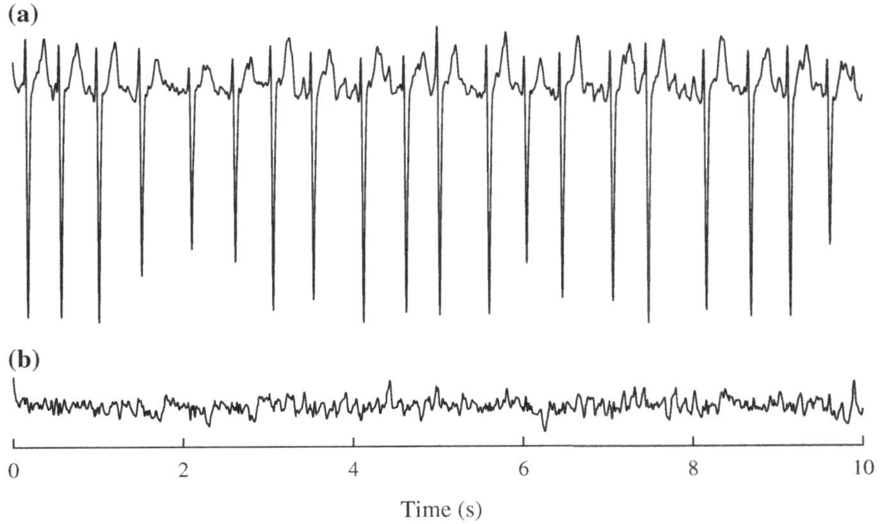

Fig. 5.4 f wave extraction using residual constrained QRS templates in combination with exponential averaging [23]. **a** Original ECG and **b** extracted f wave signal

accurate estimation of the f wave signal. However, a shortcoming of this technique, as well as with some other variants of ABS, is the lack of a mechanism ensuring that the amplitude of the estimated f waves in the QRS interval does not differ significantly from the amplitude of the enclosing f waves; a similar observation applies to the spectral content. The presence of large-amplitude QRS residuals is particularly problematic when detailed analysis of the f wave characteristics is of interest, e.g., short-term (second-to-second) variation in the dominant atrial frequency (DAF) or the morphology of f waves. This problem may be addressed by replacing the estimated f wave signal in the QRS interval I_{QRS} with samples obtained from interpolation based on the f wave signal in the two intervals I_{JQ_0} and I_{JQ_1} which enclose I_{QRS} [37, 48]. Figure 5.5 illustrates the main steps of interpolation performed in a single-lead ECG, which in this particular example is accomplished by autoregressive modeling of the samples in I_{JQ_0} and I_{JQ_1}, see Sect. 5.3.2. During the first second (encircled), the interpolated f wave signal provides a poor estimate because of the insufficient number of samples in the enclosing intervals—a problem which holds true for any interpolation method.

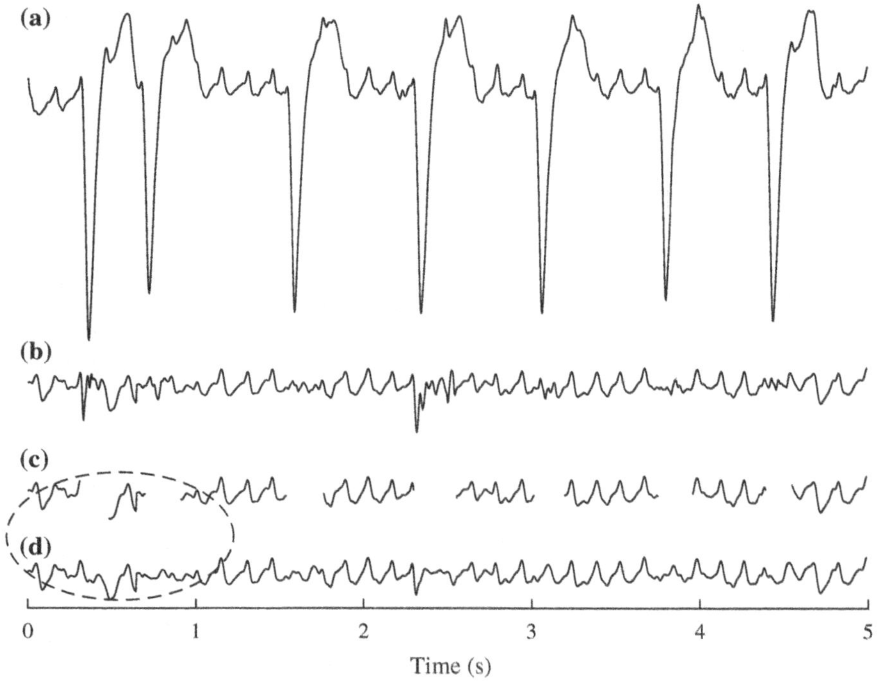

Fig. 5.5 Illustration of the main steps of QRS interval interpolation to reduce the influence of large-amplitude QRS-related residuals. **a** Original ECG, **b** extracted f wave signal using average beat subtraction, **c** same signal as in (**b**) except that the QRS intervals are replaced by gaps, and **d** extracted f wave signal with interpolated samples filling the gaps

While interpolation may be performed in every beat, its use is primarily motivated in beats for which the amplitude of the QRS-related residuals turns out to be much larger than the amplitudes of the f waves in I_{JQ_0} and I_{JQ_1}. Hence, amplitude-conditioned replacement of residuals has been explored where samples are replaced only if the amplitude exceeds a certain threshold [48]. The threshold value may be related to an index especially developed for quantifying the amplitude of QRS-related residuals [22].

Interpolation was encountered already in connection with STC (Sect. 5.2.3), used for computing the intermediate TQ-based f wave signal $\tilde{\mathbf{D}}$. In that case, however, the interpolated signal only served as a stepping stone for achieving more accurate parameter estimates of $\mathbf{A}$, $\mathbf{Q}$, and τ, not for replacing large-amplitude QRS-related residuals. In its simplest form, interpolation for replacement of residuals has been based on zero-, first-, or third-order polynomials, using only the very few samples immediately bordering to the QRS interval to compute the polynomial coefficients [49]. As noted in [3], however, polynomial interpolation is not recommended as it is plagued by a number of limitations, including the need for QRS delineation, as discussed above, and the lack of means to accurately cancel the ventricular repolarization component.

Since polynomial-based interpolation does not attempt to replicate the spectral properties of the enclosing JQ intervals, various model-based approaches to interpolation have been explored, where the model mimics certain properties of the f wave signal. The model parameters are first estimated from I_{JQ_0} and I_{JQ_1}, and then used to compute the "missing" samples in I_{QRS}. A common assumption is that the f wave signal can be modeled as a stationary process—an assumption which may be questioned since time–frequency analysis has demonstrated that the DAF may vary as much as 0.5 Hz in less than a second [50], see also [51]. Before the parameters of the interpolation model are estimated, the extracted signal should be prefiltered to ensure that its spectral content is confined to frequencies which are representative of the f wave signal.

5.3.1 *Sine/Cosine-Based Interpolation*

The first approach to model-based interpolation rests on the assumption that f waves can be described by a linear combination of sines and cosines [37]. Before interpolation, the "single beat" technique, described in Sect. 5.6.3, is used for estimating the T wave; the resulting estimate is denoted $\hat{s}_{\mathrm{T}}(n)$. The samples of the original ECG in I_{QRS} are replaced by interpolated samples. The following model describes the estimated f wave signal, beginning at the J point of the previous beat and ending at the onset of the Q wave of the subsequent beat:

$$\hat{d}(n) = \begin{cases} (s(n) - \hat{s}_{\mathrm{T}}(n)) + d(n) + v(n), & n \in \{I_{\mathrm{JQ}_0}, I_{\mathrm{JQ}_1}\}, \\ \displaystyle\sum_{p=1}^{P} (\hat{a}_{1,p}\cos(\omega_p n) + \hat{a}_{2,p}\sin(\omega_p n)), & n \in I_{\mathrm{QRS}}, \end{cases} \tag{5.47}$$

where P is the model order (set to as high as 50 in [37]). The normalized frequencies $f_p = \omega_p/2\pi$ are uniformly spaced in an interval corresponding to [0, 10] Hz. Since the duration of an f wave usually exceeds that of the QRS interval, the sine/cosine model needs to account for frequencies which are well below the DAF to provide adequate interpolation. In contrast to the samples in I_{JQ_0} and I_{JQ_1}, the interpolated samples are not influenced by noise, and therefore $\hat{d}(n)$ as a whole may, incorrectly, stand out as a signal with a locally varying noise level.

Least squares (LS) estimation is used to find $\hat{a}_{1,1}, \ldots, \hat{a}_{1,P}$ and $\hat{a}_{2,1}, \ldots, \hat{a}_{2,P}$ in (5.47) by analyzing the samples in I_{JQ_0} and I_{JQ_1}, together defining the column vector

$$\check{\mathbf{d}} = \left[\hat{d}(0)\ \hat{d}(1)\ \cdots\ \hat{d}(n_{\mathrm{Q}})\ \hat{d}(n_{\mathrm{J}})\ \cdots\ \hat{d}(N-1)\right]^T, \tag{5.48}$$

where $\check{}$ denotes that the samples $\hat{d}(n_{\mathrm{Q}}+1), \ldots, \hat{d}(n_{\mathrm{J}}-1)$ of I_{QRS} are left out since they are determined by interpolation once the estimates of the $a_{i,p}$ coefficients in (5.47) are available. Thus, the vector $\check{\mathbf{d}}$ contains $\check{N} = N - (N_{\mathrm{J}} - N_{\mathrm{Q}} + 1)$ elements. In a similar way, the cosine and sine functions define the columns of the matrix $\check{\mathbf{H}}_p$, where the samples in I_{QRS} are again left out,

$$\check{\mathbf{H}}_p = \begin{bmatrix} \cos(0) & \sin(0) \\ \cos(\omega_p) & \sin(\omega_p) \\ \vdots & \vdots \\ \cos(\omega_p n_{\mathrm{Q}}) & \sin(\omega_p n_{\mathrm{Q}}) \\ \cos(\omega_p n_{\mathrm{J}}) & \sin(\omega_p n_{\mathrm{J}}) \\ \vdots & \vdots \\ \cos(\omega_p(N-1)) & \sin(\omega_p(N-1)) \end{bmatrix}. \tag{5.49}$$

The two model parameters define the vector

$$\mathbf{a}_p = \begin{bmatrix} a_{1,p} \\ a_{2,p} \end{bmatrix} \quad p = 1, \ldots, P. \tag{5.50}$$

Using these matrix notations, the problem can be compactly formulated as one of minimizing the LS error

$$\varepsilon(\mathbf{a}_p) = \|\check{\mathbf{d}} - \check{\mathbf{H}}_p\mathbf{a}_p\|^2 \tag{5.51}$$

with respect to $a_{1,p}$ and $a_{2,p}$ for a certain frequency ω_p. Minimization of $\varepsilon(\mathbf{a}_p)$ is accomplished by calculating the gradient with respect to $\mathbf{a}_p$,

$$\nabla_{\mathbf{a}_p}\varepsilon(\mathbf{a}_p) = -2\check{\mathbf{H}}_p^T(\check{\mathbf{d}} - \check{\mathbf{H}}_p\mathbf{a}_p), \tag{5.52}$$

and setting the gradient equal to zero. The LS estimator is given by

$$\hat{\mathbf{a}}_p = (\check{\mathbf{H}}_p^T \check{\mathbf{H}}_p)^{-1} \check{\mathbf{H}}_p^T \check{\mathbf{d}}, \quad p = 1, \ldots, P, \tag{5.53}$$

where $(\check{\mathbf{H}}_p^T \check{\mathbf{H}}_p)^{-1}$ is invertible since $\check{\mathbf{H}}_p$ in (5.49) has full rank. It should be emphasized that since $\check{\mathbf{H}}_p$ is a function of ω_p, so will $\hat{\mathbf{a}}_p$. The minimum LS error is [4]

$$\varepsilon_{\min} = \check{\mathbf{d}}^T \check{\mathbf{d}} - \check{\mathbf{d}}^T \check{\mathbf{H}}_p \hat{\mathbf{a}}_p. \tag{5.54}$$

Finally, the desired interpolated samples in I_{QRS} are obtained by inserting $\hat{\mathbf{a}}_p$ in (5.47).

Since the minimum error $\varepsilon_{\min}$ of the estimator $\hat{\mathbf{a}}_p$ is closely related to Lomb's periodogram—a method briefly considered in Chap. 6 for spectral characterization of f waves—some light is shed here on this relation. The term $\check{\mathbf{d}}^T \check{\mathbf{H}}_p \hat{\mathbf{a}}_p$ in (5.54) describes the projection of $\check{\mathbf{d}}$ on the subspace spanned by $\check{\mathbf{H}}_p$, and, therefore, the energy of the component of $\check{\mathbf{d}}$ in that subspace. The interpretation in terms of energy can be used to define a spectral measure, which, for unevenly sampled signals such as the ones considered here with missing samples, is exactly the Lomb periodogram, which, for an arbitrary ω, is defined by [4]

$$\hat{S}_{\text{Lomb}}(\omega) \stackrel{\text{def}}{=} \frac{1}{\check{N}} \check{\mathbf{d}}^T \check{\mathbf{H}} \hat{\mathbf{a}}. \tag{5.55}$$

5.3.2 *Autoregressive Interpolation*

Another model-based approach assumes that the f wave signal $d(n)$ can be characterized by an autoregressive (AR) model [48],

$$d(n) = \sum_{p=1}^{P} a_p d(n-p) + e(n), \quad n = P, \ldots, N-1, \tag{5.56}$$

where P is the model order (set to 25 in [48]), $a_1, \ldots, a_P$ are the unknown model parameters, $e(n)$ is white noise, and N is the total number of samples in the three adjacent intervals I_{JQ_0}, I_{QRS}, and I_{JQ_1}. Alternatively, the model in (5.56) can be expressed as

$$\mathbf{d} = \mathbf{D}\mathbf{a} + \mathbf{e}, \tag{5.57}$$

where $\mathbf{d}$ and $\mathbf{e}$ are column vectors containing the samples $d(P), \ldots, d(N-1)$ and $e(P), \ldots, e(N-1)$, respectively, $\mathbf{D}$ is an $(N-P) \times P$ Toeplitz matrix whose structure reflects, in each row, that $d(n)$ depends on the past samples $d(n-1), \ldots,$ $d(n-P)$, and $\mathbf{a}$ is a column vector containing $a_1, \ldots, a_P$.

To find the missing samples $\mathbf{d}_{\text{QRS}}$, being a subsequence in $\mathbf{d}$ whose location is defined by I_{QRS}, the following LS problem is solved:

$$\min_{\mathbf{a},\mathbf{d}_{\mathrm{QRS}}} \|\mathbf{d} - \mathbf{D}\mathbf{a}\|^2. \tag{5.58}$$

However, joint minimization with respect to $\mathbf{a}$ and $\mathbf{d}_{\mathrm{QRS}}$ leads to a set of nonlinear equations with nontrivial solution [52], and, therefore, the problem is usually solved suboptimally in two successive steps. First, the model parameters $\mathbf{a}$ are estimated from the samples in I_{JQ_0} and I_{JQ_1} using the LS solution [53],

$$\hat{\mathbf{a}} = \left(\check{\mathbf{D}}^T\check{\mathbf{D}}\right)^{-1}\check{\mathbf{D}}^T\check{\mathbf{d}}, \tag{5.59}$$

where $\check{\mathbf{D}}$ is identical to $\mathbf{D}$, except that the rows containing the samples in I_{QRS} have been left out; this leave-out operation applies equally to the vector $\check{\mathbf{d}}$, cf. (5.48). Second, the missing samples in $\mathbf{d}_{\mathrm{QRS}}$ are obtained by solving another system of linear equations, and then making use of $\hat{\mathbf{a}}$ which results from (5.59); for a detailed description of the method, see [52, Chap. 10].

A central question in this context is "Why bother at all with f wave extraction in the QRS interval when interpolation circumvents the problem of large residuals?". The interpolation approach may work satisfactorily when the QRS interval is enclosed by JQ intervals from which the interpolation parameters can be reliably estimated. However, at higher heart rates, this approach is bound to fail when the enclosing JQ intervals are short, cf. the example in Fig. 5.5. Based on experimental findings, a minimum JQ interval length of at least one complete f wave was recommended in [44]; the length of one f wave was determined by the DAF, estimated from a signal extracted using ABS. If any of the enclosing JQ intervals was shorter than one complete f wave, interpolation was not carried out [44]. This recommendation seems to be rather optimistic for methods based on a linear combination of sines/cosines or autoregressive modeling since several f waves are required to produce reasonably accurate estimates of the model parameters. It should be noted that shorter JQ intervals imply increased vulnerability to noise—a problem which is exacerbated for shorter TQ intervals.

5.4 Extended Kalman Filtering

Average beat subtraction and variants are based on assumptions describing atrial and ventricular activity in broad terms. For example, it is assumed that the two activities are decoupled, whereas specific assumptions on f wave and QRST morphology are not introduced. A significant departure from this, rather minimalistic approach to modeling is to formulate a statistical signal model for generation of realistic-looking f waves and QRST complexes. Provided that the family of modeled f wave morphologies is sufficiently large, such an approach may lead to better f wave extraction than what is achieved by ABS and variants, especially when noisy ECG signals are analyzed. On the other hand, if the family of generated signals is too restrictive, there is an imminent risk that certain f wave morphologies are not accurately extracted.

Assuming that both f waves and QRST complexes are adequately modeled by a linear, time-varying dynamical system driven by noise, the state-space representation is attractive to consider, defined by

$$\mathbf{x}(n+1) = \mathbf{F}(n)\mathbf{x}(n) + \mathbf{G}(n)\mathbf{w}(n), \tag{5.60}$$
$$\mathbf{y}(n) = \mathbf{H}(n)\mathbf{x}(n) + \mathbf{v}(n), \tag{5.61}$$

where $\mathbf{x}(n)$ is the state vector, $\mathbf{F}(n)$ is the state transition matrix describing the system dynamics, $\mathbf{w}(n)$ is the system input noise, weighted by the matrix $\mathbf{G}(n)$. The model output vector $\mathbf{y}(n)$ is the sum of the modeled signal $\mathbf{H}(n)\mathbf{x}(n)$ and the measurement noise $\mathbf{v}(n)$, where $\mathbf{H}(n)$ is the observation matrix. The noise processes $\mathbf{w}(n)$ and $\mathbf{v}(n)$ are both assumed to be white, mutually uncorrelated, and completely characterized by their respective time-varying covariance matrices $\mathbf{Q}(n)$ and $\mathbf{R}(n)$. For the signal model defined by (5.60) and (5.61), the minimum MSE estimator of $\mathbf{x}(n)$ is the well-known *discrete-time Kalman filter*, operating recursively so that the estimate $\hat{\mathbf{x}}(n)$ is only based on $\hat{\mathbf{x}}(n-1)$ and the current observation $\mathbf{y}(n)$. An elegant and comprehensive account of the Kalman filter theory can be found in [54]; see also [55] for an accessible introduction.

The simulator of ECG signals described in Sect. 3.4, comprised of different mathematical models for the generation of f waves and QRST complexes, may serve as a starting point for f wave extraction using a special type of the Kalman filter [56]. In developing this approach, the question arises whether these signal models can be fitted into the above linear, state-space framework—a question which is addressed in the following.

The sawtooth model in (3.1) for generating f waves is considered, but with the important difference that the modulation of amplitude and frequency is omitted to make the model more manageable. Thus, the f wave signal $d(n)$ is modeled by

$$d(n) = \sum_{k=1}^{K} a_k \cos(k\omega_0 n + \varphi_k), \tag{5.62}$$

where ω_0 is the (unknown) DAF, K is a the number of harmonics, and a_k and φ_k are the amplitude and phase of the k-th harmonic, respectively. Each harmonic can be generated by a linear, time-invariant system, defined by a conjugate pole pair located on the unit circle at $\pm k\omega_0$,

$$\begin{aligned} H_k(z) &= \frac{1}{(1-e^{k\omega_o}z^{-1})(1-e^{-k\omega_o}z^{-1})} \\ &= \frac{1}{1-2\cos(k\omega_0)z^{-1}+z^{-2}}, \quad k = 1, \ldots, K. \end{aligned} \tag{5.63}$$

In the time domain, the system function $H_k(z)$ corresponds to the following second-order difference equation:

$$d_k(n+1) = 2\cos(k\omega_0)d_k(n) - d_k(n-1) + w_{d,k}(n), \quad k = 1, \ldots, K, \qquad (5.64)$$

where the input noise $w_{d,k}(n)$ is assumed to be white, and, to some extent, accounting for changes in a_k and φ_k. It is straightforward to express the K difference equations in (5.64) in terms of the above linear state-space representation, however, ω_0 needs to be estimated beforehand.

The harmonic f wave model has been employed for the analysis of short, second-long ECG segments [57]. Considering, though, that the DAF is known to fluctuate over short time periods [50], the following first-order, dynamical model is employed to account for such fluctuations [56]:

$$\omega_0(n+1) = \omega_0(n) + w_{\omega_0}(n), \qquad (5.65)$$

where the input noise $w_{\omega_0}(n)$ is assumed to be white. Another, related approach to modeling of fluctuations in ω_0 is described in [58]. Together with (5.64) and (5.65), the complete f wave signal model is defined by

$$d(n) = \sum_{k=1}^{K} d_k(n). \qquad (5.66)$$

It is noted that the models in (5.62) and (5.56) both generate f waves through filtering of white noise, however, the spectrum associated with (5.62) is much more constrained since the resonance frequencies are harmonically related, while not so for the spectrum associated with the AR model in (5.56). While the equations in (5.64) and (5.65) are both linear, a joint state-space representation is not possible due to the multiplication of $d_k(n)$ with $2\cos(k\omega_0(n))$, now depending on the state $\omega_0(n)$, which makes the combined system nonlinear.

A popular approach to the modeling of QRST complexes is to use a linear combination of Gaussian functions. In the continuous-time case, a QRST complex is modeled by [59, 60]

$$s(\tau) = \sum_{l=1}^{L} \alpha_l \exp\left[-\frac{(\tau - \tau_l)^2}{2\sigma_l^2}\right], \qquad (5.67)$$

where L is the number of Gaussians, α_l is the amplitude, σ_l is the width, and τ_l is the location. Provided that L is sufficiently large, a wide range of QRST morphologies can be modeled. In the context of f wave extraction, four Gaussians ($L = 4$) have been suggested for basic modeling of the QRST complex—three Gaussians assigned to the QRS complex and one to the T wave [56].[4]

[4]In general, it is desirable to use an orthogonal set of basis functions for signal representation, so that the signal component associated with a certain basis function do not interfere with the

To model an ECG with recurrent beats, the linearly growing time τ in (5.67) is replaced by an artificial, wrapped phase function $\tau(t)$ defining the local time of a complete cardiac cycle [60, 63],

$$s(t) = \sum_{l=1}^{L} \alpha_l \exp\left[-\frac{(\tau(t) - \tau_l)^2}{2\sigma_l^2}\right], \tag{5.68}$$

where $\tau(t)$ is obtained by the following 2π-modulo operation:

$$\tau(t) = \Omega_r t \bmod 2\pi. \tag{5.69}$$

The local time begins at $-\pi$ and ends at π, where the onset may be defined by the time instant of some "landmark ECG feature," such as the time instant of the R wave peak. The parameter Ω_r is an angular frequency set to $2\pi/T$, where the period length T may be related to the length of the current RR interval, the average length of several RR intervals, or replaced by a time-varying function $\Omega_r(t)$ which accounts for changes in heart rate [60].

A dynamical formulation of the QRST model in (5.68) and (5.69) is obtained by differentiating $s(t)$ and $\tau(t)$. Together with discretization at a sampling rate of F_s, differentiation of the QRST model results in

$$\tau(n+1) = \left(\tau(n) + \frac{\Omega_r}{F_s}\right) \bmod 2\pi, \tag{5.70}$$

$$s(n+1) = s(n) - \frac{\Omega_r}{F_s} \sum_{l=1}^{L} \alpha_l \frac{(\tau(n) - \tau_l)}{\sigma_l^2} \exp\left[-\frac{(\tau(n) - \tau_l)^2}{2\sigma_l^2}\right]. \tag{5.71}$$

The complete ECG signal model is defined by (5.64)–(5.62), (5.70), and (5.71). Since the observed samples are noisy, the signal model is extended so that the cardiac activity, i.e., $s(n) + d(n)$, as well as the phase function $\tau(n)$, are corrupted by additive, white noise,

$$y_1(n) = \tau(n) + v_1(n), \tag{5.72}$$

$$y_2(n) = s(n) + d(n) + v_2(n), \tag{5.73}$$

where the noise terms $v_1(n)$ and $v_2(n)$ are assumed to be mutually uncorrelated. Thus, the observation vector $\mathbf{y}(n)$ is composed of $y_1(n)$ and $y_2(n)$. The state vector $\mathbf{x}(n)$ is defined by the variables used for the modeling of f waves and QRST complexes,

$$\mathbf{x}(n) = \left[\omega_0(n)\ d_1(n)\ d_1(n-1) \cdots d_K(n)\ d_K(n-1)\ \tau(n)\ s(n)\right]^T. \tag{5.74}$$

---

components associated with the other basis functions. For example, the Hermite functions are well-suited for modeling of the QRST complex [61, 62]. However, orthogonality is of less importance to the described simulation model, and, therefore, Gaussian functions, being nonorthogonal basis functions, are considered.

It should be noted that the second-order difference equation in (5.64) is represented by two state variables, and an estimate of f wave signal $d(n)$ is obtained by summing the estimates of $d_1(n), \ldots, d_K(n)$. The system input noise vector is given by

$$\mathbf{w}(n) = \left[w_{\omega_0}\ w_{d,1}\ \cdots\ w_{d,K}\ \Omega_r\ \alpha_1\ \cdots\ \alpha_L\ \sigma_1\ \cdots\ \sigma_L\ \tau_1\ \cdots\ \tau_L,\right]^T, \tag{5.75}$$

where all the parameters depend on the time n, although, for convenience, this is not explicitly indicated.

The equations defining the complete ECG model, including the observation noise $\mathbf{v}(n)$, form a nonlinear, dynamic system, thus requiring a nonlinear state-space model, defined by

$$\mathbf{x}(n+1) = f(\mathbf{x}(n), \mathbf{w}(n)), \tag{5.76}$$

$$\mathbf{y}(n) = h(\mathbf{x}(n), \mathbf{v}(n)), \tag{5.77}$$

where the functions $f(\cdot)$ and $h(\cdot)$ describe the model nonlinearities. Since the discrete Kalman filter is no longer optimal for a nonlinear model, the *extended Kalman filter* (EKF) is preferred, which, prior to operation, requires that the state-space model is linearized about a working point [54, 64]. Here, "linearization" implies that the first-order terms of the respective Taylor series expansion of $f(\cdot)$ and $h(\cdot)$ are retained. Although the EKF is not an optimal estimator, it is nonetheless the standard technique used in nonlinear state estimation. For the above dynamical ECG model, the details related to the linearization have been worked out in [56], together with a complete description of the related EKF. Other applications of the EKF in ECG analysis can be found in [63, 65–69].

An interesting property of the Kalman filter approach is that the estimation of the f wave signal and the QRST complex is performed jointly, because the state vector $\mathbf{x}(n)$ in (5.74) contains information on both types of activity. This property stands in sharp contrast to ABS and variants, where the QRST complex is first estimated and then subtracted from the ECG to produce an estimate of the f wave signal. From a theoretical viewpoint, joint estimation is to be preferred over a two-step estimation procedure with respect to optimality in the MSE sense. From a practical viewpoint, however, this advantage may turn out to be less significant since the accuracy of the estimated f wave signal depends on how accurately the QRST complexes can be modeled. Another interesting property is that the EKF is one of the very few methods which, in addition to separating the ventricular and atrial activities, extracts an f wave signal whose noise level is most likely lower than the noise level of the original ECG [56].

A major difference between the EKF-based approach and ABS and variants, with practical implications, is the number of model parameters involved. While ABS involves only one single parameter, i.e., the number of beats for averaging, the EKF-based approach involves a huge number of parameters which have to be determined prior to operation, and possibly also updated during operation. The extraction method described in [56] makes use of four Gaussians to model the QRST complex, thus requiring a total of 12 parameters. Assuming that the color of the system noise and the observation noise is white, the diagonal elements, i.e., the variances, of the corresponding covariance matrices need to be determined, amounting to another 17 parameters; a colored noise assumption would obviously increase the number of parameters even further, and, therefore, this assumption is rarely pursued in the practice. It should be emphasized that these two covariance matrices play a crucial role in the design of the Kalman filter, as they inform the filter to what degree the observations can be trusted, and to what degree the observed signal is expected to differ from the assumed model. A procedure for determining the model parameters from the observed ECG signal has been proposed in [63], and implemented in many subsequent studies. Moreover, the elements of the initial state vector $\mathbf{x}(0)$ need to be determined, where $\omega_0(0)$ is a crucial parameter in f wave extraction. In [56], $\omega_0(0)$ is estimated from the f waves in successive TQ segments, i.e., an approach whose limitations have already been discussed in this chapter. Bearing in mind the huge number of model parameters, it would be of particular interest to perform sensitivity analysis to gain insight into how sensitive the extraction method is to small changes in the model parameter values.

While the irregular rhythm typical of AF is not explicitly accounted for in the ECG model, the degree of irregularity can be controlled by the variance related to the angular frequency Ω_r. It remains to be established, however, whether this type of implicit modeling is sufficient to produce the desired extraction performance. Earlier results obtained by the same ECG model and EKF, though used for denoising of fetal ECG signals, suggest that changes in the length of successive RR intervals which is less than 20% from one interval to the next have no significant influence on performance [63]. For larger changes, often observed in AF, the phase error of the model may lead to errors in the location of the Gaussian functions, which in turn may translate to reduced extraction performance.

Since the EKF can track only one QRST morphology, typically that of the dominant beat, it is not possible to accurately extract the f wave signal in the presence of occasional VPBs with deviating morphology. Hence, the EKF-based method suffers from the same limitation as, for example, the methods based on ensemble averaging. While the handling of multiple beat morphologies has yet to find its way into EKF-based f wave extraction, such a technique may very well draw on the idea proposed in [67], namely to use multiple state-space models for representing the dominant beat, ectopic beats (following morphologic clustering), and beats with unknown morphologies. The model providing the best fit to the observed beat is determined using a statistical likelihood test on the error ("innovations") associated with each of the state-space models.[5]

Since the EKF can track slow changes in QRST morphology, it is less likely that respiratory-induced changes should lead to large-amplitude QRS-related residuals

[5]An early precursor to the idea of using multiple state-space models was explored for the identification of certain persistent ECG rhythms [70], although none of them were AF. In that study, the proposed rhythm models were linear in nature, and, therefore, the discrete-time Kalman filter could be used.

as it would when, for example, exponential averaging is used. The main reason is that the EKF relies on a priori information on QRST morphology, translating to better tracking [4, Chap. 4], provided that the Gaussian model is adequate. With respect to T wave modeling, adequacy may be called in question when only one Gaussian is used [56], as neither biphasic nor asymmetric T waves can be modeled, although such morphologies are commonly encountered. The significance of coarse T wave modeling on extraction performance remains to be investigated, however, fewer Gaussians are needed to model the T wave than the QRS complex thanks to its low-frequency content.

5.5 Adaptive Filtering

Adaptive filters of varying complexity have been considered for f wave extraction, including the classical least mean square (LMS) filter (Sect. 5.5.1) and nonlinear filters based on the recurrent neural network (RNN), i.e., a network whose hidden neurons have feedback connections to the input as well as to the hidden layers of the filter [71]. Either all weights of the network (Sect. 5.5.2) or only the weights of the output layer (Sect. 5.5.3) are subject to continuous training during network operation. The block diagram in Fig. 5.6 shows the general structure of methods for f wave extraction based on adaptive filtering, applying to all filtering techniques described in this section.

The extraction of f waves is performed in the ECG lead taken as the primary input $x(n)$ to the adaptive filter, whereas another lead with negligible atrial activity, denoted $s_r(n)$, serves as reference input. Thus, the reference lead has to be acquired with an electrode positioned far away from the atria. Although at least two leads are required in adaptive filtering, multi-lead ECGs are nowadays acquired in most applications where f wave analysis is performed, e.g., continuous, long-term ambu-

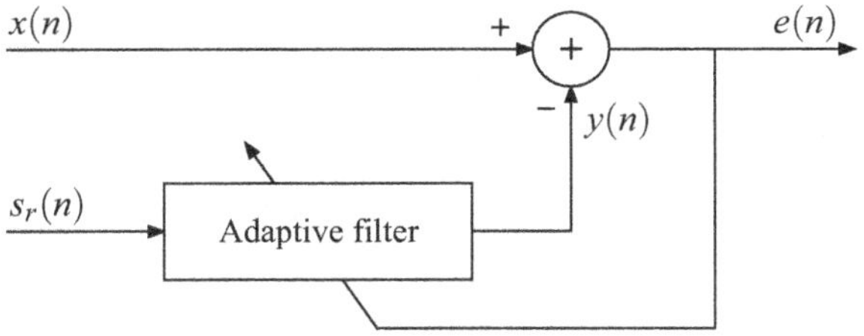

Fig. 5.6 Adaptive filtering for f wave extraction. The primary input $x(n)$ is an ECG lead for extraction, whereas the reference input $s_r(n)$ is a lead with negligible atrial activity. The output $y(n)$ of the adaptive filter provides an estimate of the ventricular activity in $x(n)$, and the error $e(n)$ between $x(n)$ and $y(n)$ provides an estimate of the f wave signal. The error $e(n)$ is fed back to the adaptation algorithm of the filter for weight update. This particular filter structure is sometimes referred to as a dual-input adaptive noise canceller [53]

latory monitoring based on either traditional electrodes or intelligent textiles with integrated conductive electrodes. The choice of reference lead is not particularly critical when analyzing the standard 12-lead ECG, since leads V_5 and V_6 typically contain much less atrial activity than leads V_1 and V_2.

Since the ventricular activity contained in $x(n)$ and $s_r(n)$ is correlated with each other, the weights of the adaptive filter are adjusted so that the error between the two input signals is minimized with respect to a cost function. The output of the adaptive filter, denoted $y(n)$, is an estimate of the ventricular activity contained in $x(n)$, and, therefore, the error $e(n)$ between $y(n)$ and $x(n)$ provides an estimate of the f wave signal.

Prior to f wave extraction, $s_r(n)$ may be preprocessed by linear filtering so that the ventricular activity is further accentuated. Another means to improve performance is to augment $s_r(n)$ with additional leads with negligible atrial activity. Yet another means is to mathematically derive a number of "support leads" from an existing lead with negligible atrial activity [72]. For the latter two approaches, the scalar reference input $s_r(n)$ becomes a vector reference input $\mathbf{s}_r(n)$.

In adaptive filtering, the input signals are processed sequentially, implying that jumps at the boundaries between successive beats is less of an issue than when the ECG is processed on a beat-by-beat basis.

5.5.1 Least Mean Square Linear Filtering

In the LMS-based approach to f wave extraction, the primary input signal $x(n)$ is described by the model

$$x(n) = s(n) + d(n), \tag{5.78}$$

where $d(n)$ is the f wave signal. This model is identical to the one in Sect. 5.2.2 considered for deriving signal- and noise-dependent weighted averaging, apart from that the beat index is dropped in (5.78) since LMS-based extraction, in its general form, does not require knowledge on the occurrence times of the beats. The reference input $s_r(n)$ contains ventricular activity which is correlated to $s(n)$. Moreover, both $s(n)$ and $s_r(n)$ are uncorrelated to $d(n)$,

$$E\left[d(n)s(n)\right] = 0, \tag{5.79}$$

$$E\left[d(n)s_r(n)\right] = 0. \tag{5.80}$$

The adaptive filter has a finite impulse response of the direct form, defined by the time-varying weights $h(l, n)$, where l denotes the local time within the impulse response and n the time at which the impulse response is valid. The output of the filter is given by the convolution

$$y(n) = \sum_{l=0}^{L-1} h(l, n) s_r(n - l), \tag{5.81}$$

where L is the number of filter weights. The filter weights $h(l, n)$ are determined so that the MSE $E\left[e^2(n)\right]$ is minimized, where $e(n) = x(n) - y(n)$. Using (5.79) and (5.80), the MSE can alternatively be expressed as

$$E\left[e^2(n)\right] = E\left[(s(n) - y(n))^2\right] + E\left[d^2(n)\right], \tag{5.82}$$

underlining the fact that the filter output $y(n)$ is an estimate of the ventricular activity $s(n)$. The LMS algorithm results from minimizing the MSE using the steepest descent technique [53],

$$h(l, n+1) = h(l, n) + 2\mu e(n) s_r(n - l), \quad l = 0, \ldots, L - 1, \tag{5.83}$$

where μ is a positive-valued step size parameter. Thus, $e(n)$ represents the estimate of the f wave signal, i.e., $\hat{d}(n) \equiv e(n)$.

The performance achieved by LMS-based filtering is inferior to that of ABS because the extracted f wave signal initially contains a considerable amount of large-amplitude QRS-related residuals, especially for a small μ, see Fig. 5.7. As a result, this approach has not received much attention in the literature, except being mentioned in a few review papers [22, 73]. A description of the LMS algorithm is nevertheless well-motivated as it is central to the filter with *impulse-correlated reference*

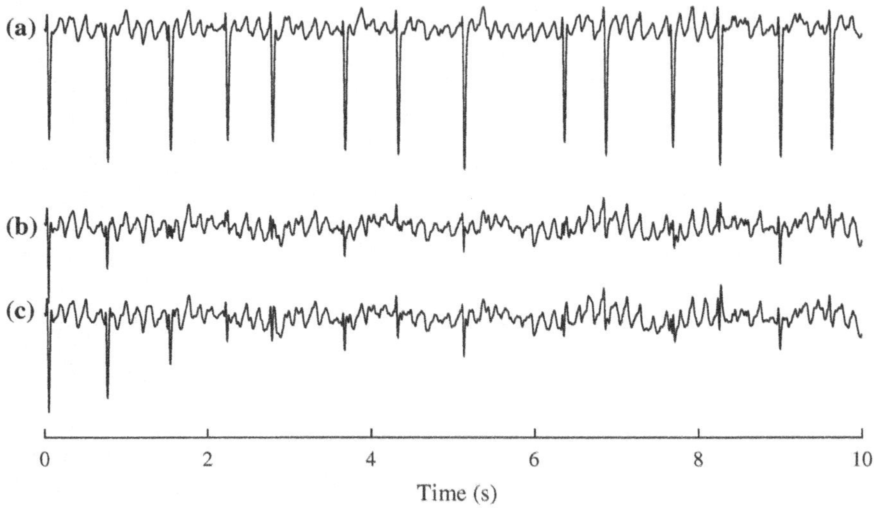

Fig. 5.7 f wave extraction based on adaptive filtering using the LMS algorithm. **a** Original ECG, and extracted f wave signal obtained for a step size set to **b** $\mu = 0.2$ and **c** $\mu = 0.05$. A larger value of μ produces smaller QRS-related residuals due to faster convergence, but also increases the risk of filter instability

input [74]; see also [75] for a detailed analysis of this filter. Rather than using a separate ECG lead as reference input, a train of impulses positioned at the occurrence times θ_i of the beats serves as input,

$$s_r(n) = \sum_i \delta(n - \theta_i). \tag{5.84}$$

Consequently, $s_r(n)$ does not convey any information on atrial activity, suggesting that f wave extraction can be performed in single-lead ECGs, with θ_i determined from $x(n)$.

The idea to use an impulse-correlated reference input was introduced to facilitate not only the analysis of signals recorded during AF, but also arrhythmias with P waves [74]. Based on one single ECG example, the performance was labelled as "excellent", although this superlative was not corroborated by quantitative results. Much later, the same approach was investigated anew for f wave extraction, but then accompanied with a quantitative performance evaluation on a small data set consisting of 10 1-min ECGs [76]. The results suggested that the LMS-based filtering approach performed slightly better than ABS, a result which probably can be attributed to better tracking of slow changes in QRST morphology.

Initially, the adaptive filter with impulse-correlated reference input was viewed as a time-variant filter [74, 75]. However, a few years later, this filter was shown to be identical to the exponential averager defined in (5.9), the reason being that time-aligned beats represent a perfectly periodic input to the filter. Hence, the filter is not adaptive in the well-established sense in which a gradient search is performed based on the error at the current time [77], but, indeed, it is the exponential averager, i.e., a linear, time-invariant comb filter whose passbands have a width proportional to the step size μ in (5.83). Thus, the results presented in [76] should be viewed in light of what adaptive filtering with impulse-correlated reference input stands for.

5.5.2 Nonlinear Filtering Based on a Simple Recurrent Network

Adaptive nonlinear filtering for f wave extraction was first considered using a simplified version of the dynamically driven RNN [78], also known as the time delay Elman network [79]. The main motivation for using this type of network was its architecture, which was claimed to be better suited for learning the rapid changes which characterize ventricular activity than would the linear, transversal structure of the LMS filter. Therefore, it was argued that better performance with respect to estimating the QRST complex could be expected. The simplified RNN network consists of the following main building blocks: an input layer with unit time delays and a nonlinear hidden layer with neurons whose output is fed forward to the output layer, and then fed backwards to the input layer, see Fig. 5.8; the output layer consists of only one neuron.

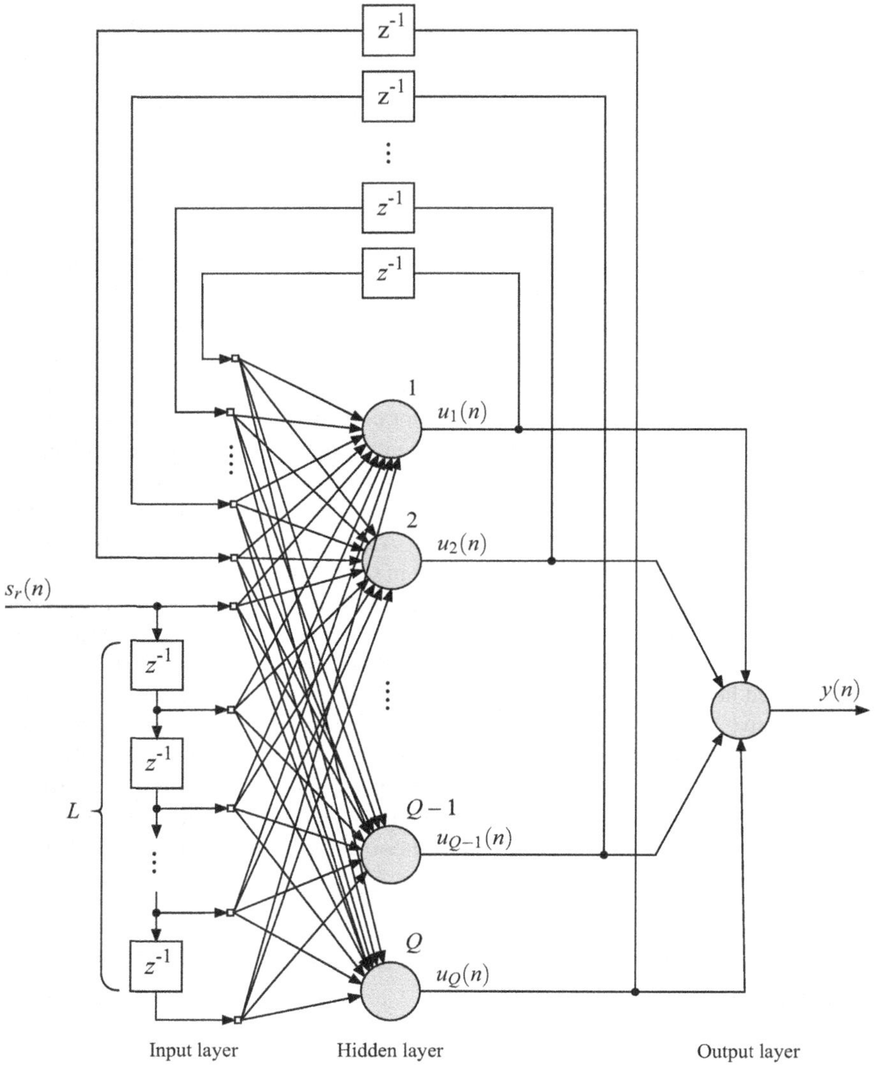

Fig. 5.8 Simplified recursive neural network with time delayed input. The hidden layer consists of Q neurons (circles filled with grey), where each neuron is described by a weighted sum of the input signals, followed by an activation function limiting the output signal of the neuron. The feedback paths are also weighted. The output layer consists of only one (linear) neuron. This network structure is similar to that of the echo state network, see Sect. 5.5.3, with the important difference that the large hidden layer is sparsely connected and only the weights of the output layer are updated during operation

The simplified RNN filter has a dynamic behavior described by the following nonlinear equations [71, Chap. 15]:

$$\mathbf{u}(n+1) = \varphi(\mathbf{W}_u \mathbf{u}(n) + \mathbf{W}_s \mathbf{s}_r(n)), \tag{5.85}$$

$$y(n) = \mathbf{w}_o^T \mathbf{u}(n), \tag{5.86}$$

where $\mathbf{u}(n)$ is a $Q \times 1$ state vector describing the output of the neurons of the hidden layer, $\mathbf{s}_r(n)$ is an $L \times 1$ input vector with time delayed samples,

$$\mathbf{s}_r(n) = \begin{bmatrix} s_r(n) \\ s_r(n-1) \\ \vdots \\ s_r(n-L+1) \end{bmatrix}, \tag{5.87}$$

and $y(n)$ is the scalar output signal. The feedback paths originating from the hidden neurons have weights defined by the elements of the matrix $\mathbf{W}_u$, whereas the paths of the input layer have weights defined by the elements in $\mathbf{W}_s$; these two matrices have the dimensions $Q \times Q$ and $Q \times L$, respectively. The output layer, consisting of only one neuron, is linear and defined by the $Q \times 1$ weight vector $\mathbf{w}_o$. The simplified RNN involves feedback from the hidden neurons, but no global feedback from the output neuron. The memoryless activation function $\varphi(\cdot)$ is associated with each neuron in the hidden layer, defined by the sigmoidal function

$$\varphi(x) = \frac{1}{1+e^{-x}}. \tag{5.88}$$

Similar to the LMS-based adaptive filter, the simplified RNN assumes that the primary input $x(n)$ is the ECG lead which is subject to extraction, and the reference input $s_r(n)$ is a lead with negligible atrial activity, see Fig. 5.6. Based on the quadratic error between $x(n)$ and the network output signal $y(n)$, the weights of the connections, i.e., $\mathbf{W}_u$, $\mathbf{W}_s$, and $\mathbf{w}_o$, are adjusted sequentially ("online training") using the well-known backpropagation algorithm performing gradient descent minimization [71, 80].

The number of time delays L of the input layer and the number of hidden neurons Q are the two main design parameters of the simplified RNN [78]. From a learning set consisting of randomly selected beats from the MIT–BIH Arrhythmia Database, see Sect. 3.1, the choice $L = 9$ and $Q = 20$ was found to provide good performance in terms of attenuating the ventricular activity in the f wave signal. The performance was studied on a tiny data set consisting of three recordings from the same database, of which only one was from a patient in AF. When compared to exponential averaging, the simplified RNN filter performed better in signals with noise or with changing QRS morphology. On the other hand, no improvement was observed in low noise signals with stable rhythm, but f waves were rather attenuated and T waves were poorly estimated, both aspects having repercussions on extraction performance. Considering the size of the data set, it was not possible to conclude whether the simplified RNN is well-suited for f wave extraction.

5.5.3 Nonlinear Filtering Based on an Echo State Network

A well-known problem with RNNs is related to the fact that weight adjustment is usually based on the gradient descent method, providing exceedingly slow learning since all weights, i.e., input, recurrent, and output, have to be adjusted. Moreover, it is well-known that RNNs are prone to bifurcations during the learning process which may cause the network to become unstable [81, 82]. In the context of f wave extraction, bifurcations would manifest themselves by oscillations in the output signal of the network, caused by infrequently occurring waveforms such as premature VPBs and spike artifacts.

Both these problems can be circumvented by using the *echo state network* (ESN) [83, 84], which is a large, sparsely connected RNN serving as a "reservoir" of recurrently connected neurons. Depending on the size of the reservoir, about 5–20% of all neurons are connected to each other. In general, the reservoir is driven by the input signal as well as the output signal through feedback; however, the output feedback has not been considered in f wave extraction [72]. The sparse connections of the reservoir are generated randomly, with weights that remain fixed throughout operation. The desired output signal is obtained as a linear combination of the nonlinear response signals. The weights of the linear combination are the only ones of the ESN which are subject to training, performed continuously as the input signal is being processed, see Fig. 5.9. As a result, training turns out to be simple and fast, never getting stuck in local minima.

The ESN has been found to outperform the much more complex, fully trained RNN in many different applications [86], and is particularly well-suited for processing of

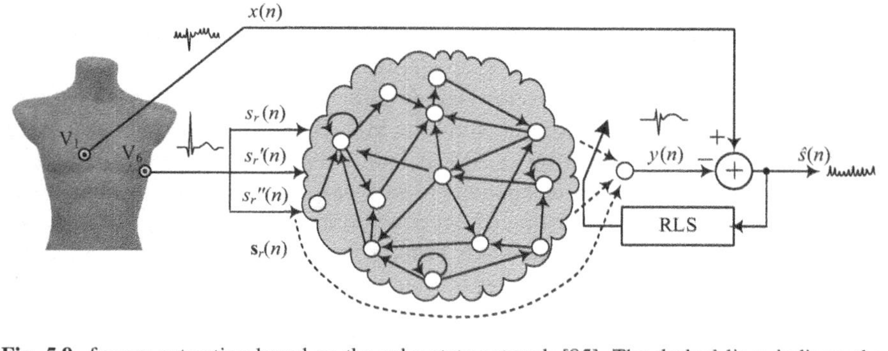

Fig. 5.9 f wave extraction based on the echo state network [85]. The dashed lines indicate the weights updated during network operation. The weights of the reservoir (grey area) are fixed after initialization. Weight adaptation is performed using the recursive least squares (RLS) algorithm

signals with time-varying dynamics such as the ECG. The term "echo state" relates to the fact that previous states of the reservoir can "echo" after they have been once passed, i.e., an output signal is generated which resembles some earlier encountered signal segment. It should be emphasized, though, that f wave extraction can be performed even if a beat morphology has never before been encountered by the ESN.

The dynamic behavior of the ESN is described by essentially the same equation as the simplified RNN in (5.85),

$$\mathbf{u}(n+1) = \varphi(\mathbf{W}_u\mathbf{u}(n) + \mathbf{W}_s\mathbf{s}_r(n)), \tag{5.89}$$

where $\mathbf{W}_s$ is the input weight matrix, $\mathbf{W}_u$ is the reservoir weight matrix in which most of the elements are set to zero, i.e., a sparse matrix, and $\mathbf{u}(n)$ is the reservoir state vector. The reference input vector $\mathbf{s}_r(n)$ contains the sample at time n of the input signals, thus differing from the vector in (5.85) which instead contains the time delayed samples of a scalar input signal. Another important difference is that the output signal $y(n)$ is computed as the scalar product of the time-varying weight vector $\mathbf{w}_o(n)$, i.e., weights requiring an update at each time n, and the state vector $\mathbf{z}(n)$, augmented with $\mathbf{s}_r(n)$,

$$y(n) = \mathbf{w}_o^T(n)\mathbf{z}(n), \tag{5.90}$$

where

$$\mathbf{z}(n) = \begin{bmatrix} \mathbf{u}(n) \\ \mathbf{s}_r(n) \end{bmatrix}. \tag{5.91}$$

For the previously described methods based on adaptive filtering, a scalar reference signal $s_r(n)$ was analyzed for f wave extraction. However, it has been demonstrated that better performance can be achieved when $s_r(n)$ is supplemented with information on its first and second derivatives, denoted $s_r'(n)$ and $s_r''(n)$, respectively [72]. Hence, the reference input vector to the ESN is defined by

$$\mathbf{s}_r(n) = \begin{bmatrix} s_r(n) \\ s_r'(n) \\ s_r''(n) \end{bmatrix}. \tag{5.92}$$

The primary input $x(n)$ may also be used to supplement $\mathbf{s}_r(n)$, provided that all the samples of $x(n)$ are set to zero, except those which are contained in a small interval centered around each QRS complex [85]. The resulting signal can be viewed as a variant of the impulse-correlated reference input, stripped of atrial information, cf. (5.84). Assuming that the hidden layer contains Q neurons, see Fig. 5.8, the dimensions of $\mathbf{W}_s$ and $\mathbf{W}_u$ are $Q \times 3$ and $Q \times Q$, respectively.

When initializing the ESN, the weight matrices $\mathbf{W}_s$ and $\mathbf{W}_u$ are randomly generated using a uniform probability density function (symmetric around zero and invariant to training). To further ensure network stability, the equation describing the dynamic behavior of the reservoir is commonly modified to include an artificial white noise term $\mathbf{v}(n)$ [87],

$$\mathbf{u}(n+1) = \varphi\left(\mathbf{W}_u\mathbf{u}(n) + \mathbf{W}_s\mathbf{s}_r(n)\right) + \mathbf{v}(n). \tag{5.93}$$

Another modification of the network is due to the property that the activation function $\varphi(\cdot)$, defined by a hyperbolic tangent function in [72], is memoryless, thereby causing $\mathbf{u}(n)$ to be impulsive in nature. As a result, the output signal $y(n)$ is also impulsive, which is a highly undesirable property. This problem can be solved by smoothing $\mathbf{u}(n)$ using exponential averaging, implying that the update equation in (5.93) is altered to [88]

$$\mathbf{u}(n+1) = \alpha\mathbf{u}(n) + (1-\alpha)(\varphi(\mathbf{W}_u\mathbf{u}(n) + \mathbf{W}_s\mathbf{s}_r(n)) + \mathbf{v}(n)), \tag{5.94}$$

where the forgetting factor α (also known as the "leakage rate") is a positive constant less than 1. While smoothing is not an explicit part of the simplified RNN, the filtering introduced by delaying and weighting the input samples $\mathbf{s}_r(n)$, cf. (5.87), may be viewed as a counterpart to the smoothing in (5.94).

Since offline, supervised training of the ESN is inefficient for signals with rapid changes in morphology, i.e., the QRS complex, it is necessary to perform continuous, online training to ensure fast adaptation of $\mathbf{w}_o(n)$. While adaptation can be accomplished with the LMS algorithm, the recursive least squares (RLS) algorithm offers superior convergence rate for highly correlated input signals—a property which obviously applies to $s_r(n)$ as well as to its first and second derivatives. The RLS algorithm minimizes the weighted LS error between the "desired signal" $x(n)$ and the output signal $y(n)$ through recursive computation of both $\mathbf{w}_o(n)$ and the inverse of the exponentially weighted, deterministic correlation matrix $\mathbf{R}_z(n)$ of the state vector $\mathbf{z}(n)$ in (5.91),

$$\mathbf{P}_z(n) = \mathbf{R}_z^{-1}(n) = \left(\sum_{i=0}^{n} \xi^{n-i}\mathbf{z}(i)\mathbf{z}^T(i)\right)^{-1}, \tag{5.95}$$

where ξ is a weighting factor usually chosen to be between 0.95 and 1. The RLS algorithm is defined by [55]

$$\mathbf{g}(n) = \frac{\xi^{-1}\mathbf{P}_z(n-1)\mathbf{z}(n)}{1+\xi^{-1}\mathbf{z}^T(n)\mathbf{P}_z(n-1)\mathbf{z}(n)}, \tag{5.96}$$

$$\mathbf{P}_z(n) = \xi^{-1}\mathbf{P}_z(n-1) - \xi^{-1}\mathbf{g}(n)\mathbf{z}^T(n)\mathbf{P}_z(n-1), \tag{5.97}$$

$$\mathbf{w}_o(n) = \mathbf{w}_o(n-1) + \mathbf{g}(n)(x(n) - \mathbf{w}_o^T(n-1)\mathbf{z}(n)), \tag{5.98}$$

and initialized by $\mathbf{w}_o(0) = \mathbf{0}$ and $\mathbf{P}_z(0) = \delta^{-1}\mathbf{I}$, where δ is a small positive constant. A number of techniques have been proposed for improving the numerical instability that may occur when computing $\mathbf{P}_z(n)$ [53], one attractive technique being LS prewhitening of $\mathbf{z}(n)$ [89], implemented in the ESN proposed for f wave extraction [72].

Given that the large-size matrices defining the reservoir, i.e., $\mathbf{W}_s$ and $\mathbf{W}_u$, are randomly generated, only a few parameters need to be determined before operation of the ESN. The reservoir size Q, the forgetting factor α in (5.94), and the RLS weighting factor ξ are the parameters which significantly influence performance, whereas the reservoir connectivity and δ have less influence on performance [72]. Using the RMS error between the simulated f wave signal and the corresponding estimate as performance measure, the following values were found to provide good performance: $Q = 100$, $\alpha = 0.8$, and $\xi = 0.999$. Compared to the simplified RNN, involving only 20 neurons in the hidden layer, the ESN offers, thanks to a much larger number of neurons, better ability to learn aberrant signal patterns such as those composed of VPBs.

As already noted, an essential feature of the ESN is the sparse connectivity of the reservoir, implying a substantial reduction in the computational complexity over the fully connected reservoir. In [72], a connectivity of 20% was found to provide good performance, though a lower percentage may be used if desired. Other design aspects, including normalization of the input signals for faster training of the network, the relationship between the spectral radius of $\mathbf{W}_u$ and network stability, and the distribution of the nonzero elements in the weight matrices, are discussed at length in [88], see also [90]. Based on experience from analyzing thousands of hours with ECG data, network instability has never been encountered.

The extraction performance of the ESN has been studied from different perspectives, especially in relation to ABS and STC. Considerable amplitude variation in the QRS complex as well as the T wave is handled well by the ESN, considering that large-amplitude QRS-related residuals are absent in the estimated f wave signal, see Fig. 5.10a. Ventricular premature beats, whose morphologies are vastly different from that of the dominant beat, neither cause much residuals, see Fig. 5.10b.

Initialization of the ESN introduces a transient in the f wave signal which may be problematic when processing brief ECG signals as well as brief AF episodes. This issue was investigated in terms of the RMS error [72]. The results showed that the ESN converges in about one second—a result which should be contrasted with the much longer "convergence time" required by ABS and variants, where a transient behavior typically would last for about 20–30 beats until the QRS-related residuals reaches an acceptable level (depending on f wave amplitude and prevailing noise level).

The convergence aspect is closely related to the handling of rhythm transitions, i.e., from sinus rhythm to AF and vice versa, see Fig. 5.11. This example demonstrates that the ESN can extract f waves accurately immediately after the transitions to AF, without any noticeable transient behavior. Interestingly, the same observation also applies to P waves which are accurately extracted immediately after the transitions from AF to sinus rhythm, in spite of the fact that the ESN was not designed to extract P waves. For ABS and variants, the f waves are inaccurate immediately after the transition to AF, and then become increasingly better as the AF episode progresses. If, on the other hand, a QRST template has been determined from an earlier episode, it may be used to extract f waves provided that the QRST morphology has not changed.

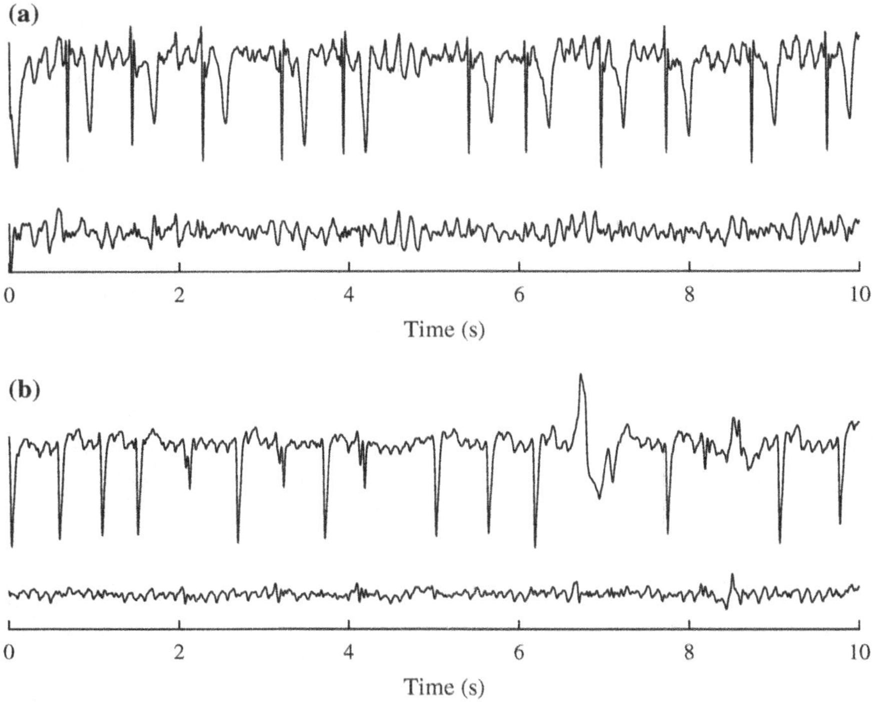

Fig. 5.10 f wave extraction based on an echo state network, performed in the presence of **a** considerable variation in QRS amplitude and large-amplitude T waves, and **b** multiform ventricular premature beats. The upper signal is the original ECG and the lower signal the extracted f wave signal

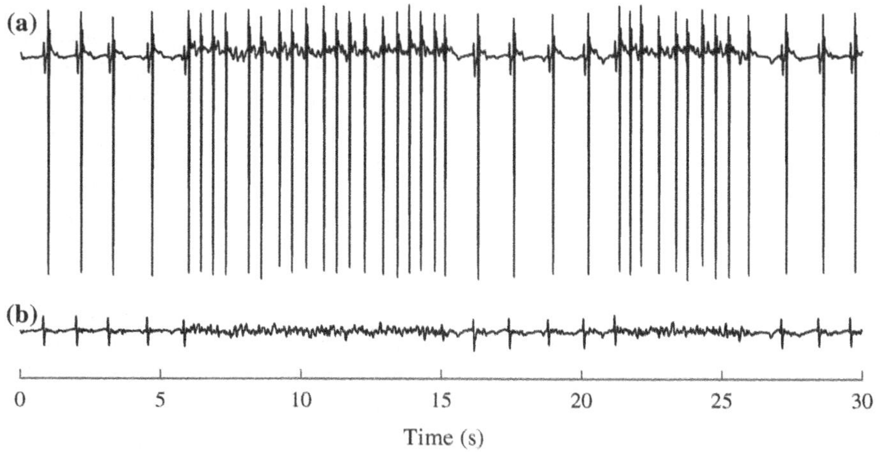

Fig. 5.11 f wave extraction performed in the presence of rhythm transitions from sinus rhythm to AF, and vice versa, using the echo state network. **a** Original ECG and **b** extracted f wave signal

5.6 Principal Component Analysis

Principal components analysis (PCA) is a statistical technique which performs an orthogonal linear transformation of the observed signal for the purpose of decorrelating the samples of the signal and maximizing the variance of the transformed vector, i.e., the principal components [91]. The first axis of the transformed coordinate system corresponds to the maximal variance, the second axis to the maximal variance in the direction orthogonal to the first axis, and so on. The emphasis on variance stems from the observation that larger variance is usually associated with the more interesting dynamics of the signal, whereas lower variance is usually associated with noise. The subspace defined by the principal components with the largest variances usually receive the most attention as that subspace offers optimal dimensionality reduction in the LS sense. However, other subspaces may also be of interest, to be discussed below. Principal component analysis has a long and rich history in computerized ECG analysis—the first paper was published already in 1964 [92]—and has been instrumental in the development of many methods for data compression, waveform classification, tracking of waveform changes related to myocardial ischemia, and noise reduction [93].

In the context of f wave extraction, some PCA-based methods have been designed for single-lead ECG analysis, whereas others for multi-lead ECG analysis, described in Sects. 5.6.1 and 5.6.2, respectively. While different signal properties are explored by PCA-based methods, they share the idea that an "atrial subspace" needs to be identified onto which the observed signal can be projected for extraction of f waves.

Methods based on PCA do not involve assumptions on various signal properties, but are confined to exploring either intralead (sample-to-sample) correlation or interlead correlation, depending on whether single- or multi-lead ECGs are subject to analysis. The correlation matrix required for finding the transformation can be easily estimated from the observed signal. No assumptions have to be made on decoupling between the ventricular and atrial activities, rhythm irregularity, dominant beat morphology, noise properties in terms of variance and color, and so on. Neither does any assumption have to be made on the statistical distribution of the observed signal samples.

5.6.1 *Single-Lead PCA*

In single-lead PCA [94], each beat is segmented by selecting the samples of an interval centered around a QRS-related fiducial point. The samples of the k-th beat are contained in the vector

$$\mathbf{x}_k = \begin{bmatrix} x(n_k) \\ x(n_k+1) \\ \vdots \\ x(n_k+N-1) \end{bmatrix}, \quad k = 1, \ldots, M, \tag{5.99}$$

where n_k is the onset of the k-th beat, N is the number of samples per beat, and M is the number of beats. The ensemble of beats is represented by the $N \times M$ data matrix

$$\mathbf{X} = \begin{bmatrix} \mathbf{x}_1 \ \mathbf{x}_2 \ \cdots \ \mathbf{x}_M \end{bmatrix}. \tag{5.100}$$

While $\mathbf{X}$ may contain beats with widely different morphology, it is often desirable to only include beats with similar morphology as an homogenous ensemble implies that a smaller value of M is needed.

The transformation producing the principal components $\mathbf{w} = \begin{bmatrix} w_1 \ w_2 \ \cdots \ w_N \end{bmatrix}^T$ rests on the assumption that the observed signal $\mathbf{x}$ can be treated as a zero-mean random process, where $\mathbf{x}_1, \ldots, \mathbf{x}_M$ are different realizations of $\mathbf{x}$. This process is characterized by the intralead correlation matrix $\mathbf{R}_x = E[\mathbf{x}\mathbf{x}^T]$, i.e., the correlation between different samples in $\mathbf{x}$. The principal components $\mathbf{w}$ result from applying an orthogonal linear transformation to $\mathbf{x}$, defined by the $N \times N$ matrix

$$\boldsymbol{\Phi} = \begin{bmatrix} \boldsymbol{\varphi}_1 \ \boldsymbol{\varphi}_2 \ \cdots \ \boldsymbol{\varphi}_N \end{bmatrix}, \tag{5.101}$$

and

$$\mathbf{w} = \boldsymbol{\Phi}^T \mathbf{x}. \tag{5.102}$$

This transformation rotates $\mathbf{x}$ so that the elements of $\mathbf{w}$ become mutually uncorrelated. The first principal component is obtained as the scalar product $w_1 = \boldsymbol{\varphi}_1^T \mathbf{x}$, where the vector $\boldsymbol{\varphi}_1$ is chosen so that the variance of w_1,

$$E[w_1^2] = E[\boldsymbol{\varphi}_1^T \mathbf{x}\mathbf{x}^T \boldsymbol{\varphi}_1] = \boldsymbol{\varphi}_1^T \mathbf{R}_x \boldsymbol{\varphi}_1, \tag{5.103}$$

is maximized subject to the constraint that $\boldsymbol{\varphi}_1^T \boldsymbol{\varphi}_1 = 1$. The maximal variance is obtained when $\boldsymbol{\varphi}_1$ is chosen as the normalized eigenvector corresponding to the largest eigenvalue of $\mathbf{R}_x$, denoted λ_1. The resulting variance is

$$E[w_1^2] = \boldsymbol{\varphi}_1^T \mathbf{R}_x \boldsymbol{\varphi}_1 = \lambda_1 \boldsymbol{\varphi}_1^T \boldsymbol{\varphi}_1 = \lambda_1. \tag{5.104}$$

Subject to the constraint that w_1 and the second principal component w_2 should be uncorrelated, w_2 is obtained by choosing $\boldsymbol{\varphi}_2$ as the eigenvector corresponding to the second largest eigenvalue of $\mathbf{R}_x$, and so on until the variance of $\mathbf{x}$ is completely represented by $\mathbf{w}$. Accordingly, to obtain the whole set of N different principal components, the eigenvector equation for $\mathbf{R}_x$ needs to be solved,

$$\mathbf{R}_x \boldsymbol{\Phi} = \boldsymbol{\Phi} \boldsymbol{\Lambda}, \tag{5.105}$$

where $\boldsymbol{\Lambda}$ is a diagonal matrix defined by the eigenvalues $\lambda_1, \ldots, \lambda_N$. Since $\mathbf{R}_x$ is not known in practice, the $N \times N$ sample correlation matrix $\hat{\mathbf{R}}_x$, defined by

$$\hat{\mathbf{R}}_x = \frac{1}{M}\mathbf{X}\mathbf{X}^T, \tag{5.106}$$

replaces $\mathbf{R}_x$ when computing the eigenvectors in (5.105).

The number of beats M should be large enough to produce a useful estimate of $\mathbf{R}_x$, where "large enough" may be translated to about 40 beats [94] or to the number of beats occurring during one minute [22]. In the standard implementation of PCA, the mean of the realizations $\mathbf{x}_1, \ldots, \mathbf{x}_M$ is removed before $\hat{\mathbf{R}}_x$ is computed. Since mean removal is not performed in this context, thus making PCA identical to the Karhunen–Loève transform [93], the eigenvector related to the largest eigenvalue, i.e., $\boldsymbol{\varphi}_1$, is virtually identical to a scaled version of the mean, i.e., the ensemble average.[6] The seven first eigenvectors are illustrated in Fig. 5.12 for five different ECGs; the related, normalized eigenvalues are presented in Fig. 5.13.

For the purpose of extracting the f wave signal, the following decomposition of $\mathbf{x}$ was proposed in [94]:

$$\mathbf{x} = \sum_{k=1}^{N} w_k \boldsymbol{\varphi}_k = \sum_{k=1}^{N_v} w_k \boldsymbol{\varphi}_k + \sum_{k=N_v+1}^{N_a} w_k \boldsymbol{\varphi}_k + \sum_{k=N_a+1}^{N} w_k \boldsymbol{\varphi}_k, \tag{5.107}$$

where the weights are determined by $w_k = \boldsymbol{\varphi}_k^T \mathbf{x}$. The ventricular subspace is spanned by the first N_v eigenvectors, i.e., the eigenvectors corresponding to the N_v largest eigenvalues, the atrial subspace is spanned by the next $N_a - N_v$ eigenvectors, and the "noise subspace" is spanned by the remaining eigenvectors. From the decomposition in (5.107), it is evident that an estimate of the f wave signal may be obtained by subtracting the two sums which produce estimates of the QRST complex and the noise from $\mathbf{x}$,

$$\begin{aligned} \hat{\mathbf{d}} &= \mathbf{x} - \sum_{k=1}^{N_v} w_k \boldsymbol{\varphi}_k - \sum_{k=N_a+1}^{N} w_k \boldsymbol{\varphi}_k \\ &= \sum_{k=N_v+1}^{N_a} w_k \boldsymbol{\varphi}_k, \end{aligned} \tag{5.108}$$

[6] For the eigenvector $\boldsymbol{\varphi}_1$ to be virtually identical to a scaled version of the ensemble average, the ensemble with similar-shaped beats should be reasonably well-aligned in time and the noise level should not be so high that f waves are completely obscured, i.e., two conditions which are easily met in practice.

(a)

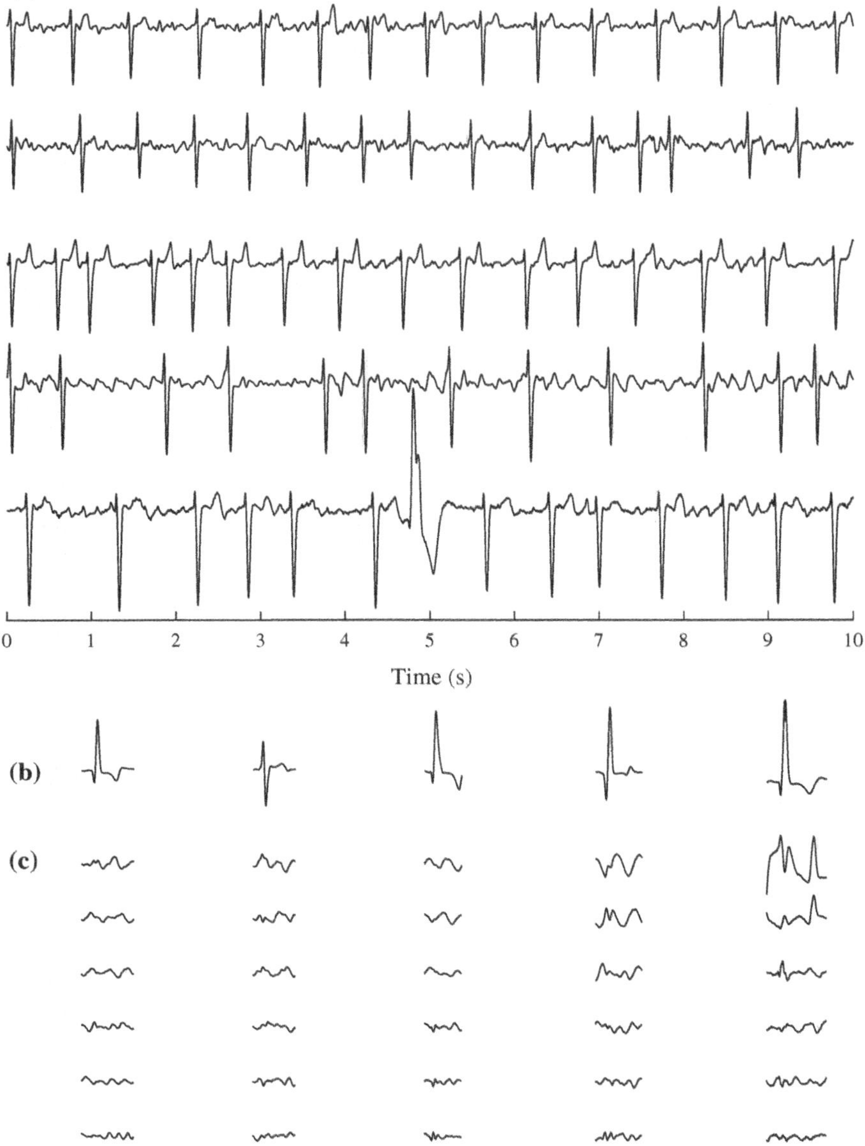

(b)

(c)

Fig. 5.12 **a** Single-lead ECGs obtained from five different patients, **b** the first eigenvector φ_1 and the ensemble average of the dominant beats (the two waveforms coincide so they cannot be distinguished from one another), and **c** the second until the seventh most significant eigenvectors. The eigenvectors are plotted using a time scale zoomed by a factor of two relative to the scale used in (**a**)

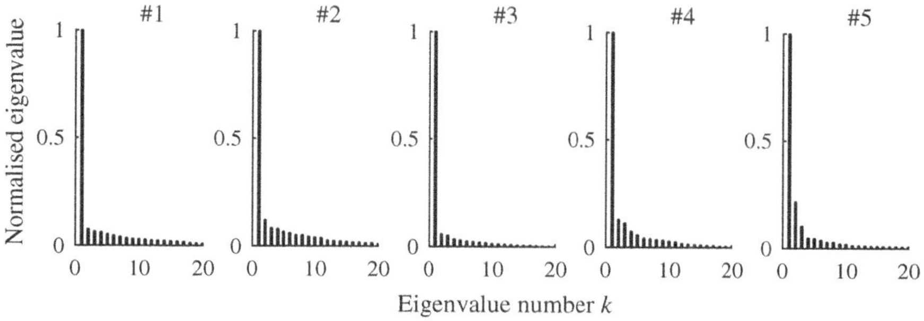

Fig. 5.13 The normalized eigenvalues $\lambda'_k = \lambda_k/\lambda_1, k = 1, \ldots, 20$, of the five single-lead ECGs displayed in Fig. 5.12

provided that the two dimensionality parameters N_v and N_a have first been properly identified.

When all beats in $\mathbf{X}$ have similar morphology, i.e., the situation usually encountered in practice, single-lead PCA is actually closely related to ABS. Since $\boldsymbol{\varphi}_1$ is virtually identical to a scaled version of the ensemble average, and all other eigenvectors are associated with much smaller eigenvalues, it is natural to assume that this eigenvector defines the ventricular space, i.e., $N_v = 1$. Ignoring the noise-related terms in (5.108), the f wave signal is simply estimated by $\hat{\mathbf{d}} = \mathbf{x} - w_1\boldsymbol{\varphi}_1$. Amplitude scaling is also performed in the single-lead version of STC, implying that single-lead PCA and STC are closely related when homogenous ensembles are processed. It was claimed that the estimator in (5.108) can be used to remove some of the noise in the ECG [94], however, it remains to be demonstrated to what extent such noise removal can improve f wave extraction.

When $\mathbf{X}$ becomes increasingly heterogenous, the ventricular subspace is spanned by two or more eigenvectors, i.e., $N_v > 1$, where the second and higher-order eigenvectors are related to morphologic variability. In such situations, single-lead PCA can be viewed as a generalization of ABS, where the template beat, defined as a linear combination of eigenvectors with data-dependent weights, replaces the ensemble average. With the additional degrees of freedom, the template beat may offer better handling of beat-to-beat variation in QRST morphology due to respiration and other extracardiac factors [94].

The vital point of single-lead PCA is the identification of N_v and N_a—inaccurate identification of these two parameters leads to that the extracted activities, both of ventricular and atrial origin, contain unwanted signal components. In many applications, PCA is employed as a technique for concentrating the information in $\mathbf{x}$ into a subset of principal components, i.e., $w_1, \ldots, w_K$, where $K \ll N$. In that case, the choice of K can be guided by various statistical performance indices, measuring, for example, how well the subset of principal components approximates the ensemble with respect to energy [91, 95]. However, such indices are unsuitable in f wave extraction since the problem is to find the dimensionality of two subspaces (where each subspace is rather loosely defined), rather than to concentrate the signal energy into as few components as possible.

Unfortunately, no method for subspace identification has been devised in the literature, and, consequently, it is unclear how a PCA-based method can be implemented in an automated system for AF analysis. Based on a small data set with AF patients, using visual subspace identification, the dimensionality of the atrial subspace, i.e., $N_a - N_v$, was found to range from 4 to 10 [94]. Thus, the use of fixed values of N_v and N_a does not seem to be a feasible alternative. As illustrated in Fig. 5.13, the falling-off pattern of λ_k does not exhibit any clearcut feature which can be explored for identification of these two parameters.

The decomposition in (5.107) is based on the assumption that an atrial subspace can always be identified for ECGs in AF. However, the variance maximization performed by PCA does not necessarily imply that a "pure" atrial subspace exists, since some eigenvector(s) may contain a mixture of atrial and ventricular activity. Neither does the orthogonality constraint necessarily facilitate the identification of an atrial subspace.

A simplified version of single-lead PCA rests on the assumptions that the ventricular subspace is one-dimensional, i.e., $N_v = 1$, and that the noise term in (5.108) can be ignored [22, 96], thereby avoiding the identification of an atrial subspace. As already noted, these two assumptions imply that f wave extraction is closely related to ABS when homogenous ensembles are processed, with the difference that amplitude scaling is not performed in ABS. While the method described in [22] was labeled "SVD-based," it can just as well be labeled "PCA-based" since the eigenvectors associated with PCA can be determined directly from $\mathbf{X}$ by means of SVD, rather than by diagonalization of the sample correlation matrix $\hat{\mathbf{R}}_x$. The SVD means that an $N \times M$ matrix can be decomposed as [34]

$$\mathbf{X} = \mathbf{U}\boldsymbol{\Sigma}\mathbf{V}^T, \tag{5.109}$$

where $\mathbf{U}$ is an $N \times N$ orthogonal matrix whose columns are the left singular vectors, and $\mathbf{V}$ is an $M \times M$ orthogonal matrix whose columns are the right singular vectors. The matrix $\boldsymbol{\Sigma}$ is an $N \times M$ non-negative diagonal matrix containing the singular values $\sigma_1, \ldots, \sigma_N$,

$$\boldsymbol{\Sigma} = \begin{bmatrix} \sigma_1 & 0 & \cdots & 0 & \cdots & 0 \\ 0 & \sigma_2 & \cdots & 0 & \cdots & 0 \\ \vdots & \vdots & \ddots & \vdots & \ddots & \vdots \\ 0 & 0 & \cdots & \sigma_N & \cdots & 0 \end{bmatrix}, \tag{5.110}$$

for $N < M$. If $N > M$, $\boldsymbol{\Sigma}$ has instead more rows than columns and contains the singular values $\sigma_1, \ldots, \sigma_M$.

Using SVD, $\hat{\mathbf{R}}_x$ in (5.106) can be expressed in terms of $\mathbf{U}$ and a diagonal matrix $\boldsymbol{\Lambda}$ whose entries are the normalized and squared singular values $\sigma_1^2/M, \ldots, \sigma_M^2$/M,

$$\hat{\mathbf{R}}_x = \frac{1}{M}\mathbf{X}\mathbf{X}^T = \frac{1}{M}\mathbf{U}\boldsymbol{\Sigma}\mathbf{V}^T\mathbf{V}\boldsymbol{\Sigma}^T\mathbf{U}^T = \mathbf{U}\boldsymbol{\Lambda}\mathbf{U}^T. \tag{5.111}$$

Comparing (5.111) with (5.105), it is obvious that the eigenvectors associated with PCA are obtained as the left singular vectors of $\mathbf{U}$, i.e., $\boldsymbol{\Phi} = \mathbf{U}$, and the eigenvalues λ_k as σ_k^2/M. In a similar way, the right singular vectors of $\mathbf{V}$ contain information on interbeat correlation as they are associated with $\hat{\mathbf{R}}_x$, see below.

The method described in [22] includes various means for avoiding jumps at the beat boundaries. Rather than using fixed positions of the onset and end relative to a fiducial point, the positions are adjusted so that the magnitude of the jumps is minimized. Gaussian windowing is applied to the resulting f wave signal to further reduce the influence of jumps.

5.6.2 *Multi-Lead PCA*

Another approach to f wave extraction is to explore the redundant information in multi-lead ECGs [97, 98]. Similar to single-lead PCA, the idea with multi-lead PCA is to decompose the multi-lead ECG into a ventricular subspace, an atrial subspace, and a noise subspace. Since multi-lead PCA differs in certain respects from single-lead PCA, it is described in the following. The $L \times 1$ vector $\mathbf{x}(n)$ contains the L available leads at the time n,

$$\mathbf{x}(n) = \begin{bmatrix} x_1(n) \\ x_2(n) \\ \vdots \\ x_L(n) \end{bmatrix}, \quad n = 0, \ldots, N_t - 1, \tag{5.112}$$

where N_t denotes the total number of samples, typically so many that several beats are included. In contrast to single-lead PCA, where beat segmentation is required before forming the data matrix in (5.100), the data vector in (5.112) does not require segmentation since the ECG is processed on a sample-by-sample basis. The principal components are obtained by

$$\mathbf{w}(n) = \boldsymbol{\Phi}^T\mathbf{x}(n), \tag{5.113}$$

where the columns of the matrix $\boldsymbol{\Phi}$ are defined by the eigenvectors of the $L \times L$ matrix $\mathbf{R}_x$ describing interlead correlation. This matrix is estimated from the observed samples $\mathbf{x}(n)$ by

$$\hat{\mathbf{R}}_x = \frac{1}{N_t}\sum_{n=0}^{N_t-1}\mathbf{x}(n)\mathbf{x}^T(n). \tag{5.114}$$

The transformation in (5.113) redistributes the energy of the original ECG leads so that the most significant principal components contain information on the ventricular activity. Then, the next few components usually contain information on the atrial activity, although there is no guarantee that the ventricular and atrial activities are well-separated into different components. The remaining principal components contain noise, and are therefore discarded. Hence, the counterpart to finding N_v and N_a in single-lead PCA is to find the principal component(s) related to the atrial activity. Again, no automated method has been devised for the identification of the atrial component(s), although spectral analysis may be considered for finding the principal component(s) most likely of atrial origin, cf. Sect. 5.9.

With multi-lead PCA, a global f wave signal is extracted, defined as a linear combination of all the L leads, whereas single-lead PCA produces a lead-specific f wave signal. From a clinical perspective, it is unclear how the global f wave signal should be interpreted since its amplitude has been scaled by $\boldsymbol{\Phi}$. However, lead-specific f wave signals $\hat{\mathbf{d}}(n)$ may be extracted using a transformation which is inverse to the one given in (5.113) [4, 99],

$$\hat{\mathbf{d}}(n) = \boldsymbol{\Phi}\tilde{\mathbf{w}}(n), \tag{5.115}$$

where $\tilde{\mathbf{w}}(n)$ denotes that all components of $\mathbf{w}(n)$ are set to zero except the one which contains f waves. If f waves are identified in more than one principal component, the vector $\tilde{\mathbf{w}}(n)$ needs to be modified accordingly.

The standard 12-lead ECG was analyzed in [98], meaning the eight leads $V_1, \ldots, V_6$, I, and II since the other four leads can be determined as linear combinations of I and II. The minimum number of leads required for successful extraction of an atrial component was not investigated. Therefore, it is unclear whether multi-lead PCA can be applied to long-term, ambulatory ECGs which usually are recorded with only three leads.

Figure 5.14 illustrates multi-lead PCA when applied to a standard 12-lead ECG. Visually, the atrial activity is most easily identified in the fourth principal component, although the third and the fifth components also contain some activity. These three components also contain various degrees of QRS-related residuals. The three first components contain mostly ventricular activity, whereas the fifth and higher-order components contain noise.

5.6.3 Multi-Lead PCA of Single Beats

The *dominant T wave* is a concept introduced to explain and characterize T wave morphology in different leads, inspired by the observation that the T wave morphology is strikingly similar in different leads [100, 101]. In each lead, the T wave may be viewed as the projection of a main waveform, i.e., the dominant T wave, and a weighted sum of its derivatives. Singular value decomposition can then be used to find the dominant T wave, involving the data matrix $\mathbf{X}$ whose columns are defined by

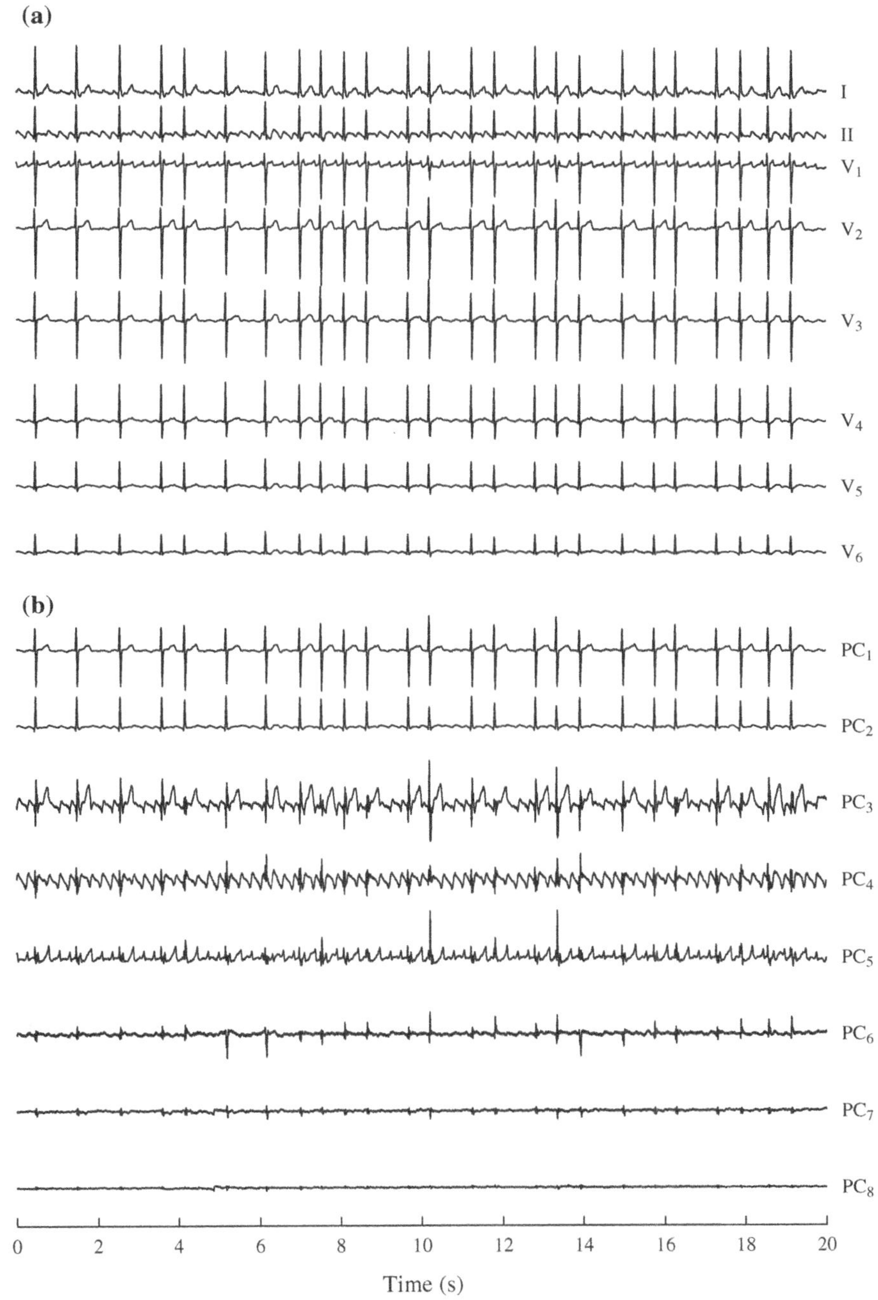

Fig. 5.14 f wave extraction using multi-lead PCA. **a** Original eight-lead ECG and **b** the corresponding principal components (PCs), displayed in order of decreasing variance. All PCs, except the first, have been scaled by a factor of 10

the samples of the JQ interval in the different leads. The dominant T wave is taken as the eigenvector, i.e., the column of $\mathbf{U}$ in (5.109), corresponding to the largest singular value.

Since the dominant T wave concept is closely related to multi-lead PCA, it is hardly surprising that this concept has been considered when developing a method for f wave extraction [37]. An important difference, though, is that the JQ interval of a single beat is analyzed, not the JQ interval of many beats as done in single-lead PCA, motivating *single beat cancellation* as the name of the method. In each lead, the T wave model, based on the most significant eigenvector, is scaled to fit to the observed T wave for subtraction. Then, the atrial activity within the QRS interval is estimated by interpolation between the two enclosing JQ intervals (with cancelled T waves), using the harmonic model in (5.47). Thus, no attempt is made to retrieve the f waves concealed inside the QRS complex. This procedure is repeated for all beats in the ECG, including ectopic beats.

The dominant T wave is defined as a linear combination of the most significant eigenvector and its derivatives [37, 100], see also [102, 103]. Since the eigenvector may contain noise which affects the derivatives quite considerably, a smooth mathematical function is fitted to the eigenvector from which the derivatives are computed. The following function was employed for modeling the T wave, defined as the product of two logistic functions:

$$f_{\mathrm{T}}(t;\boldsymbol{\theta}_{\mathrm{T}}) = p_1\left(p_2 + \frac{1}{1+e^{p_3(t-p_5)}}\cdot\frac{1}{1+e^{p_4(t-p_5)}}\right), \tag{5.116}$$

where p_1 is a scale factor, p_2 defines the wave offset, p_3 and p_4 define the positive and negative slope, respectively, and p_5 defines the timing of the T wave apex. These five parameters constitute together the vector $\boldsymbol{\theta}_{\mathrm{T}}$. The sporadically occurring U wave is modeled by a Gaussian function,

$$f_{\mathrm{U}}(t;\boldsymbol{\theta}_{\mathrm{U}}) = p_6 e^{-(t-p_8)^2/p_7^2}, \tag{5.117}$$

where p_6 is a scale factor, p_7 defines the width, and p_8 defines the timing of the U wave apex. These three parameters together define the vector $\boldsymbol{\theta}_{\mathrm{U}}$. The T-plus-U wave model, given by $f_{\mathrm{T}}(t;\boldsymbol{\theta}_{\mathrm{T}}) + f_{\mathrm{U}}(t;\boldsymbol{\theta}_{\mathrm{U}})$, is fitted to the most significant eigenvector, using nonlinear optimization to find $\hat{\boldsymbol{\theta}}_{\mathrm{T}}$ and $\hat{\boldsymbol{\theta}}_{\mathrm{U}}$. The two waves of the l-th lead are modeled by the following linear combination [37]:

$$\hat{x}_l(t) = \hat{a}_{0,l} f_{\mathrm{T}}(t;\hat{\boldsymbol{\theta}}_{\mathrm{T}}) + \hat{a}_{1,l} f_{\mathrm{T}}'(t;\hat{\boldsymbol{\theta}}_{\mathrm{T}}) + \hat{a}_{2,l} f_{\mathrm{T}}''(t;\hat{\boldsymbol{\theta}}_{\mathrm{T}}) + \hat{a}_{3,l} f_{\mathrm{U}}(t;\hat{\boldsymbol{\theta}}_{\mathrm{U}}), \tag{5.118}$$

where the four weight estimates $\hat{a}_{0,l}, \ldots, \hat{a}_{3,l}$ are obtained using LS estimation. Thus, a total of 12 parameters are needed to model the T-plus-U wave in each lead.

It should be noted that the estimation of the U wave model parameters in individual beats is challenged by the fact that the U wave is usually concealed by f waves, since the U wave amplitude is usually about 10% of the T wave amplitude [104]. The U wave has actually been studied in the presence of f waves, but then only after ensemble averaging of several QRST-U complexes [105]. The results showed that the U wave amplitude in lead V_1, on average, is 0.17 times the amplitude of the f wave (ranging from 0.1 to 0.4).

In the JQ interval, the f wave signal is obtained by subtracting $\hat{x}_l(t)$ from the original ECG, whereas, in the QRS interval, it is obtained by interpolation. The desired f wave signal is obtained by concatenation of JQ and QRS intervals. Since concatenation may introduce jumps at the interval boundaries, lowpass filtering was applied to reduce their influence.

Single beat cancellation was found to produce considerably smaller QRS-related residuals than ABS, a result which applied to both dominant and ectopic beats [37]. This result is hardly unexpected since the f wave signal in the QRS interval is obtained by interpolation using the samples of the enclosing JQ intervals.

5.7 Singular Spectral Analysis

The identification of ventricular and atrial subspaces represents, whether single- or multi-lead, the Achilles heel of the PCA-based methods, probably explaining why these methods have not been much used in clinical studies. By instead focusing on the samples of the TQ intervals, the problem of identifying an (unwanted) ventricular subspace is transformed into a simpler problem where only the atrial subspace has to be identified. The f wave signal in the QRST interval is estimated by projecting the QRST samples on the atrial subspace [106]. The method for subspace identification is based on *singular spectral analysis* (SSA) [107, 108], a method which is related to single-lead PCA since the eigenvectors and eigenvalues of a correlation matrix plays a central role in both methods. However, whereas single-lead PCA involves a correlation matrix estimated from an ensemble of time-aligned QRST complexes, the correlation matrix of SSA is estimated from the samples of consecutive TQ intervals. The motivation for pursuing SSA was to extract an f wave signal suitable for estimation of the DAF, whereas other characteristics of the extracted signal, such as amplitude and morphology, were not the focus of interest [106].

The f waves are observed in the TQ intervals (ignoring the presence of U waves and noise), whereas they are considered missing in the QRST intervals. In order to produce a reliable estimate of the correlation matrix, the ECG signal needs to be long enough so that it includes sufficiently many TQ intervals. Assuming that the locations of the QRST intervals have previously been determined, the binary indicator function $g(n)$ is introduced to describe whether the f waves, here denoted $x(n), n = 0, \ldots, N-1$, are missing,

$$g(n) = \begin{cases} 1, & n \in \text{TQ intervals}, \\ 0, & \text{otherwise}. \end{cases} \quad (5.119)$$

Moreover, $x(n)$ is assumed to be a wide-sense stationary process characterized by M different lags of the correlation function $r_x(k)$. The following estimator handles the fact that some observations are missing [109], see also [110, 111],

$$\hat{r}_x(k) = \frac{\sum_{n=0}^{N-k-1} x(n)g(n)x(n+k)g(n+k)}{\sum_{n=0}^{N-k-1} g(n)g(n+k)}, \quad |k| = 0, \ldots, M-1, \quad (5.120)$$

where $\hat{r}_x(k) = \hat{r}_x(-k)$. The estimated correlation function $\hat{r}_x(k)$ is used to form the $M \times M$ correlation matrix, defined by

$$\mathbf{R}_x = \begin{bmatrix} r_x(0) & r_x(-1) & r_x(-2) & \cdots & r_x(-M+1) \\ r_x(1) & r_x(0) & r_x(-1) & \cdots & r_x(-M+2) \\ r_x(2) & r_x(1) & r_x(0) & \cdots & r_x(-M+3) \\ \vdots & \vdots & \vdots & \ddots & \vdots \\ r_x(M-1) & r_x(M-2) & r_x(M-3) & \cdots & r_x(0) \end{bmatrix}. \quad (5.121)$$

It should be noted that $\mathbf{R}_x$ differs in structure from the correlation matrix in (5.106) since it is related to a wide-sense stationary process. Thus, the matrix is not only symmetric, but it is also Toeplitz.

Similar to single-lead PCA, the eigenvectors $\varphi_1(n), \ldots, \varphi_M(n)$ and the eigenvalues $\lambda_1, \ldots, \lambda_M$ of $\mathbf{R}_x$ are computed for the purpose of identifying the atrial subspace. The eigenvectors associated with the M_a largest eigenvalues are considered to span the atrial subspace, whereas the eigenvectors associated with the remaining eigenvalues span the noise subspace. Thus, the crucial step of this extraction method is to determine the dimensionality $M_a (\ll M)$ so that the atrial subspace is composed of only wanted signal components. Figure 5.15 displays the 10 most significant eigenvectors of $\mathbf{R}_x$, where $r_x(k)$ is estimated from the TQ intervals of a 1-min ECG recorded in AF. Most eigenvectors have a sinusoidal shape, where the frequency of the first three eigenvectors is approximately equal to the DAF. For increasingly smaller eigenvalues, the frequency of the eigenvectors becomes increasingly higher. As noted in [106], and illustrated in Fig. 5.15, an oscillatory mode is typically characterized by a pair of sinusoidal eigenvectors with the same frequency, but with the phase shifted about 90°.

Rather than determining the weights w_i, i.e., the i-th principal component, by correlating $x(n)$ with $\varphi_i(n)$ in a fixed-location interval as done in (5.102), the weights are now determined as a function of time by letting $\varphi_i(n)$ slide on a sample-by-sample basis across $x(n)$, using a window length of M samples,

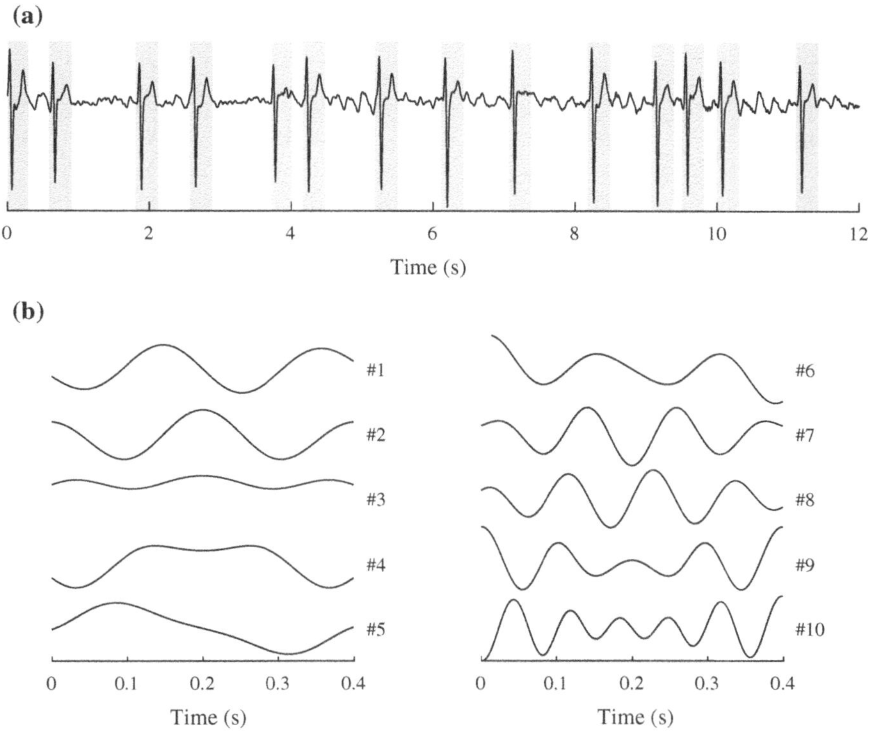

Fig. 5.15 **a** Single-lead ECG (grey areas indicating the QRST intervals) and **b** the corresponding 10 most significant eigenvectors of the correlation matrix in (5.121), determined using the estimator in (5.120). The DC level was removed before estimation

$$w_i(n) = \sum_{j=0}^{M-1} x(n+j)\varphi_i(j), \quad n = 0, \dots, N-M, \quad i = 1, \dots, M_a. \tag{5.122}$$

It should be emphasized that $w_i(n)$ is computed in both QRST and TQ intervals, whereas $\varphi_i(n)$ is estimated from TQ intervals.

In single-lead PCA, the f wave signal estimate $\hat{d}(n)$ is obtained as an orthogonal expansion of the eigenvectors, cf. (5.108). However, since $w_i(n)$ is estimated for every n and shorter than $x(n)$, i.e., $(N - M + 1)$ instead of N, the f wave signal is estimated by means of the following two steps, which together are optimal in the LS sense [107]. First, the i-th signal component $\hat{d}_i(n)$, associated with $\varphi_i(n)$, is obtained ("reconstructed") by

$$\hat{d}_i(n) = \begin{cases} \dfrac{1}{n+1}\displaystyle\sum_{j=0}^{n} w_i(n-j)\varphi_i(j), & n = 0, \ldots, M-2, \\ \dfrac{1}{M}\displaystyle\sum_{j=0}^{M-1} w_i(n-j)\varphi_i(j), & n = M-1, \ldots, N-M, \\ \dfrac{1}{N-n}\displaystyle\sum_{j=n-N+M}^{M-1} w_i(n-j)\varphi_i(j), & n = N-M+1, \ldots, N-1, \end{cases} \tag{5.123}$$

where $\hat{d}_i(n), i = 1, \ldots, M_a$, may be viewed as a result of filtering $w_i(n)$ using the impulse response $\varphi_i(n)$. Another interpretation of (5.123) is that $\hat{d}_i(n)$ is obtained as a weighed reconstruction of each eigenvector, sampled at different lags. Second, the desired estimate of the f wave signal is obtained by summing the M_a signal components $\hat{d}_i(n)$,

$$\hat{d}(n) = \sum_{i=1}^{M_a} \hat{d}_i(n). \tag{5.124}$$

The number of lags M and the number of components M_a constitute the two main design parameters of SSA. The choice of M determines the lowest frequency which can be captured by the reconstructed signal. With respect to f wave signals, this implies that M should be chosen so that frequencies down to 3–4 Hz can be captured. The number of components M_a should either be set to a fixed value, or determined by means of an iterative version of SSA. In the latter case, new components $\hat{d}_i(n)$ are added until the reconstructed signal $\hat{d}(n)$ no longer improves significantly; this version is known as *iterative singular spectral analysis* [112]. In [106], the reconstruction of $\hat{d}(n)$ was performed using at least $M_a = 7$ components.

The iterative SSA method rests on the assumption that $x(n)$ is a wide-sense stationary process, implying that the DAF must be constant within the analyzed interval $[0, N-1]$. Accordingly, the extracted f wave signal cannot reflect the time-varying spectral structure commonly observed in f wave signals [50]. As noted earlier, the iterative SSA method was introduced to fill in samples in the QRST intervals so that the DAF could be more reliably estimated, while characterization of the spectral harmonics was not addressed.

Figure 5.16 illustrates f wave extraction using iterative SSA, as well as two-template ABS for comparison. As expected, the f wave signal extracted by the former method is characterized by more pronounced narrowband oscillations than the f wave signal of the latter method. Moreover, it is noted that the f wave signal extracted by iterative SSA exhibits less variation in amplitude in the QRST intervals than in the TQ intervals.

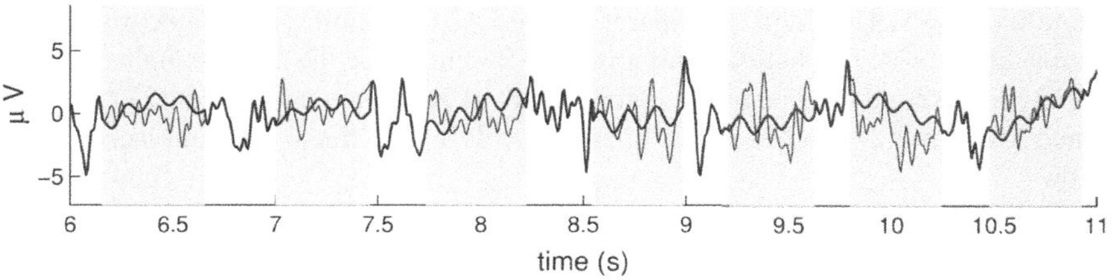

Fig. 5.16 f wave estimation using iterative SSA (thick line) at a heart rate of 75 beats per minute. For comparison, the estimate produced by two-template ABS is displayed (thin line). Grey areas indicate the QRST intervals. Note that the two signals coincide in the TQ intervals. (Reprinted from [106] with permission)

5.8 Autoregressive Modeling and Prediction Error Analysis

Rather than exploring the second-order statistics in a nonparametric fashion for separating the mixture of signal sources of atrial, ventricular, and extracardiac origin, as done in multi-lead PCA, the sources can be separated by modeling the f wave signal as an AR process [113], described in this section. The model-based approach also explores the second-order statistics, but in a parametric fashion. The following linear observation model is assumed:

$$\mathbf{x}(n) = \mathbf{A}\mathbf{s}(n), \tag{5.125}$$

where $\mathbf{A}$ is an $L \times L$ instantaneous, unknown mixing matrix.[7] The vector $\mathbf{s}(n)$ contains L different "source" signals,

$$\mathbf{s}(n) = \begin{bmatrix} s_1(n) \\ s_2(n) \\ \vdots \\ s_L(n) \end{bmatrix}, \tag{5.126}$$

whereas $\mathbf{x}(n)$ contains the L "sensor" signals, i.e., the ECG leads. For the situation when the number of sensors and sources differ, the number of sensors is usually assumed to be at least as many as the number of sources. Assuming that $s_1(n)$ is the atrial source, the goal is to find an $L \times 1$ "demixing" vector $\mathbf{w}$ such that the desired f wave signal $d(n)$ is expressed as

$$d(n) = s_1(n) = \mathbf{w}^T\mathbf{x}(n), \tag{5.127}$$

[7]The observation model in (5.125) is also central to the methods exploring higher-order statistics for independent component analysis, see Sect. 5.9. The mixing matrix $\mathbf{A}$ is here constrained to be orthogonal, whereas not so in Sect. 5.9.

where $\mathbf{w}^T\mathbf{w} = 1$. The demixing operation in (5.127) is similar to the linear transformation in (5.113), but with the important difference that the "demixing" vector in (5.113) corresponds to one of the eigenvectors of $\hat{\mathbf{R}}_x$, whereas $\mathbf{w}$ in (5.127) is determined from optimization of a cost function involving the linear prediction errors, see below.

Modeling of $d(n)$ as an AR process was already considered in Sect. 5.3, where AR-based interpolation was used to replace large-amplitude QRS residuals; the model parameters were estimated in the enclosing JQ intervals. Autoregressive modeling is closely related to the linear prediction problem in which $d(n)$ is predicted from the P preceding samples $d(n-1), \ldots, d(n-P)$, using a finite impulse response (FIR) filter structure of the predictor,

$$\hat{d}(n) = \sum_{p=1}^{P} a_p d(n-p). \tag{5.128}$$

Initially, the model parameters $a_1, \ldots, a_P$ and the model order P are assumed to be known; however, to make the method practical, the parameters need, at a later stage, to be estimated from the observed signal $\mathbf{x}(n)$. The prediction error $e_d(n)$ of the demixed signal $d(n)$,

$$e_d(n) = d(n) - \hat{d}(n), \tag{5.129}$$

can, by use of the demixing equation in (5.127), be expressed in terms of the observed signal $\mathbf{x}(n)$,

$$e_d(n) = \mathbf{w}^T\mathbf{e}_x(n), \tag{5.130}$$

where

$$\mathbf{e}_x(n) = \mathbf{x}(n) - \sum_{p=1}^{P} a_p \mathbf{x}(n-p). \tag{5.131}$$

Similar to the standard approach to estimating the AR model parameters, the demixing vector $\mathbf{w}$ can be found by minimizing the mean square prediction error [114]

$$E[e_d^2(n)] = \mathbf{w}^T E\left[\mathbf{e}_x(n)\mathbf{e}_x^T(n)\right]\mathbf{w}, \tag{5.132}$$

subject to the constraint that $\mathbf{w}^T\mathbf{w} = 1$. Unfortunately, such an approach may not always lead to the extraction of the source signal $s_1(n)$, because the minimum value of $E[e_d^2(n)]$ may correspond to another source signal, even when the AR parameters modeling $s_1(n)$ are known a priori [113]. This serious limitation can be addressed by noting that the prediction error related to $s_1(n)$, i.e., $e_d(n)$, should be a white process, whereas the prediction errors related to the other source signals should not (this observation assumes that the model order P is large enough). Introducing the mean cross prediction error $E[e_d(n)e_d(n-q)]$ for studying the whiteness of $e_d(n)$,

$$E[e_d(n)e_d(n-q)] = \mathbf{w}^T \mathbf{E}_x(q)\mathbf{w}, \quad q = 1, \ldots, P, \tag{5.133}$$

where

$$\mathbf{E}_x(q) = E\left[\mathbf{e}_x(n)\mathbf{e}_x^T(n-q)\right], \tag{5.134}$$

the goal is to find that particular $\mathbf{w}$ which sets $\mathbf{w}^T\mathbf{E}_x(q)\mathbf{w}$ equal to zero, reflecting that $e_d(n)$ is uncorrelated at the time lag q. Instead of finding the roots which solve $\mathbf{w}^T\mathbf{E}_x(q)\mathbf{w} = 0$, minimization of the following cost function, the *mean square cross prediction error* (MSCPE), is proposed for finding $\mathbf{w}$ [113]:

$$J(\mathbf{w}; q) = \mathbf{w}^T \mathbf{E}_x(q)\mathbf{E}_x^T(q)\mathbf{w}, \quad 0 < q \leq P, \tag{5.135}$$

which again is subject to the constraint that $\mathbf{w}^T\mathbf{w} = 1$. It is well-known that the minimization of $J(\mathbf{w}; q)$ is equivalent to finding the eigenvector which corresponds to the smallest eigenvalue of the symmetric matrix $\mathbf{E}_x(q)\mathbf{E}_x^T(q)$ [115]. The "square" factor appearing in the MSCPE is related to squaring of the mean cross prediction error in (5.133),

$$\left(\mathbf{w}^T\mathbf{E}_x(q)\mathbf{w}\right)^2 = \mathbf{w}^T\mathbf{E}_x(q)\mathbf{w}\mathbf{w}^T\mathbf{E}_x^T(q)\mathbf{w}. \tag{5.136}$$

Minimizing this expression implies that the projection of the vector $\mathbf{w}^T\mathbf{E}_x(q)$ onto $\mathbf{w}$ is minimized. Since $E[e_d(n)e_d(n-q)]$ should be equal to zero due to the whiteness property, this process is equivalent to minimizing the norm of $\mathbf{w}^T\mathbf{E}_x(q)$ itself, as obtained by minimizing the cost function $J(\mathbf{w}; q)$.[8]

To speed up the determination of $\mathbf{w}$, the input signal $\mathbf{x}(n)$ is whitened, i.e., its components are decorrelated and scaled to have unit variance. Whitening represents a standard preprocessing technique in blind source separation [95, 116]. Recalling that the correlation matrix $\mathbf{R}_x$ can be decomposed into its eigenvectors $\boldsymbol{\Phi}$ and eigenvalues $\boldsymbol{\Lambda}$,

$$\mathbf{R}_x = \boldsymbol{\Phi}\boldsymbol{\Lambda}\boldsymbol{\Phi}^T, \tag{5.137}$$

the following linear transformation whitens $\mathbf{x}(n)$,

$$\mathbf{x}_w(n) = \boldsymbol{\Lambda}^{-\frac{1}{2}}\boldsymbol{\Phi}^T\mathbf{x}(n), \tag{5.138}$$

since

$$E[\mathbf{x}_w(n)\mathbf{x}_w^T(n)] = \boldsymbol{\Lambda}^{-\frac{1}{2}}\boldsymbol{\Phi}^T E[\mathbf{x}(n)\mathbf{x}^T(n)]\boldsymbol{\Phi}\boldsymbol{\Lambda}^{-\frac{1}{2}} = \mathbf{I}. \tag{5.139}$$

[8] Alternatively, the mean cross prediction error $E[e_d(n)e_d(n-q)]$ in (5.133) may be minimized by introducing the cost function $J'(\mathbf{w}; q) = \mathbf{w}^T(\mathbf{E}_x(q) + \mathbf{E}_x^T(q))\mathbf{w}$. The eigenvector corresponding to the smallest eigenvalue of the symmetric matrix $\mathbf{E}_x(q) + \mathbf{E}_x^T(q)$ minimizes $J'(\mathbf{w}; q)$, and therefore taken as an estimate of $\mathbf{w}$.

Thus, the whitened signal $\mathbf{x}_w(n)$ replaces $\mathbf{x}(n)$ in the algorithm for determining $\mathbf{w}$, defined by (5.128)–(5.135).

The model parameters $\mathbf{a} = \left[a_1 \; \cdots \; a_P\right]^T$ are estimated using an iterative procedure. The initial estimate of $\mathbf{a}$ is coarse in the sense that it is determined from an f wave signal $\hat{d}^{(0)}(n)$, formed by concatenation of the samples contained in the latter half of consecutive RR intervals (rather than by concatenation of successive TQ intervals as done in, e.g., [25, 117]).[9] Cubic spline interpolation is employed for reducing the jumps which tend to occur at the boundaries of the TQ intervals. Then, the initial estimate $\hat{\mathbf{a}}^{(0)}$ is used for determining $\mathbf{w}^{(0)}$, which in turn is used for producing the f wave signal estimate $\hat{d}^{(1)}(n)$ computed using (5.127). Since $\hat{d}^{(1)}(n)$ is a connected signal, i.e., defined for all samples, rather than a concatenated signal as $\hat{d}^{(0)}(n)$, the new estimate $\hat{\mathbf{a}}^{(1)}$ is expected to be more accurate. The demixing vector $\mathbf{w}^{(1)}$ is then used for producing the f wave signal estimate $\hat{d}^{(2)}(n)$, and so on. The iterative procedure is terminated when

$$\|\hat{\mathbf{a}}^{(i)} - \hat{\mathbf{a}}^{(i-1)}\| < \eta_a, \tag{5.140}$$

where η_a is a preset tolerance.

A weak point of this iterative procedure is the initial estimate $\hat{d}^{(0)}(n)$, which may not be representative as the lengths of concatenated intervals can be very short, thus jeopardizing convergence. The performance evaluation presented in [113], involving only eight 8-s ECGs from AF patients, does not shed much light on convergence properties. Therefore, the influence of a nonrepresentative $\hat{d}^{(0)}(n)$ on overall performance remains to be investigated. Another aspect to be investigated is how often the AR parameter estimates need to be updated when signals much longer than 8 s are analyzed. Yet another aspect relates to how critical the choice of η_a in (5.140) is to performance.

While the cost function $J(\mathbf{w}; q)$ may be evaluated for any time lag $q = 1, \ldots, P$, only $J(\mathbf{w}; q = 1)$ was considered in [113]. One reason for this particular choice is that minimization of a short time lag is likely to convey more important information on whiteness than does minimization of larger time lags.[10] The definition of a cost function involving all P lags has not been considered in the context of f wave extraction.

Figure 5.17 illustrates f wave extraction in five patients with persistent AF. Several interesting observations can be made from the extracted f wave signals. For beats with dominant morphology, large-amplitude QRS-related residuals do not seem to be much of an issue, whereas VPBs remain uncanceled in the extracted signal (patients #2 and #4). As can be observed from the TQ intervals, the f wave amplitude of the extracted signal is sometimes much larger than that of the original ECG

[9] An overview of methods for AR model parameter estimation, as well as for model order estimation, can be found in, e.g., [4, 118].

[10] The analysis of lagged covariance matrices for the purpose of separating different signal sources with a temporal structure was first explored in a nonparametric setting, leading to the AMUSE algorithm [119], see also [95, Chap. 18]. Using that algorithm, the time lag is usually chosen to be $q = 1$, just like in [113].

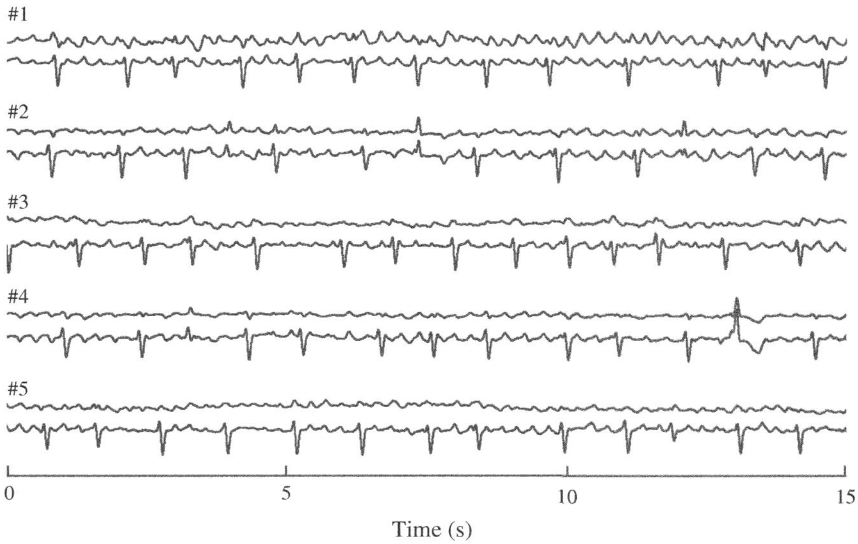

Fig. 5.17 f wave extraction in single-lead ECGs using autoregressive modeling and prediction error analysis. The ECG was acquired from five different patients with persistent AF. (Reprinted from [113] with permission)

(patients #1 and #5), whereas, in other situations, f waves which are clearly visible in the original ECG are barely visible in the extracted signal (patient #3). Moreover, the f wave morphology observed in the TQ intervals of the original ECG is not always preserved in the extracted signal. Thus, the method appears to be unsuitable for characterization of f wave amplitude and morphology, see, e.g., [57, 58].

5.9 Independent Component Analysis

Independent component analysis (ICA) has the following linear observation model at its core [95, 120],

$$\mathbf{x}(n) = \mathbf{A}\mathbf{s}(n), \tag{5.141}$$

where the observed signals $\mathbf{x}(n)$ are a mixture of the source signals $\mathbf{s}(n)$. This model has already been considered for blind source separation based on AR modeling and prediction error analysis, cf. (5.125). The mixing matrix $\mathbf{A}$ is assumed to be square ($L \times L$), i.e., the number of sources is assumed to be equal to the number of ECG leads, and invertible. Since neither $\mathbf{s}(n)$ nor $\mathbf{A}$ are known a priori, they have to be estimated from the observations, contained in the matrix

$$\mathbf{X} = \left[\mathbf{x}(0)\ \mathbf{x}(1)\ \cdots\ \mathbf{x}(N-1)\right]. \tag{5.142}$$

The source signals $\mathbf{s}(n)$ are assumed to be statistically independent of each other, characterized by the joint probability density function

$$p(\mathbf{s}(n)) = \prod_{l=1}^{L} p_l(s_l(n)), \tag{5.143}$$

where the l-th source signal $s_l(n)$ is characterized by an (unknown) non-Gaussian probability density function $p_l(s_l(n))$. Relying on the central limit theorem, it is well-known that mixing with $\mathbf{A}$ causes $\mathbf{x}(n)$ to become more Gaussian than $\mathbf{s}(n)$. The assumptions of statistical independence and non-Gaussianity are general in nature as they do not impose any constraint on the spectral properties of the source signals, whereas source separation based on AR modeling does. It should be emphasized that the mixing is instantaneous, meaning that the model does not involve any filtering of the source signals (models involving filtering are known as "convolutive mixtures"). Another important observation is that the model in (5.141) does not explicitly account for the presence of noise. However, when the noise originates from a separate source, it should end up in one of the independent components, thus being separated from the other signal sources.

The goal of ICA is to find a demixing matrix $\mathbf{W}$ producing statistically independent estimates $\hat{\mathbf{s}}(n)$ of the source signals $\mathbf{s}(n)$,

$$\hat{\mathbf{s}}(n) = \mathbf{W}\mathbf{x}(n), \tag{5.144}$$

commonly referred to as "independent components". Owing to the two assumptions of statistical independence and non-Gaussianity of the source signals, it is always possible to identify a demixing matrix $\mathbf{W}$ [121]. An obvious solution is to chose the demixing matrix as $\mathbf{W} = \mathbf{A}^{-1}$, however, this choice is made difficult since the estimation of $\mathbf{W}$ involves ambiguity with respect to permutation, amplitude, and sign of the source signals. The order of the independent components is ambiguous as they may be permuted; thus application-specific techniques are required to identify the independent component(s) of special interest. Moreover, the amplitudes and signs of the independent components cannot be uniquely determined since $\mathbf{x}(n)$ is the product of two unknown quantities, i.e., $\mathbf{A}$ and $\mathbf{s}(n)$. In matrix terms, these ambiguities may be described by a product of the permutation matrix $\mathbf{P}$, the diagonal amplitude scaling matrix $\mathbf{B}$ with positive elements, and the diagonal sign matrix $\mathbf{S}$ with elements ± 1; all three matrices with dimension $L \times L$. Then, if $\mathbf{W}$ is a demixing matrix producing statistically independent estimates $\hat{\mathbf{s}}(n)$, so is also the matrix product $\mathbf{PBSW}$.

The observed signal $\mathbf{x}(n)$ is usually preprocessed before being used in the computation of the independent components. Preprocessing includes centering, i.e., subtraction of the mean in each independent component, and whitening, i.e., decorrelation and variance normalization. Whitening is usually accomplished with PCA, cf. (5.138), as it reduces the dimensionality of the problem and speeds up the maximization of non-Gaussianity, see below.

A description of the many methods developed for estimating $\mathbf{W}$ is far beyond the scope of this text; the interested reader is referred to the comprehensive literature on this topic, see, e.g., [95] and references therein. Here, two aspects central to ICA are briefly considered: how to quantify non-Gaussianity and how to chose $\mathbf{W}$ so that non-Gaussianity is maximized.

When decomposing ECG leads into independent components, the source signals should be non-Gaussian, otherwise ICA will fail. Therefore, a measure needs to be defined quantifying the degree of non-Gaussianity. Kurtosis is a well-known measure for describing the peakedness of a distribution in relation to the Gaussian distribution. For a continuous-valued, scalar random variable y with zero-mean and unit variance, kurtosis is defined by

$$\kappa_y = E\left[y^4\right] - 3. \tag{5.145}$$

The kurtosis is equal to zero for Gaussian random variables, whereas it is negative for sub-Gaussian ("flat") distributions and positive for super-Gaussian ("spiky") distributions. Therefore, the absolute value of the kurtosis may serve as a measure of non-Gaussianity.

Another, more commonly used measure of non-Gaussianity is based on the differential entropy of the scalar random variable y,

$$H(y) = -\int_{-\infty}^{\infty} p(y) \ln p(y) dy, \tag{5.146}$$

where $p(y)$ is its probability density function. For random variables with equal variance, $H(y)$ reaches its maximum when y is Gaussian. Based on $H(y)$, a normalized measure, named *negentropy*, is defined by

$$J(y) = H(y_G) - H(y), \tag{5.147}$$

where y_G is a zero-mean, unit variance Gaussian random variable. The negentropy $J(y)$ is positive for all random variables, except for the Gaussian ones when $J(y)$ is equal to zero. Due to difficulties with estimating $p(y)$, $J(y)$ is not computed directly from its definition, but can be well-approximated by the use of expectations of general nonquadratic functions

$$J(y) \approx \left(E\left[g(y)\right] - E\left[g(y_G)\right]\right)^2, \tag{5.148}$$

where $g(x) = -\exp(-x^2/2)$ is a common choice. This approach to approximation turns out to be more robust to outliers than are approximations involving "polynomial" moments such as $E\left[y^4\right]$ in (5.145) [95].

The other central aspect of ICA relates to the maximization of the non-Gaussianity measure, where the goal is to find the weight (column) vectors $\mathbf{w}_1, \ldots, \mathbf{w}_L$, together defining $\mathbf{W}$. Similar to the f wave extraction method described in Sect. 5.8, whitening of $\mathbf{x}(n)$, defined by

$$\mathbf{x}_w(n) = \boldsymbol{\Lambda}^{-\frac{1}{2}}\boldsymbol{\Phi}^T\mathbf{x}(n), \tag{5.149}$$

is performed before the estimation of $\mathbf{W}$, implying that the demixing equation in (5.144) may be rewritten as

$$\hat{\mathbf{s}}(n) = \mathbf{W}\mathbf{x}(n) = \mathbf{U}\mathbf{x}_w(n). \tag{5.150}$$

The goal is then to find an estimator of the new demixing matrix $\mathbf{U} = \begin{bmatrix}\mathbf{u}_1 & \cdots & \mathbf{u}_L\end{bmatrix}^T$, whereas the two matrices related to whitening, i.e., $\boldsymbol{\Lambda}$ and $\boldsymbol{\Phi}$, are estimated as described in Sect. 5.8.

The FastICA algorithm offers fast, iterative maximization of an approximation of the negentropy, and is one of the most popular methods for estimating $\mathbf{U}$ [95, 122]. With this algorithm, described by pseudocode in Table 5.1, the weight vectors $\mathbf{u}_k$ are found successively subject to the constraint that the estimate $\hat{\mathbf{u}}_k$ is orthonormal to the previously found estimates $\hat{\mathbf{u}}_1, \ldots, \hat{\mathbf{u}}_{k-1}$. The weight update equation finds a vector $\mathbf{u}_k$ such that the projection $\mathbf{u}_k^T\mathbf{x}_w(n)$ maximizes non-Gaussianity. This equation involves the first and second derivatives of the function $g(x)$, denoted $g'(x)$ and $g''(x)$, respectively, and may be defined as exemplified above. To make the algorithm practical, the expectations in the weight update equation are replaced by sample averages, computed from the observations in $\mathbf{X}$. It should be noted that the FastICA algorithm is particularly simple to use since no step size parameter is involved, which otherwise is required in gradient-based algorithms such as the LMS algorithm.

Table 5.1 The FastICA algorithm for estimating the independent component $\hat{\mathbf{u}}_k$

$\mathbf{x}_w(n) = \boldsymbol{\Lambda}^{-\frac{1}{2}}\boldsymbol{\Phi}^T\mathbf{x}(n),\ n = 0, \ldots, N-1,$ (whitening)
$\mathbf{u}_k = \mathbf{u}_{k,\text{init}},$ (initialization as random vector)
$\varepsilon = \infty,$
while $\varepsilon > \eta,$ (η is a preset convergence tolerance)
  $\tilde{\mathbf{u}}_k \leftarrow E[\mathbf{x}_w(n)g'(\mathbf{u}_k^T\mathbf{x}_w(n))] - E[g''(\mathbf{u}_k^T\mathbf{x}_w(n))]\mathbf{u}_k,$ (weight update)
  $\tilde{\mathbf{u}}_k \leftarrow \tilde{\mathbf{u}}_k - \sum_{j=1}^{k-1}(\tilde{\mathbf{u}}_k^T\hat{\mathbf{u}}_j)\hat{\mathbf{u}}_j,$ (orthogonalization)
  $\tilde{\mathbf{u}}_k \leftarrow \tilde{\mathbf{u}}_k/\|\tilde{\mathbf{u}}_k\|,$ (weight normalization)
  $\varepsilon = \|\tilde{\mathbf{u}}_k - \mathbf{u}_k\|,$ (error)
  $\mathbf{u}_k \leftarrow \tilde{\mathbf{u}}_k,$
end
$\hat{\mathbf{u}}_k = \tilde{\mathbf{u}}_k.$

5.9.1 *f Waves and Modeling Assumptions*

Independent component analysis has been considered for f wave extraction in several papers [25, 123–126], see also the reviews [73, 127]. Similar to multi-lead PCA, ICA requires that the independent component with atrial activity is identified before the f wave signal can be analyzed. In practice, the "atrial" component is rarely a pure f wave signal, but, to various degrees, mixed with activity originating from other sources.

The identified atrial component represents a global f wave signal, with contributions from all analyzed leads, whose amplitude is not easily translated to clinical terms since the signal variance has been normalized. Hence, lead-specific information is lost as the components derive from one or several signal sources. The independent components with atrial activity may, however, be transformed back to the individual leads using a linear transformation [99].

Overall, ICA has been found robust to modeling errors, owing to that the underlying signal model involves just a few assumptions [128]; notably, none of the assumptions concern rhythm irregularity or beat morphology. Nevertheless, the few statistical assumptions involved with ICA need to be reasonably valid. While the assumptions cannot be easily verified since the source signals cannot be observed, different arguments have been put forward to make it plausible that the ICA model is suitable for f wave extraction.

The main signal sources are related to ventricular activity, atrial activity, and extracardiac noise, e.g., muscular activity, respiration, and electrode movement. These sources are assumed to be *statistically independent*. While the ventricular activity is coupled to the atrial activity during normal sinus rhythm, these two activities may be treated as independent statistical processes during AF as the atrial wavefronts initiate ventricular depolarization at highly irregular time instants. The assumption of statistical independence between the two cardiac sources and the extracardiac sources seems to be largely accepted, although respiration and ventricular activity are dependent since QRS amplitude, to a certain degree, is mechanically modulated by respiration (cf. page 141). While it has been shown that controlled respiration, through the autonomic nervous system, modulates DAF in certain patients with permanent AF [111], no one has reported on mechanical modulation of spontaneous respiration on f wave amplitude.

The assumption of *non-Gaussian source signals* is valid for ventricular activity, because histogram analysis of the ECG samples, at least at reasonably good SNRs, shows that the kurtosis is much larger than zero, i.e., the ventricular activity is clearly super-Gaussian [5, 123]. On the other hand, this assumption is more questionable for atrial activity as the kurtosis may approach zero [25], although it was initially believed that the atrial activity could be described by a sub-Gaussian distribution [123]. The statistical distributions of the extracardiac noise sources have not received much attention in the literature, although it has been noted that muscular activity can range from sub-Gaussian [129] to approximately Gaussian [5], implying that the independent component(s) with atrial activity may contain a certain amount of muscular noise.

The validity of assuming *linear, instantaneous mixing*, modeled by $\mathbf{W}$, may be motivated by the structure of the solution to the forward problem [123], see also [130]. In this solution, the electrical potential on the body surface results from adding partial contributions of the potentials on the epicardial surface, where each point is weighted by a linear, instantaneous transfer coefficient. The coefficients account for the conductivity of the human torso when approximated as an isotropic, homogeneous volume conductor. The validity of this assumption may be questioned though because the cardiac source changes its position over time due to, e.g., respiration [5], thus calling for more advanced ICA methods which allow the instantaneous mixture matrix to change over time.

A convolutive mixture model has been investigated for f wave extraction [131]. In that model, being more complex than the instantaneous mixture model in (5.141), each element in $\mathbf{W}$ is a linear, time-invariant filter instead of a scalar. Consequently, memory is introduced in the model. However, the results, deriving from a set of simulated ECG signals, showed that the instantaneous mixture model yields better performance than methods based on a convolutive mixture model with finite impulse response filters.

5.9.2 f Wave Identification

Identification of the f wave component(s) is a critical step in ICA. The first algorithm proposed for this purpose made use of kurtosis-based reordering of the components, exploring the assumption that f waves are associated with sub-Gaussian components, whereas noise and artifacts are associated with approximately Gaussian components and ventricular activity with super-Gaussian sources [123]. The following estimator of kurtosis was used:

$$\hat{\kappa}_i = \frac{1}{N}\sum_{n=0}^{N-1}\left(\frac{\hat{s}_i(n) - \hat{\mu}_i}{\hat{\sigma}_i}\right)^4 - 3, \tag{5.151}$$

where $\hat{s}_i(n)$ is the i-th independent component, and $\hat{\mu}_i$ and $\hat{\sigma}_i$ are the related sample mean and sample variance, respectively. Figure 5.18 illustrates the outcome of ICA when processing a 12-lead ECG, with the independent components ordered according to their kurtosis. The first component contains most of the f waves, and is relatively free of large-amplitude QRS residuals; some of the other components are also influenced by atrial activity. The amplitude histograms in Fig. 5.19 show that the first component is sub-Gaussian due to its negative kurtosis, whereas the last component is super-Gaussian. Ordering according to kurtosis was combined with power spectral analysis of the sub-Gaussian components in order to validate whether a dominant spectral peak exists in the interval [4, 12] Hz, i.e., the interval where the DAF is contained [123].

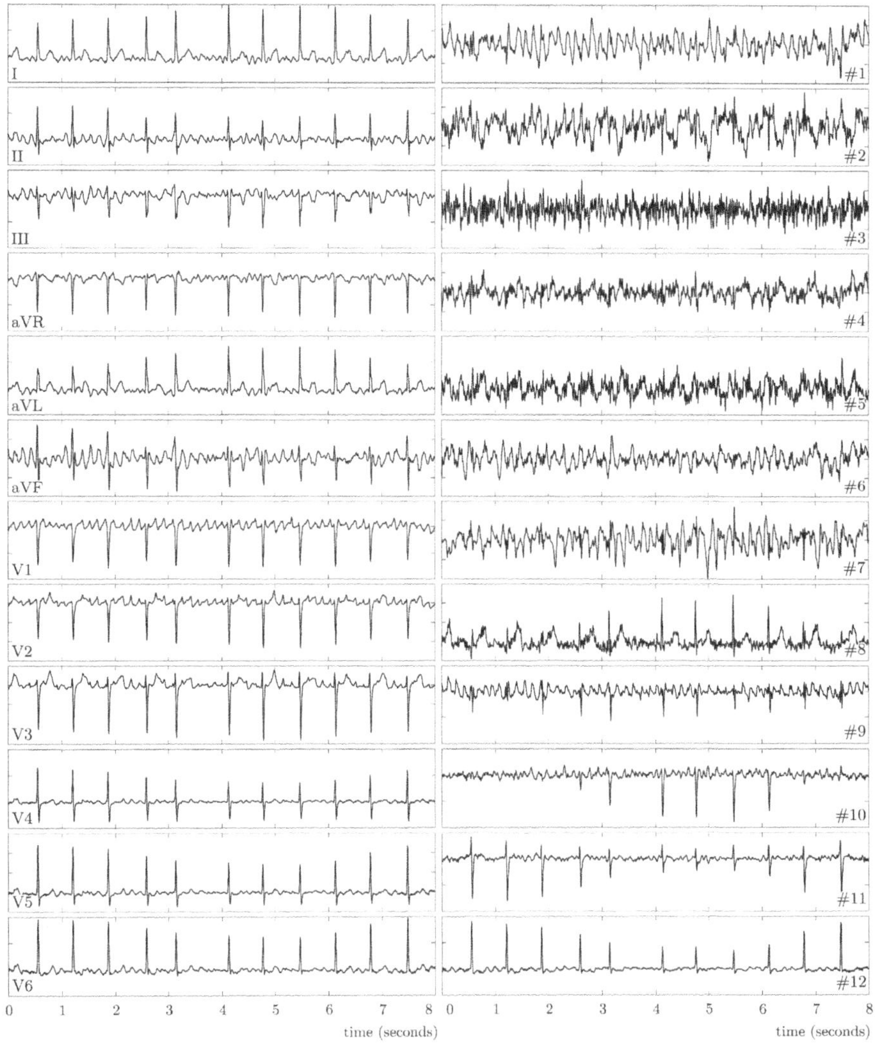

Fig. 5.18 f wave extraction using independent component analysis. **a** Original 12-lead ECG and **b** resulting independent components, displayed for increasing values of kurtosis (top to bottom). (Reprinted from [123] with permission)

Kurtosis alone is insufficient for accurate identification of the f wave component. One important reason is that the f wave characteristics change as the disease progresses, from almost flutter-like, well-organized, and clearly visible f waves to disorganized and often waxing and waning, making it increasingly difficult to distinguish the f wave component from noise components using kurtosis. Even worse, the f waves turn out to be approximately Gaussian in patients with persistent or perma-

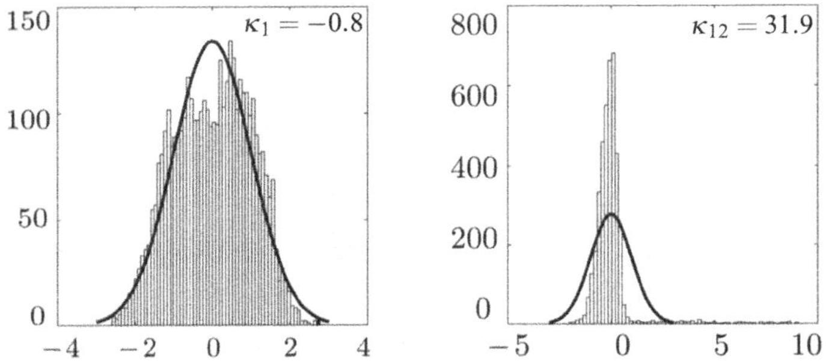

Fig. 5.19 Amplitude histogram of the samples in **a** component #1 containing atrial activity, and **b** component #12 containing ventricular activity, both displayed in Fig. 5.18. A fitted Gaussian probability density function is superimposed for comparison. Kurtosis is denoted k in the figure. (Reprinted from [123] with permission)

nent AF, causing problems for ICA which cannot separate more than one Gaussian source [127].

An improved approach to atrial component identification is to supplement kurtosis ordering and spectral analysis with a technique which retains the f wave components and excludes the ventricular components [25], see also [132]. Since the ventricular components usually have high kurtosis, they can be excluded by a simple threshold test (using a threshold of 1.5, most ventricular components can be excluded accurately). The nonventricular components, i.e., atrial activity, noise, and artifacts, whose kurtosis is close to zero, are separated using second-order blind identification (SOBI). This technique aims at separating a mixture of uncorrelated sources with different spectral content by analyzing second-order statistics which account for temporal information of the sensor signals [133], see also [95, Chap. 18]. To achieve this aim, SOBI finds a transformation that simultaneously diagonalizes several correlation matrices at different lags. Since, in general, no transformation exists which can handle such a strict condition, a function is introduced which measures the degree of joint (approximate) diagonalization at different lags. Similar to other ICA-based techniques, the component with a spectral peak in the interval [4, 12] Hz is selected as the one containing f waves. It may be noted that SOBI bears some resemblance to the AR-based method in Sect. 5.8, since both methods involve the diagonalization of one or several correlation matrices.

5.10 Performance Measures

A wealth of measures have been proposed for evaluating the performance of methods for f wave extraction. Depending on whether the ECG signal is real or simulated, different types of performance measures are usually employed. Performance evaluation based on real ECGs relies on indirect measures which, for example, reflect disproportionately large changes in f wave amplitude in the QRS interval. On the other hand, performance evaluation based on simulated ECGs allows direct quantification of the error between the extracted and true f wave signals, so that well-established measures can be employed, e.g., the MSE or the crosscorrelation coefficient. In several studies, more than one measure is employed to provide a more detailed description of performance. Unfortunately, since the measures, as well as the data sets used for evaluation, differ between studies, it is rarely possible to make a fair comparison of performance.

In the following, the simulated f wave signal in single-lead ECGs is denoted $d(n)$, and the extracted f wave signal is denoted $\hat{d}(n)$. For multi-lead ECGs, the corresponding matrix notations are $\mathbf{D}$ and $\hat{\mathbf{D}}$, respectively. All signals are assumed to have a length of N samples, which either refers to the length of a beat or a signal.

5.10.1 Real Signals

An important aim of performance measures designed for real signals is to quantify the amount of large-amplitude QRS residuals, as it would evidence inadequate f wave extraction. A simple approach is to compare the f wave amplitude which is representative of the QRST interval with that which is representative of the TQ interval—a comparison based on the assumption that good performance is manifested by a insignificant difference in amplitude between the two intervals [99]. Amplitude may be defined in various ways, for example, as the mean peak-to-peak f wave amplitude, see Sect. 6.2. A nonparametric statistical test, such as the Mann–Whitney test, can be used to determine whether the f wave amplitudes in the QRST interval and the TQ interval are significantly different. The two statistical populations are created by measuring the amplitude in several beats. This test provides a rather coarse description of performance as the results are expressed in terms of statistical significance of differences between the two populations (expressed as a p value), assuming that the statistical description is representative of performance.

Another approach to quantifying large-amplitude QRS-related residuals is to analyze the amplitude distribution of $\hat{d}(n)$ [37]. First, its median $m_{\hat{d}}$ and interquartile range Q_{50}, i.e., the difference between 75-th and 25-th percentiles, are computed. Then, rather than determining the standard deviation $\sigma_{\hat{d}}$ directly from the samples, it can be robustly estimated from Q_{50} through $\sigma_{\hat{d}} = Q_{50}/0.6745$, where the scale factor derives from the properties of a normal distribution, assumed to describe f wave amplitude. The percentage of samples in the QRS interval, whose absolute value exceeds the threshold $\eta = m_{\hat{d}} + 2\sigma_{\hat{d}}$, defines the performance measure. A high percentage suggests that the extracted f wave signal contains large-amplitude residuals, and vice versa.

The *ventricular residue* (VR) is a time domain measure describing the extent by which the f wave amplitude in the QRS interval deviates from the overall f wave amplitude in $\hat{d}(n)$ [22], see also [44, 76]. This measure involves the maximum absolute amplitude, defined by

$$A_{\max,i} = \max_{n=n_i-N_Q,\ldots,n_i+N_Q} (|\hat{d}(n)|), \tag{5.152}$$

where n_i is the occurrence time of the i-th QRS complex, and $2N_Q + 1$ is the length of the QRS window centered around n_i. The overall power of $\hat{d}(n)$, determined in the interval $[N_0, N_1]$, is obtained as

$$P_{\hat{d}}(N_0, N_1) = \frac{1}{N_1 - N_0 + 1} \sum_{n=N_0}^{N_1} \hat{d}^2(n). \tag{5.153}$$

For the i-th QRS complex, the ventricular residue measure is defined by

$$P_{\mathrm{VR},i} = A_{\max,i} \cdot \frac{\sqrt{P_{\hat{d}}(n_i - N_Q, n_i + N_Q)}}{P_{\hat{d}}(0, N-1)}. \tag{5.154}$$

When an f wave signal is accurately extracted, $P_{\mathrm{VR},i}$ assumes a value close to one, whereas it becomes increasingly larger as the amplitude of the QRS residuals becomes increasingly larger, illustrated in Fig. 5.20. An elaborated version of $P_{\mathrm{VR},i}$ has been proposed in [134], with the aim to better characterize both QRS-related residuals as well as the accuracy of extracted f waves.

While the above-mentioned performance measures provide useful information on the relationship between the local and the overall amplitude of $\hat{d}(n)$, they are blind to perturbations in the spectral properties of $\hat{d}(n)$. An accurately extracted f wave signal is usually characterized by a spectral peak which is narrower than the peak of a signal with large-amplitude residuals; the location of the spectral peak is determined by the DAF. This observation was the main motivation for introducing the measure *spectral concentration* P_{SC}, defined as the normalized spectral power in a small interval centered around the dominant spectral peak located at ω_p [25], see also [106, 113, 131, 135–137],

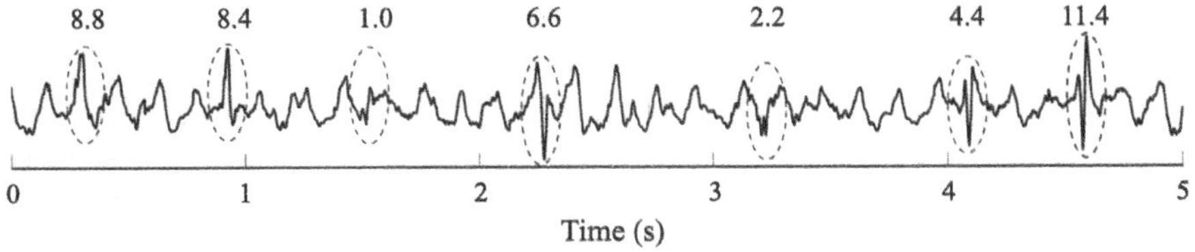

Fig. 5.20 Illustration of the ventricular residue performance measure $P_{\mathrm{VR},i}$, presented for each QRS complex (encircled), when applied to an f wave signal extracted using average beat subtraction. (Modified from [22] with permission)

$$P_{\mathrm{SC}} = \frac{\displaystyle\int_{\alpha_0\omega_p}^{\alpha_1\omega_p} S_{\hat{d}}(\omega)d\omega}{\displaystyle\int_0^{\pi} S_{\hat{d}}(\omega)d\omega}, \tag{5.155}$$

where $S_{\hat{d}}(\omega)$ is the power spectrum of $\hat{d}(n)$, and $0 < \alpha_0 < 1$ and $\alpha_1 > 1$ define the length of the integration interval; the interval is usually defined by $\alpha_0 = 0.82$ and $\alpha_1 = 1.17$. The spectral concentration measure has mostly been used for evaluating blind source separation methods, where the output is a global f wave signal, but also to find the demixing vector $\mathbf{w}$ in ICA-based f wave extraction [135, 136]. Higher spectral concentration has been considered synonymous with better extraction performance. However, the relevance of P_{SC} may be questioned since it is well-known that the DAF often varies over time [50, 51], implying that the dominant spectral peak may be broadened for physiological reasons. Neither does P_{SC} provide much information on the magnitude of QRST-related residuals. If higher spectral concentration is synonymous with better performance, then a very narrow bandpass filter, whose center frequency is defined by the DAF, would offer excellent performance. However, such an approach to f wave extraction is not particularly convincing.

Wavelet entropy [138] measures the "degree of organization" of a signal by decomposing $\hat{d}(n)$ into different scales (frequency bands) and analyzing the temporal distribution of the wavelet coefficient energy of each scale [139], see also [140] and page 110.[11] The use of wavelet entropy as a performance measure is motivated by the observation that QRS-related residuals are associated with wavelet energies which are unevenly distributed over time (though with a "periodicity" related to the heart rate), whereas the wavelet energies of the f waves are much more evenly distributed over time. Accordingly, the entropy of a scale with f waves is higher than the entropy of a scale containing QRS-related residuals. Since the spectral content of the atrial and ventricular activities differ, each type of activity is characterized by the wavelet entropy of certain, relevant scales; see [139] for details on the scales used for analyzing the wavelet entropy. Since both the time and scale domains are explored, performance measures based on wavelet entropy should provide a more detailed characterization of $\hat{d}(n)$ than, for example, the above-mentioned amplitude measures. It was claimed that "optimum suppression of ventricular activity would be expected to maximize the wavelet entropies at scales corresponding to ventricular and atrial activities [139]." While lower amplitude of the QRS-related residuals certainly implies higher entropy, it is unclear why the maximization of entropy implies optimum suppression.

Another, much more indirect approach to evaluating performance is to analyze whether ECG-derived measurements, such as the DAF, are in agreement with measurements obtained from intracardiac, simultaneously recorded signals, where the latter type of signal is treated as reference [1, 76].

[11] The wavelet coefficients are obtained by correlating the observed signal with the selected mother wavelet at different dilations and time shifts.

5.10.2 Simulated Signals

The MSE was the measure first employed for quantifying the performance of methods for f wave extraction, defined such that the overall noise level in a multi-lead, simulated ECG signal is also taken into account [30]. For the i-th beat, the MSE is defined by

$$P_{\text{MSE},i} = \frac{1}{N}\|\mathbf{D}_i - \hat{\mathbf{D}}_i\|_F^2 - \hat{\sigma}^2, \tag{5.156}$$

where $\mathbf{D}_i$ and $\hat{\mathbf{D}}_i$ relate to the i-th beat. First, an estimate of the noise level σ_l^2 is provided by the mean of the ensemble variance, defined in (5.23), of the l-th lead. The ensemble variance is preferably determined from the ventricular signal before the simulated f waves have been added. Hence, it is assumed that the noise level is identical in all beats of a lead. The overall noise level $\hat{\sigma}^2$ of the ECG is obtained by summing $\hat{\sigma}_l^2$ across all the leads. The MSE P_{MSE} is obtained by averaging the errors $P_{\text{MSE},i}$ for all beats. Among the performance measures used for simulated signals, P_{MSE} is the only one which handles multi-lead ECG signals and accounts for the noise level, although the latter is done in a crude way since one single noise variance estimate $\hat{\sigma}^2$ applies to all beats and leads.

For single-lead extraction, the MSE in (5.156) may be simplified to

$$P_{\text{MSE}} = \frac{1}{N}\sum_{n=0}^{N-1}(d(n) - \hat{d}(n))^2, \tag{5.157}$$

where the correction of noise variance and information on beat occurrence, introduced by the index i in (5.156), have been left out.

Due to its ease of interpretation, the NMSE has become a popular performance measure [21, 22, 37, 56, 76, 141],

$$P_{\text{NMSE}} = \frac{\sum_{n=0}^{N-1}(d(n) - \hat{d}(n))^2}{\sum_{n=0}^{N-1} d^2(n)}. \tag{5.158}$$

In most studies, the NMSE is defined as the square root of P_{NMSE}, although such a definition is incorrect from a formal viewpoint. A disadvantage with P_{NMSE} is its dependence on absolute amplitude, meaning that a certain error $(d(n) - \hat{d}(n))$ becomes less significant when $d(n)$ becomes larger—an undesirable property in performance evaluation. Indeed, proper evaluation of the accuracy of amplitude estimation is essential, since, for example, large-amplitude f waves have been found to predict AF termination in patients with persistent AF who have undergone catheter ablation [142].

The Pearson's *correlation coefficient* (CC) measures the similarity between $d(n)$ and $\hat{d}(n)$ [22, 76, 94, 136, 137, 141],

$$P_{\text{CC}} = \frac{\sum_{n=0}^{N-1} d(n)\hat{d}(n)}{\sqrt{\sum_{n=0}^{N-1} d^2(n)}\sqrt{\sum_{n=0}^{N-1} \hat{d}^2(n)}}, \tag{5.159}$$

where the mean values of $d(n)$ and $\hat{d}(n)$ have been first removed. Identical morphology of $d(n)$ and $\hat{d}(n)$ is reflected by P_{CC} being equal to one, whereas P_{CC} drops to zero as the two morphologies become increasingly different. The signed correlation index is robust variation on P_{CC} to measure morphologic similarity [143], see also Sect. 6.4.3. An important limitation of P_{CC}, as well as of P_{NMSE}, is their invariance to changes in amplitude.

The *improvement in SNR*, denoted ΔSNR, may be defined as the ratio between the power of the observed signal $x(n)$ but without f waves, and the MSE of the extracted f wave signal, i.e., P_{MSE} [56],

$$\Delta\text{SNR} = 10 \cdot \log_{10} \frac{\sum_{n=0}^{N-1} (d(n) - x(n))^2}{\sum_{n=0}^{N-1} (d(n) - \hat{d}(n))^2}. \tag{5.160}$$

In contrast to the earlier performance measures for simulated signals, all being functions of $d(n)$ and $\hat{d}(n)$, ΔSNR involves the power of the ventricular activity and the noise. As a consequence, the accuracy of the extracted f wave signal is evaluated correctly when the power of the ventricular activity and the noise, i.e., the numerator in (5.160), are held fixed from one realization of the simulated signal to the next. If not, an improved SNR may be due to changes in the properties of the ventricular activity and the noise.

Since the DAF f_0 is known for a simulated f wave signal, the mean absolute error (MAE) of the corresponding estimate $\hat{f}_0$ may serve as an indirect performance measure [106],

$$P_{\text{MAE}} = \frac{1}{M} \sum_{m=1}^{M} |f_{0,m} - \hat{f}_{0,m}|, \tag{5.161}$$

where averaging is performed over an ensemble of M different f wave signals, for example, created by adding different noise realizations to the ventricular activity. The RMS error, being closely related to the MAE, has been used to evaluate how well changes in the DAF can be tracked on a sample-to-sample basis [56]. When the f wave morphology is the target of the subsequent analysis, the performance measures in (5.156)–(5.160) are preferred over P_{MAE}.

5.11 Extraction Performance

Performance evaluation has been approached in many different ways in the literature. The data set used for evaluation may be composed of just a handful of signals or a huge number of real ECGs complemented with simulated signals. Sometimes, the performance of the proposed method is compared to some existing method, typically ABS despite the fact that a great number of methods have been proposed since ABS was conceived. Hence, the advancement in performance of a new method is not easily established, especially since the data sets, as well as their sizes, vary considerably from one study to another. The lack of comparative results, other than those involving ABS, may be explained by the difficulties associated with implementing a method from its description in the original publication.

The purpose of this section is not to rank the performance of the extraction methods described in this chapter, but to point out various aspects of performance in relation to these methods. The aspects highlighted in the beginning of this chapter, including the number of available leads, the presence of VPBs, and the duration of the analyzed ECG, are here supplemented with some additional aspects with particular relevance to performance evaluation.

QRS detection. Average beat subtraction and variants require that QRS complexes are detected before any ensemble operation can be performed. If the QRS detector operates at a low sampling rate, i.e., much below 1 kHz, it is necessary to improve the accuracy of the occurrence times through time alignment of all QRS complexes of the ensemble relative to a template QRS [4].

In general, methods based on adaptive filtering and multi-lead, blind source separation do not require QRS detection, although the ESN requires detection for construction of a spike-type reference signal [85]. The computational advantage of a method operating independently of a QRS detector is relatively modest though, since f wave extraction is usually preceded by several building blocks performing ECG signal processing, where QRS detection is the most fundamental block.

Number of leads. The methods described in this chapter can extract f waves in single-lead ECGs, except those which need two or more leads to separate the ventricular source(s), the atrial source(s), and the extracardiac noise sources. In fact, results describing the performance of ICA-based methods and multi-lead PCA have always been based on the standard 12-lead ECG. Thus, this family of signal separation methods appears to be less suitable for analysis of long-term, ambulatory ECGs where only two or three leads are recorded. It has been argued that only two leads are sufficient for ICA-based methods, since electrode positioning can be optimized on a patient-to-patient basis by analyzing information derived from body surface potential mapping [144]. However, the clinical acceptance of a tailored, two-lead ECG configuration is unclear.

Duration of ECG recording. As already noted, some 30–40 beats are at least needed by ABS and variants to produce a QRST template in which the f waves are reasonably well-suppressed, provided that the noise level is low. When the ECG recording is as short as 10 s, multi-lead PCA, ICA and the ESN constitute the more interesting alternatives. In general, the recommended minimum signal duration required for extraction has not been specified, although such information has practical significance.

Transitions from non-AF rhythms to AF deserve special attention, especially when brief AF episodes occur. Ideally, an f wave signal should be extracted immediately after the transition. However, since any method requires a certain "convergence time," brief AF episodes lasting less than 30 s may not be possible to analyze with respect to f wave characteristics. Provided that the characteristics of successive, brief AF episodes do not change much from one episode to the next, the convergence time may be shortened by simply suspending operation during non-AF rhythms so that the need to relearn the parameters for each new episode is bypassed.

Ventricular premature beats. Specific information on extraction performance in the presence of VPBs is rarely reported, although such beats can be abundant, and, consequently, with considerable influence on performance. In one of the few comparative studies [99], the data set was composed of ECGs completely free of VPBs, thus providing no information on how VPBs were handled. When simulated signals are used for evaluation, produced by adding simulated f waves to ECGs in sinus rhythm, see, e.g., [21, 30, 56, 135], only sporadic VPBs are likely to be present. When ECGs in AF are analyzed, the percentage of VPBs is not reported, nor is any information on whether f wave signals extracted during VPBs are included in the performance results. Thus, it is desirable to complement the above-mentioned performance measures with some other measure reflecting performance in the presence of VPBs.

The original design of most methods need to be modified to properly handle VPBs. For example, ABS and variants need additional beat templates to accomplish f wave extraction [96], whereas single-lead PCA would need a larger ventricular subspace to accommodate different VPB morphologies. As already noted, the EKF would need an expanded state-space model to handle VPBs with morphologies substantially different from that of the dominant beat. With respect to f wave extraction based on multi-lead PCA or ICA, it is unclear how VPBs are handled since such information is lacking in the literature, including the lack of illustrative examples. Using the ESN in its original design, sporadic VPBs were found to rarely cause large-amplitude residuals, as illustrated in Fig. 5.10b. When large-amplitude residuals were observed, the reservoir size was likely too small to remember VPB morphologies.

The interpolation-based methods and the methods based on SSA and AR modeling and prediction error analysis all rely on the TQ interval. Accordingly, VPB-related problems are avoided since the samples of the QRST interval are not analyzed. In cases of premature VPBs, however, the TQ interval preceding the VPB can be too short for these approaches to work satisfactorily.

Processing of non-AF rhythms. It is evident that methods for f wave extraction are designed to process ECGs in AF. However, the output of the AF detector may not always be accurate, implying that extraction has to be performed in non-AF rhythms. In such cases, it would be desirable to still have the atrial activity extracted, irrespective of whether atrial flutter or P waves are present. Since performance evaluation is almost always confined to extraction in signals which are known to contain f waves, it would be of great interest to establish what are the characteristics of the extracted signal in non-AF rhythms.

Average beat subtraction was originally developed for the detection of rhythms with P waves dissociated from the QRS complexes, but later applied to f wave extraction. This method is useful also for rhythms where P waves are associated with the QRS complexes. In such cases, the aim of averaging is rather to suppress the noise than the dissociated P waves. The same observation applies to the variants of ABS since they are based on the same principle as ABS. For STC, it is noted that the TQ-based f wave signal, introduced to improve parameter estimation in the presence of f waves, is superfluous when non-AF rhythms are processed.

Methods which build on the assumption that f waves exist in the JQ/TQ interval, allowing a coarse estimate of the DAF, will no longer work in non-AF rhythms. This observation applies to SSA, AR modeling and prediction error analysis, and the EKF. Moreover, the residual-constrained QRS template method involves certain general assumptions on the f wave properties which make that approach less suitable for processing of non-AF rhythms.

Typically, atrial flutter is not handled gracefully by methods for f wave extraction. One important reason is that the fundamental assumption of decoupled atrial and ventricular rates, explored in ABS and variants, is invalidated when atrial flutter is present. One of the very few papers describing a method which from the bottom is designed to extract both AF and atrial flutter explores the principles of blind source separation [137]. In that method, however, an estimate of the DAF is essential, making the method unsuitable for operation in P wave rhythms.

Signal-to-noise ratio. The SNR may vary quite markedly over time in an ECG recording, either due to physiologically mediated variation in f wave amplitude or variation in noise level. The waxing and waning of the f wave amplitude becomes increasingly more pronounced as the disease progresses to permanent AF, manifested by f waves which may vanish for several seconds due to progressive fibrosis of the atria, as well as other complicating factors. Obviously, f waves may vanish completely at higher noise levels. In both situations, the low SNR precludes reliable f wave analysis in the JQ and TQ intervals, implying that f wave extraction may have to be discontinued until the SNR becomes acceptable again. The problems arising at low SNRs bring up the issue of how to quantify f wave signal quality—a critical issue which barely has received any attention in the literature. It should be noted that methods less suitable for handling of non-AF rhythms are also less suitable for operation at lower SNRs.

Crucial parameters. Any extraction method involves at least one parameter whose setting is crucial to performance. For example, the number of beats M used for ensemble averaging is such a parameter in ABS, as well as the only, whose influence on performance has been investigated in [21, 96]. The size of the reservoir with recurrently connected neurons in the ESN is another parameter which is crucial to performance, especially when only 20–30 neurons are used [72]. In PCA- and ICA-based methods, the automated procedure required for identifying the atrial subspace will have to involve at least one crucial parameter, thus motivating a study investigating to what degree extraction performance is sensitive to changes in the parameter setting.

Clinical applications. Given that the data set used for performance evaluation is often small and composed of short-duration ECGs with distinct f waves, lacks non-AF arrhythmias, and has a low noise level, it is desirable that the robustness of the developed method is put to the test in clinical applications. For example, f wave extraction can be performed continuously on day-long signals, recorded under conditions which are much more challenging than those of the data set used for the initial performance evaluation. ECGs recorded in patients administered an antiarrhythmic drug can serve as a powerful test bed for evaluating performance since the f wave characteristics, as well as the signal quality, often change rather dramatically during the course of the recording [145–148]. Since performance evaluation of most methods have been confined to short recordings, the analysis of long-term recordings provides important insight into the approach taken to merging the f wave signals extracted in successive segments, or the means taken to process the signal in a segment-free (sequential) fashion.

Of the many methods developed for f wave extraction, it is only ABS and STC which have been extensively used in studies published in clinical journals. While this observation should not be taken as a confirmation of excellent performance, it nonetheless indicates that these two methods have proven useful for processing ECGs whose properties are more complex than those of simulated signals, produced by adding simulated f waves to low-noise ECGs in sinus rhythm.

References

1. M. Holm, S. Pehrsson, M. Ingemansson, L. Sörnmo, R. Johansson, L. Sandhall, M. Sunemark, B. Smideberg, C. Olsson, S.B. Olsson, Non-invasive assessment of atrial refractoriness during atrial fibrillation in man–Introducing, validating, and illustrating a new ECG method. Cardiovasc. Res. **38**, 69–81 (1998)
2. A. Bollmann, N. Kanuru, K. McTeague, P. Walter, D.B. DeLurgio, J. Langberg, Frequency analysis of human atrial fibrillation using the surface electrocardiogram and its response to ibutilide. Am. J. Cardiol. **81**, 1439–1445 (1998)
3. J.L. Salinet Jr., J.P.V. Madeiro, P.C. Cortez, P.J. Stafford, G.A. Ng, F.S. Schlindwein, Analysis of QRS-T subtraction in unipolar atrial fibrillation electrograms. Med. Biol. Eng. Comput. **51**, 1381–1391 (2013)

4. L. Sörnmo, P. Laguna, *Bioelectrical Signal Processing in Cardiac and Neurological Applications* (Elsevier (Academic Press), Amsterdam, 2005)
5. G.D. Clifford, F. Azuaje, P.E. McSharry (eds.), *Advanced Methods and Tools for ECG Data Analysis* (Artech House, Boston, 2006)
6. D.S. Rosenbaum, R.J. Cohen, Frequency based measures of atrial fibrillation in man, in *Proceeding of International Conference of the IEEE Engineering in Medicine and Biology Society (EMBC)*, vol. 12 (1990), pp. 582–583
7. L. Sörnmo, M. Stridh, J.J. Rieta, Atrial activity extraction from the ECG, in *Understanding Atrial Fibrillation: The Signal Processing Contribution*, ed. by L.T. Mainardi, L. Sörnmo, S. Cerutti, ch. 3 (San Francisco: Morgan & Claypool, 2008), pp. 53–80
8. J.J. Rieta, F. Hornero, Comparative study of methods for ventricular activity cancellation in atrial electrograms of atrial fibrillation. Physiol. Meas. **28**, 925–936 (2007)
9. L. Stark, J. Dickson, G. Whipple, H. Horibe, Remote real-time diagnosis of clinical electrograms by a digital computer system. Ann. N.Y. Acad. Sci. **127**, 851–872 (1966)
10. S. Blumlein, G. Harvey, V. Murthy, J. Haywood, New technique for detection of changes in QRS morphology of ECG signals. Am. J. Physiol. **244**, H560–566 (1983)
11. J. Slocum, E. Byrom, L. McCarthy, A.V. Sahakian, S. Swiryn, Computer detection of atrioventricular dissociation from surface electrocardiograms during wide QRS complex tachycardia. Circulation **72**, 1028–1036 (1985)
12. J. Slocum, A.V. Sahakian, S. Swiryn, Diagnosis of atrial fibrillation from surface electrocardiograms based on computer-detected atrial activity. J. Electrocardiol. **25**, 1–8 (1992)
13. S. Shkurovich, A.V. Sahakian, S. Swiryn, Detection of atrial activity from high-voltage leads of implantable ventricular defibrillators using a cancellation technique. IEEE Trans. Biomed. Eng. **45**, 229–234 (1998)
14. Q. Xi, A.V. Sahakian, S. Swiryn, The effect of QRS cancellation on atrial fibrillatory wave signal characteristics in the surface electrocardiogram. J. Electrocardiol. **36**, 243–249 (2003)
15. A. Fujiki, M. Sakabe, K. Nishida, K. Mizumaki, H. Inoue, Role of fibrillation cycle length in spontaneous and drug-indcued termination of human atrial fibrillation–Spectral analysis of fibrillation waves from surface electrocardiogram. Circ. J. **67**, 391–395 (2003)
16. D.C. Shah, T. Yamane, K.J. Choi, M. Haïssaguerre, QRS subtraction and the ECG analysis of atrial ectopics. Ann. Noninvasive Electrocardiol. **9**, 389–398 (2004)
17. F. Beckers, W. Anne, B. Verheyden, C. van der Dussen de Kestergat, E. van Herk, L. Janssens, R. Willems, H. Heidbuchel, A. E. Aubert, Determination of atrial fibrillation frequency using QRST-cancellation with QRS-scaling in standard electrocardiogram leads, in *Proceedings of Computers in Cardiology*, vol. 32 (IEEE Press, 2005), pp. 339–342
18. S. Petrutiu, A.V. Sahakian, S. Swiryn, Abrupt changes in fibrillatory wave characteristics at the termination of paroxysmal atrial fibrillation in humans. Europace **9**, 466–470 (2007)
19. H. Grubitzsch, D. Modersohn, T. Leuthold, W. Konertz, Analysis of atrial fibrillatory activity from high-resolution surface electrocardiograms: evaluation and application of a new system. Exp. Clin. Cardiol. **13**, 29–35 (2008)
20. M. Sterling, D.T. Huang, B. Ghoraani, Developing a new computer-aided clinical decision support system for prediction of successful postcardioversion patients with persistent atrial fibrillation. Comput. Math. Methods Med. (2015)
21. H. Dai, S. Jiang, Y. Li, Atrial activity extraction from single lead ECG recordings: evaluation of two novel methods. Comput. Biol. Med. **43**, 176–183 (2013)
22. R. Alcaraz, J.J. Rieta, Adaptive singular value cancelation of ventricular activity in single-lead atrial fibrillation electrocardiograms. Physiol. Meas. **29**, 1351–1369 (2008)
23. V.D.A. Corino, M.W. Rivolta, R. Sassi, F. Lombardi, L.T. Mainardi, Ventricular activity cancellation in electrograms during atrial fibrillation with constraints on residuals' power. Med. Eng. Phys. **35**, 1770–1777 (2013)
24. E. Bataillou, E. Thierry, H. Rix, O. Meste, Weighted averaging using adaptive estimation of the weights. Signal Process. **44**, 51–66 (1995)
25. F. Castells, J.J. Rieta, J. Millet, V. Zarzoso, Spatiotemporal blind source separation approach to atrial activity estimation in atrial tachyarrhythmias. IEEE Trans. Biomed. Eng. **52**, 258–267 (2005)

26. A.L. Goldberger, L.A. Amaral, L. Glass, J.M. Hausdorff, P.C. Ivanov, R.G. Mark, J.E. Mietus, G.B. Moody, C.K. Peng, H.E. Stanley, PhysioBank, PhysioToolkit, and PhysioNet: components of a new research resource for complex physiologic signals. Circulation **101**, E215–220 (2000)
27. P. Laguna, L. Sörnmo, Sampling rate and the estimation of ensemble variability for repetitive signals. Med. Biol. Eng. Comput. **38**, 540–546 (2000)
28. J. Malmivuo, R. Plonsey, *Bioelectromagnetism* (Oxford University Press, Oxford, 1995)
29. G.J.M. Huiskamp, A. van Oosterom, Heart position and orientation in forward and inverse electrocardiography. Med. Biol. Eng. Comput. **30**, 613–620 (1992)
30. M. Stridh, L. Sörnmo, Spatiotemporal QRST cancellation techniques for analysis of atrial fibrillation. IEEE Trans. Biomed. Eng. **48**, 105–111 (2001)
31. L. Sörnmo, Vectorcardiographic loop alignment and morphologic beat-to-beat variability. IEEE Trans. Biomed. Eng. **45**, 1401–1413 (1998)
32. R. Goya-Esteban, F. Sandberg, Ó. Barquero-Pérez, A. García Alberola, L. Sörnmo, J.L. Rojo-Álvarez, Long-term characterization of persistent atrial fibrillation: wave morphology, frequency, and irregularity analysis. Med. Biol. Eng. Comput. **52**, 1053–1060 (2014)
33. V.D.A. Corino, F. Sandberg, L.T. Mainardi, P.G. Platonov, L. Sörnmo, Noninvasive assessment of atrioventricular nodal function: effect of rate-control drugs during atrial fibrillation. Ann. Noninvasive Electrocardiol. **20**, 534–541 (2015)
34. G.H. Golub, C.F. van Loan, *Matrix Computations*, 2nd edn. (The Johns Hopkins University Press, Baltimore, 1989)
35. J. Waktare, K. Hnatkova, C.J. Meurling, H. Nagayoshi, T. Janota, A.J. Camm, M. Malik, Optimal lead configuration in the detection and subtraction of QRS and T wave templates in atrial fibrillation, in *Proceedings of Computers in Cardiology*, vol. 25 (IEEE Press, 1998), pp. 629–632
36. L. Mainardi, M. Matteucci, R. Sassi, On predicting the spontaneous termination of atrial fibrillation episodes using linear and nonlinear parameters of ECG signal and RR series, in *Proceedings of Computers in Cardiology*, vol. 31 (IEEE Press, 2004), pp. 665–668
37. M. Lemay, J.-M. Vesin, A. van Oosterom, V. Jacquemet, L. Kappenberger, Cancellation of ventricular activity in the ECG: evaluation of novel and existing methods. IEEE Trans. Biomed. Eng. **54**, 542–546 (2007)
38. M. Åström, E. Carro, L. Sörnmo, P. Laguna, B. Wohlfart, Vectorcardiographic loop alignment and the measurement of morphologic beat-to-beat variability in noisy signals. IEEE Trans. Biomed. Eng. **47**, 497–506 (2000)
39. R. Bailón, L. Sörnmo, P. Laguna, A robust method for ECG-based estimation of the respiratory frequency during stress testing. IEEE Trans. Biomed. Eng. **53**, 1273–1285 (2006)
40. V. Jacquemet, A. van Oosterom, J.-M. Vesin, L. Kappenberger, Analysis of electrocardiograms during atrial fibrillation. A biophysical approach. IEEE Med. Biol. Eng. Mag. **25**, 79–88 (2006)
41. C. Li, C. Zheng, C. Tai, Detection of ECG characteristic points using the wavelet transform. IEEE Trans. Biomed. Eng. **42**, 21–28 (1995)
42. J.P. Martínez, R. Almeida, S. Olmos, A.P. Rocha, P. Laguna, A wavelet-based ECG delineator: evaluation on standard databases. IEEE Trans. Biomed. Eng. **51**, 570–581 (2004)
43. R. Almeida, J.P. Martínez, A.P. Rocha, P. Laguna, Multilead ECG delineation using spatially projected leads from wavelet transform loops. IEEE Trans. Biomed. Eng. **56**, 1996–2005 (2009)
44. H. Dai, L. Yin, Y. Li, QRS residual removal in atrial activity signals extracted from single lead: a new perspective based on signal extrapolation. IET Signal Process. **10**, 1169–1175 (2016)
45. X. Du, N. Rao, F. Ou, G. Xu, L. Yin, G. Wang, f-wave suppression method for improvement of locating T-wave ends in electrocardiograms during atrial fibrillation. Ann. Noninvasive Electrocardiol. **18**, 262–270 (2013)
46. B. Niu, Y. Zhu, X. He, H. Wu, MCPSO: A multi-swarm cooperative particle swarm optimizer. Appl. Math. Comput. **2**, 1050–1062 (2007)

47. F. Van den Bergh, A.P. Engelbrecht, A cooperative approach to particle swarm optimization. IEEE Trans. Evol. Comput. **8**, 225–239 (2004)
48. P. Bonizzi, M. Stridh, L. Sörnmo, O. Meste, Ventricular activity residual reduction in remainder ECGs based on short-term autoregressive model interpolation, in *Proceedings of Computers in Cardiology*, vol. 36, pp. 813–816 (2009)
49. A. Ahmad, J.L. Salinet, P.D. Brown Jr., J.H. Tuan, P.J. Stafford, G.A. Ng, F.S. Schlindwein, QRS subtraction for atrial electrograms: flat, linear and spline interpolation. Med. Biol. Eng. Comput. **49**, 1321–1328 (2011)
50. M. Stridh, L. Sörnmo, C.J. Meurling, S.B. Olsson, Sequential characterization of atrial tachyarrhythmias based on ECG time-frequency analysis. IEEE Trans. Biomed. Eng. **51**, 100–114 (2004)
51. M. Stridh, L. Sörnmo, C. Meurling, S.B. Olsson, Characterization of atrial fibrillation using the surface ECG: time-dependent spectral properties. IEEE Trans. Biomed. Eng. **48**, 19–27 (2001)
52. S.V. Vaseghi, *Advanced Digital Signal Processing and Noise Reduction*, 3rd edn. (Wiley, 2006)
53. S. Haykin, *Adaptive Filter Theory*, 5th edn. (Pearson, New Jersey, 2014)
54. B.D.O. Anderson, J.B. Moore, *Optimal Filtering* (Prentice-Hall, Englewood Cliffs, N.J., 1979)
55. M. Hayes, *Statistical Digital Signal Processing and Modeling* (Wiley, New York, 1996)
56. E.K. Roonizi, R. Sassi, An extended Bayesian framework for atrial and ventricular activity separation in atrial fibrillation. IEEE J. Biomed. Health Inform. **21**, 1573–1580 (2017)
57. M. Stridh, D. Husser, A. Bollmann, L. Sörnmo, Waveform characterization of atrial fibrillation using phase information. IEEE Trans. Biomed. Eng. **56**, 1081–1089 (2009)
58. A. Buttu, E. Pruvot, J. Van Zaen, A. Viso, A. Forclaz, P. Pascale, S.M. Narayan, J. Vesin, Adaptive frequency tracking of the baseline ECG identifies the site of atrial fibrillation termination by catheter ablation. Biomed. Signal Process. Control **8**, 969–980 (2013)
59. M.E. Nygårds, J. Hulting, An automated system for ECG monitoring. Comput. Biomed. Res. **12**, 181–202 (1979)
60. P.E. McSharry, G.D. Clifford, L. Tarassenko, L.A. Smith, A dynamical model for generating synthetic electrocardiogram signals. IEEE Trans. Biomed. Eng. **50**, 289–294 (2003)
61. L. Sörnmo, P.O. Börjesson, M.E. Nygårds, O. Pahlm, A method for evaluation of QRS shape features using a mathematical model for the ECG. IEEE Trans. Biomed. Eng. **28**, 713–717 (1981)
62. P. Laguna, R. Jané, S. Olmos, N.V. Thakor, H. Rix, P. Caminal, Adaptive estimation of QRS complex by the Hermite model for classification and ectopic beat detection. Med. Biol. Eng. Comput. **34**, 58–68 (1996)
63. R. Sameni, M.B. Shamsollahi, C. Jutten, G.D. Clifford, A nonlinear Bayesian filtering framework for ECG denoising. IEEE Trans. Biomed. Eng. **54**, 2172–2185 (2007)
64. J.V. Candy, *Bayesian Signal Processing: Classical, Modern, and Particle Filtering Methods*, 2nd edn. (Wiley, 2016)
65. O. Sayadi, M.B. Shamsollahi, ECG denoising and compression using a modified extended Kalman filter structure. IEEE Trans. Biomed. Eng. **55**, 2240–2248 (2008)
66. E. Pueyo, M. Malik, P. Laguna, A dynamic model to characterize beat-to-beat adaptation of repolarization to heart rate changes. Biomed. Signal Process. Control **3**, 29–43 (2008)
67. J. Oster, J. Behar, O. Sayadi, S. Nemati, A.E.W. Johnson, G.D. Clifford, Semisupervised ECG ventricular beat classification with novelty detection based on switching Kalman filters. IEEE Trans. Biomed. Eng. **62**, 2125–2134 (2015)
68. E.K. Roonizi, R. Sassi, A signal decomposition model-based Bayesian framework for ECG components separation. IEEE. Trans. Signal Process. **64**, 665–674 (2016)
69. M. Rahimpour, B.M. Asl, P wave detection in ECG signals using an extended Kalman filter: an evaluation in different arrhythmia contexts. Physiol. Meas. **37**, 1089–1104 (2016)
70. D.E. Gustafson, A.S. Willsky, J.Y. Wang, M.C. Lancaster, J.H. Triebwasser, ECG/VCG rhythm diagnosis using statistical signal analysis–I. Identification of persistent rhythms. IEEE Trans. Biomed. Eng. **25**, 344–353 (1978)

71. S. Haykin, *Neural Networks: A Comprehensive Foundation*, 2nd edn. (Prentice Hall, 1998)
72. A. Petrėnas, V. Marozas, L. Sörnmo, A. Lukoševičius, An echo state neural network for QRST cancellation during atrial fibrillation. IEEE Trans. Biomed. Eng. **59**, 2950–2957 (2012)
73. V. Zarzoso, Extraction of ECG characteristics using source separation techniques: Exploiting statistical independence and beyond, in *Advanced Biosignal Processing*, ed. by A. Naït-Ali (Springer, Berlin Heidelberg, 2013), pp. 15–47
74. N.V. Thakor, Z. Yi-Sheng, Applications of adaptive filtering to ECG analysis: noise cancellation and arrhythmia detection. IEEE Trans. Biomed. Eng. **38**, 785–794 (1991)
75. P. Laguna, R. Jané, O. Meste, P.W. Poon, P. Caminal, H. Rix, N.V. Thakor, Adaptive filter for event-related bioelectric signals using an impulse correlated reference input: comparison with signal averaging techniques. IEEE Trans. Biomed. Eng. **39**, 1032–1044 (1992)
76. J. Lee, M.H. Song, D.G. Shin, K.J. Lee, Event synchronous adaptive filter based atrial activity estimation in single-lead atrial fibrillation electrocardiograms. Med. Biol. Eng. Comput. **50**, 801–811 (2012)
77. P. Laguna, R. Jané, E. Masgrau, P. Caminal, The adaptive linear combiner with a periodic-impulse reference input as a linear comb filter. Signal Process. **48**, 193–203 (1996)
78. C. Vásquez, A. Hernández, F. Mora, G. Carrault, G. Passariello, Atrial activity enhancement by Wiener filtering using an artificial neural network. IEEE Trans. Biomed. Eng. **48**, 940–944 (2001)
79. J.L. Elman, Finding structure in time. Cogn. Sci. **14**, 179–211 (1990)
80. J.A. Anderson, *An Introduction to Neural Networks* (MIT Press, 1995)
81. K. Doya, Bifurcations of recurrent neural networks in gradient descent learning. IEEE Trans. Neural Netw. **1**, 75–80 (1993)
82. B.A. Pearlmutter, Gradient calculations for dynamic recurrent neural networks: a survey. IEEE Trans. Neural Netw. **6**, 1212–1228 (1995)
83. H. Jaeger, The 'echo state' approach to analysing and training recurrent neural networks, GMD Report 148 (German National Research Center for Information Technology, 2001)
84. M.C. Ozturk, D. Xu, J.C. Principe, Analysis and design of echo state networks. Neural Comput. **19**, 111–138 (2007)
85. A. Petrėnas, L. Sörnmo, A. Lukoševičius, V. Marozas, Detection of occult paroxysmal atrial fibrillation. Med. Biol. Eng. Comput. **53**, 287–297 (2015)
86. M. Lukoševičius, H. Jaeger, Reservoir computing approaches to recurrent neural network training. Comput. Sci. Rev. **3**, 127–149 (2009)
87. H. Jaeger, M. Lukoševičius, D. Popovici, U. Siewert, Optimization and applications of echo state networks with leaky integrator neurons. Neural Netw. **20**, 335–352 (2007)
88. M. Lukoševičius, A practical guide to applying echo state networks, in *Neural Networks: Tricks of the Trade*, ed. by G. Montavon, G.B. Orr, K.-R. Müller, 2nd edn. (Springer, 2012)
89. S.C. Douglas, Numerically-robust $\mathcal{O}(N^2)$ RLS algorithms using least-squares prewhitening, in *IEEE International Conference on Acoustics, Speech and Signal Processing (ICASSP)*, vol. 25 (2000), pp. 412–415
90. A. Rodan, P. Tiňo, Minimum complexity echo state network. IEEE Trans. Neural Netw. **22**, 131–144 (2011)
91. I.T. Joliffe, *Principal Component Analysis* (Springer, Berlin, 2002)
92. L.G. Horan, N.C. Flowers, D.A. Brody, Principal factor waveforms of the thoracic QRS-complex. Circ. Res. **14**, 131–145 (1964)
93. F. Castells, P. Laguna, L. Sörnmo, A. Bollmann, J. Millet Roig, Principal component analysis in ECG signal processing. J. Adv. Signal Process. **2007**, ID 74580 (2007)
94. F. Castells, C. Mora, J.J. Rieta, D. Moratal-Pérez, J. Millet, Estimation of atrial fibrillatory wave from single-lead atrial fibrillation electrocardiograms using principal component analysis concepts. Med. Biol. Eng. Comput. **43**, 557–560 (2005)
95. A. Hyvärinen, J. Karhunen, E. Oja, *Independent Component Analysis* (Wiley Interscience, 2001)
96. A. Martínez, R. Alcaraz, J.J. Rieta, Ventricular activity morphological characterization: Ectopic beats removal in long term atrial fibrillation recordings. Comput. Methods Programs Biomed. **109**, 283–292 (2013)

97. P. Langley, J.P. Bourke, A. Murray, Frequency analysis of atrial fibrillation, in *Proceedings of Computers in Cardiology*, vol. 27 (IEEE Press, 2000), pp. 65–68
98. D. Raine, P. Langley, A. Murray, A. Dunuwille, J.P. Bourke, Surface atrial frequency analysis in patients with atrial fibrillation: a tool for evaluating the effects of intervention. J. Cardiovasc. Electrophysiol. **15**, 1021–1026 (2004)
99. P. Langley, M. Stridh, J.J. Rieta, J. Millet, L. Sörnmo, A. Murray, Comparison of atrial signal extraction algorithms in 12-lead ECGs with atrial fibrillation. IEEE Trans. Biomed. Eng. **53**, 343–346 (2006)
100. A. van Oosterom, The dominant T wave and its significance. J. Cardiovasc. Electrophysiol. **14**, S180–S187 (2003)
101. A. van Oosterom, The dominant T wave. J. Electrocardiol. **37**, 193–197 (2004)
102. R. Sassi, L.T. Mainardi, An estimate of the dispersion of repolarization times based on a biophysical model of the ECG. IEEE Trans. Biomed. Eng. **58**, 3396–3405 (2011)
103. P. Laguna, J.P. Martínez, E. Pueyo, Techniques for ventricular repolarization instability assessment from the ECG. Proc. IEEE **104**, 392–415 (2016)
104. G.S. Wagner, *Marriott's Practical Electrocardiography*, 10th edn. (Lippincott Williams & Wilkins, Baltimore, 2001)
105. P. Langley, J.P. Bourke, A. Murray, The U wave in atrial fibrillation, in *Proceedings of Computing in Cardiology*, vol. 42, pp. 833–836 (2015)
106. R. Sassi, V.D.A. Corino, L.T. Mainardi, Analysis of surface atrial signals: time series with missing data? Ann. Biomed. Eng. **37**, 2082–2092 (2009)
107. R. Vautard, P. Yiou, M. Ghil, Singular-spectrum analysis: a toolkit for short, noisy chaotic signals. Phys. D **58**, 95–126 (1992)
108. N. Golyandina, A. Zhigljavsky, *Singular Spectrum Analysis for Time Series* (Springer, 2013)
109. E. Parzen, On spectral analysis with missing observations and amplitude modulation. Sankya A. **25**, 383–392 (1963)
110. D.H. Schoellhamer, Singular spectrum analysis for time series with missing data. Geophys. Res. Lett. **28**, 3187–3190 (2001)
111. M. Stridh, L. Sörnmo, C.J. Meurling, S.B. Olsson, Detection of autonomic modulation in permanent atrial fibrillation. Med. Biol. Eng. Comput. **41**, 625–629 (2003)
112. D. Kondrashov, M. Ghil, Spatio-temporal filling of missing points in geophysical data sets. Nonlinear Process. Geophys. **13**, 151–159 (2006)
113. G. Wang, N. Rao, S.J. Shepherd, C.B. Beggs, Extraction of desired signal based on AR model with its application to atrial activity estimation in atrial fibrillation. J. Adv. Signal Process. **8**, 1–9 (2008)
114. W. Liu, D.P. Mandic, A. Cichocki, Blind source extraction based on a linear predictor. IET Signal Process. **1**, 29–34 (2007)
115. T.K. Moon, W.C. Sterling, *Mathematical Methods and Algorithms for Signal Processing* (Prentice Hall, New Jersey, USA, 2000)
116. J.F. Cardoso, Blind signal separation: statistical principles. Proc. IEEE **86**, 2009–2025 (1998)
117. P. Bonizzi, M. de la Salud Guillem, A.M. Climent, J. Millet, V. Zarzoso, F. Castells, O. Meste, Noninvasive assessment of the complexity and stationarity of the atrial wavefront patterns during atrial fibrillation. IEEE Trans. Biomed. Eng. **57**, 2147–2157 (2010)
118. S.M. Kay, *Modern Spectral Estimation, Theory and Application* (Prentice-Hall, New Jersey, 1999)
119. L. Tong, R.-W. Liu, V.C. Soon, Y.-F. Huang, Indeterminacy and identifiability of blind identification. IEEE Trans. Circ. Syst. **38**, 499–509 (1991)
120. A. Hyvärinen, E. Oja, Independent component analysis: algorithms and applications. Neural Netw. **13**, 411–430 (2000)
121. P. Comon, Independent component analysis–a new concept? Signal Process. **36**, 287–314 (1994)
122. A. Hyvärinen, E. Oja, A fast fixed-point algorithm for independent component analysis. Neural Comput. **9**, 1483–1492 (1997)

123. J.J. Rieta, F. Castells, C. Sánchez, V. Zarzoso, J. Millet, Atrial activity extraction for atrial fibrillation analysis using blind source separation. IEEE Trans. Biomed. Eng. **51**, 1176–1186 (2004)
124. M. Lemay, J.-M. Vesin, Z. Ihara, L. Kappenberger, Suppression of ventricular activity in the surface electrocardiogram of atrial fibrillation, in *Proceedings of the International Conference Independent Component Analysis and Blind Signal Separation* (Springer, 2004), pp. 1095–1102
125. F. Castells, J. Igual, J. Millet, J.J. Rieta, Atrial activity extraction from atrial fibrillation episodes based on maximum likelihood source separation. Signal Process. **85**, 523–535 (2005)
126. R. Phlypo, Y. D'Asseler, I. Lemahieu, V. Zarzoso, Extraction of the atrial activity from the ECG based on independent component analysis with prior knowledge of the source kurtosis signs, in *Proceeding of International Conference of the IEEE Engineering in Medicine and Biology Society (EMBC)*, vol. 29 (2007), pp. 6499–6502
127. V. Zarzoso, O. Meste, P. Comon, D.G. Latcu, N. Saoudi, Noninvasive cardiac signal analysis using data decomposition techniques, in *Modeling in Computational Biology and Biomedicine: A Multidisciplinary Endeavor* ed. by F. Cazals, P. Kornprobst (Springer, Berlin Heidelberg, 2013), pp. 83–116
128. V. Zarzoso, R. Phlypo, P. Comon, A contrast for independent component analysis with priors on the source kurtosis signs. IEEE Signal Process. Lett. **15**, 501–504 (2008)
129. A. Mincholé, L. Sörnmo, P. Laguna, Detection of body position changes from the ECG using a Laplacian noise model. Biomed. Signal Process. Control **14**, 189–196 (2014)
130. A.J. Pullan, M.L. Buist, L.K. Cheng, *Mathematically Modelling the Electrical Activity of the Heart* (World Scientific, New Jersey, USA, 2005)
131. C. Vayá, J.J. Rieta, C. Sanchez, D. Moratal, Convolutive blind source separation algorithms applied to the electrocardiogram of atrial fibrillation: study of performance. IEEE Trans. Biomed. Eng. **54**, 1530–1533 (2007)
132. F.I. Donoso, R.L. Figueroa, E.A. Lecannelier, E.J. Pinoa, A.J. Rojas, Atrial activity selection for atrial fibrillation ECG recordings. Comput. Biol. Med. **43**, 1628–1636 (2013)
133. A. Belouchrani, K. Abed-Meraim, J.F. Cardoso, E. Moulines, A blind source separation technique using second-order statistics. IEEE Trans. Signal Process. **45**, 434–444 (1997)
134. J. Malik, N. Reed, C.-L. Wang, H.-T. Wu, Single-lead f-wave extraction using diffusion geometry. Physiol. Meas. **38**, 1310–1334 (2017)
135. R. Phlypo, V. Zarzoso, I. Lemahieu, Atrial activity estimation from atrial fibrillation ECGs by blind source extraction based on a conditional maximum likelihood approach. Med. Biol. Eng. Comput. **48**, 483–488 (2010)
136. R. Llinares, J. Igual, J. Miró-Borrás, A fixed point algorithm for extracting the atrial activity in the frequency domain. Comput. Biol. Med. **40**, 943–949 (2010)
137. R. Llinares, J. Igual, Exploiting periodicity to extract the atrial activity in atrial arrhythmias. J. Adv. Signal Process. 134–140 (2011)
138. O.A. Rosso, S. Blanco, J. Yordanova, V. Kolev, A. Figliola, M. Schürmann, E. Başar, Wavelet entropy: a new tool for analysis of short duration brain electrical signals. J. Neurosci. Meth. **105**, 65–75 (2001)
139. P. Langley, Wavelet entropy as a measure of ventricular beat suppression from the electrocardiogram in atrial fibrillation. Entropy **17**, 6397–6411 (2015)
140. J. Ródenas, M. García, R. Alcaraz, J.J. Rieta, Wavelet entropy automatically detects episodes of atrial fibrillation from single-lead electrocardiograms. Entropy **17**, 6179–6199 (2015)
141. J. Mateo, J.J. Rieta, Radial basis function neural networks applied to efficient QRST cancellation in atrial fibrillation. Comput. Biol. Med. **43**, 154–163 (2013)
142. I. Nault, N. Lellouche, S. Matsuo, S. Knecht, M. Wright, K.T. Lim, F. Sacher, P. Platonov, A. Deplagne, P. Bordachar, N. Derval, M.D. O'Neill, G.J. Klein, M. Hocini, P. Jaïs, J. Clémenty, M. Haïssaguerre, Clinical value of fibrillatory wave amplitude on surface ECG in patients with persistent atrial fibrillation. J. Interv. Card. Electrophysiol. **26**, 11–19 (2009)
143. J. Lian, G. Garner, D. Muessig, V. Lang, A simple method to quantify the morphological similarity between signals. Signal Process. **90**, 684–688 (2010)

144. J. Igual, R. Llinares, M.S. Guillem, J. Millet, Optimal localization of leads in atrial fibrillation episodes, in *International Conference on Acoustics, Speech and Signal Processing (ICASSP)*, vol. 31 (2006), pp. II:1192–II:1195
145. D. Husser, M. Stridh, L. Sörnmo, C. Geller, H.U. Klein, S.B. Olsson, A. Bollmann, Time-frequency analysis of the surface electrocardiogram for monitoring antiarrhythmic drug effects in atrial fibrillation. Am. J. Cardiol. **95**, 526–528 (2005)
146. A. Bollmann, A. Tveit, D. Husser, M. Stridh, L. Sörnmo, P. Smith, S.B. Olsson, Fibrillatory rate response to candesartan in persistent atrial fibrillation. Europace **10**, 1138–1144 (2008)
147. M. Aunes-Jansson, N. Edvardsson, M. Stridh, L. Sörnmo, L. Frison, A. Berggren, Decrease of the atrial fibrillatory rate, increased organization of the atrial rhythm and termination of atrial fibrillation by AZD7009. J. Electrocardiol. **46**, 29–35 (2013)
148. M. Aunes, K. Egstrup, L. Frison, A. Berggren, M. Stridh, L. Sörnmo, N. Edvardsson, Rapid slowing of the atrial fibrillatory rate after administration of AZD7009 predicts conversion of atrial fibrillation. J. Electrocardiol. **47**, 316–323 (2014)

Chapter 6
Characterization of f Waves

Leif Sörnmo, Raúl Alcaraz, Pablo Laguna and José Joaquín Rieta

6.1 Introduction

The diagnosis of atrial fibrillation (AF) is based on the finding of an irregular ventricular rhythm, further strengthened when f waves are discernible. Since no information beyond the presence of f waves is considered when making the diagnosis, f wave characterization has yet to find its way into clinical practice. At the same time, f wave characterization is receiving considerable attention in the scientific community, driven by the need for noninvasive information on electropathological alterations in the atria, which may facilitate patient-tailored treatment of AF.

Invasive measurements, acquired during electrophysiological examination or open thorax surgery, can be used to characterize the atrial activity. While invasive measurements obviously provide a much more local characterization of the atrial activity than the surface ECG, the acquisition of invasive signals must take place inside the hospital, the required equipment is expensive, and the procedure is associated with increased risk of patient complication. Moreover, the acquisition duration

L. Sörnmo (✉)
Department of Biomedical Engineering and Center for Integrative Electrocardiology,
Lund University, Lund, Sweden
e-mail: leif.sornmo@bme.lth.se

R. Alcaraz
Innovation in Bioengineering Research Group, University of Castilla–La Mancha,
Campus Universitario, Cuenca, Spain

P. Laguna
Biomedical Signal Interpretation and Computational Simulation (BSICoS),
Aragón Institute of Engineering Research (I3A), Centro de Investigacíon
Biomédica en Red de Bioingeniería, Biomateriales y Nanomedicina (CIBER-BBN), Zaragoza
University, Zaragoza, Spain

J. J. Rieta
Biomedical Synergy, Electronic Engineering Department,
Universidad Politécnica de Valencia, Gandía, Spain

L. Sörnmo (ed.), *Atrial Fibrillation from an Engineering Perspective*,
Series in BioEngineering, https://doi.org/10.1007/978-3-319-68515-1_6

is limited by the procedural duration, which may last for a few minutes only, while the surface ECG may be acquired over weeks or even months.

There are several aims of f wave characterization, many of them related to the prediction of treatment outcome [1–3]. For example, a low f wave amplitude predicts AF recurrence in patients with persistent AF undergoing catheter ablation [4], and, conversely, a large amplitude predicts termination of persistent AF during catheter ablation [5]. For patients with persistent AF undergoing cardioversion, a low atrial fibrillatory rate (AFR) predicts successful outcome [6], and, conversely, a fast rate predicts AF recurrence [7]. Monitoring of the effect of antiarrhythmic drug therapy is another application where f wave characterization provides valuable information, particularly in the developmental phase of the drug when the complications of invasive electrophysiological testing to some extent can be avoided [8]. For example, different f wave characteristics, including the AFR, have been studied in patients receiving either a drug under development or placebo, with the aim of determining what characterize patients converting to normal sinus rhythm, as well as patients not converting [9]. In all these applications, ECG-derived information may be considered for optimizing AF management and supporting therapeutic decisions at substantial cost savings.

Yet another, more general aim of f wave characterization is to investigate the structural changes and the electrophysiological remodeling that take place in the atria as AF progresses from self-terminating paroxysms to a more sustained or permanent state [10]. The outcome of such investigations may turn out to be instrumental in preventing the progression of AF.

From an engineering perspective, the problems of detecting AF and extracting f waves, treated in Chaps. 4 and 5, respectively, are considerably more clear-cut than the problem of characterizing f waves. The main reason is that methods for detection and extraction lend themselves to performance evaluation which can be expressed in technical terms, e.g., evaluation based on annotated or simulated ECG signals, whereas methods for f wave characterization, at least so far, rest on phenomenological observations which may link a certain f wave characteristic to the clinical issue at hand, be it related to prediction or evaluation of treatment outcome. As a result, research on f wave characterization implies more groping in the dark than does research on AF detection and f wave extraction. On the other hand, more room is available for investigating different techniques for signal characterization, with implications on clinical management.

The characterization of f waves has for many years revolved around f wave amplitude and AFR—the two fundamental signal characteristics which are relatively straightforward to determine [11]. However, as signal processing techniques have grown more sophisticated and diversified, research on f wave characterization has become increasingly more multifaceted. Different techniques have been investigated for analyzing nonstationary f wave signals with respect to spatiotemporal organization and nonlinear dynamics [2, 12, 13], as well as for analyzing the spatial distribution of different f wave characteristics on the body surface [14].

The majority of parameters proposed for f wave characterization are well-known in the realm of signal processing. Indeed, few parameters have been developed with reference to a statistical signal model accounting for certain specific electrophysiological phenomena. The lack of tailored, model-based parameters is most likely due to the difficulty to associate a particular f wave characteristic to a certain local electrophysiological property of the atria. This lack may be remedied using computational modeling and simulation to obtain a better understanding of the genesis of f waves [15–17].

Given the extensive work on f wave extraction, one would expect most studies on f wave characterization to be based on the extracted f wave signal—an expectation which remains to be fulfilled. With easy-to-implement methods such as average beat subtraction (ABS), the presence of QRS-related residuals will, to various extents, worsen the reliability of f wave characterization. For example, measurements of f wave amplitude are likely to be more vulnerable to such residuals than measurements on AFR. To evade this problem, several authors have confined characterization to f waves contained in TQ intervals [18–21]. However, as already pointed out in Chap. 5, the availability of fewer samples implies less accurate results, and, therefore, it is hoped that well-performing extraction methods will find their way into studies on f wave characterization.

This chapter reviews different approaches to f wave characterization, together forming a smorgasbord of "dishes" rather than a coherent body of methods. First, the two fundamental characteristics f wave amplitude and AFR are considered in Sects. 6.2 and 6.3, respectively, followed by a description of linear and nonlinear techniques for characterizing f wave morphology and regularity (Sect. 6.4). Techniques for quantifying f wave signal quality in individual leads are described in Sect. 6.5, needed to ensure that f wave characterization is performed on signals with sufficient quality. The analysis of spatial ECG information, manifested as a vectorcardiographic loop or a body surface potential map, is reviewed in Sect. 6.6. The chapter concludes with a brief overview of popular clinical applications where the herein described approaches to f wave characterization are explored.

6.2 f Wave Amplitude

In clinical studies, f wave amplitude has been manually analyzed after quantization into either fine or coarse, defined as less than or greater than 50 μV [22–26]. As caliper measurements of f wave amplitude now belong to history, such quantization has once and for all been shelved in favor of continuous-valued measurements. Based on the extracted f wave signal $x(n)$,[1] but with the QRS intervals excluded to avoid the influence of QRS-related residuals, a straightforward definition of f wave amplitude is the average of the four largest peak-to-peak amplitudes of individual f waves in a 10-s recording [27, 28]. Given that the f wave amplitude often varies over time, it may be necessary to average all peak-to-peak amplitudes contained in the recording to produce a representative measurement. Determination of the peak-to-peak amplitude

[1] For notational convenience, the extracted f wave signal is denoted $x(n)$ in this chapter, replacing the notation $\hat{d}(n)$ used in Chap. 5.

requires that a search interval is first delineated so that the extrema of the f wave can be located. The length of the search interval depends on the AFR, and thus the AFR needs to be estimated.

The f wave amplitude does not necessarily have to rely on the amplitude of local extrema, but can just as well be computed as a root mean square (RMS) amplitude of the extracted f wave signal, or, as in [29], without the square root to instead measure signal power. Another approach would be to employ classical envelope detection based on the Hilbert transform [30], where the f wave amplitude can be determined as an average of the envelope in the time interval of interest.

Envelope detection based on local extrema has also been proposed for the measurement of f wave amplitude [31], see also [32]. Once $x(n)$ has been centered, i.e., its mean m_x has been removed, the lower envelope $e_l(n)$ is obtained by connecting successive local minima of $x(n)$ using polynomial interpolation, and the upper envelope $e_u(n)$ by connecting successive local maxima of $x(n)$; a piecewise cubic Hermite interpolating polynomial was used in [31].[2] The sample-to-sample difference between $e_u(n)$ and $e_l(n)$ is taken as a measure of the local amplitude, which, when averaged over the entire N-sample signal,

$$A_{\mathrm{f}} = \frac{1}{N} \sum_{n=0}^{N-1} |e_u(n) - e_l(n)|, \tag{6.1}$$

is a measure of global f wave amplitude. The different signals involved with the computation of A_{f} are illustrated in Fig. 6.1.

The methods in [27, 31] require that the extrema of the f wave signal are determined before the amplitude can be measured. The method in [27] produces measurements which are more intuitive since the samples between peaks are not taken into account. However, as the noise level increases, peak-to-peak measurements become increasingly more unreliable than those obtained from the envelope [31]. To reduce the influence of baseline wander and muscular noise, the original ECG signal is bandpass filtered before the amplitude is measured, using a passband of either 1–50 Hz [27] or 0.5–30 Hz [31].

None of the two methods in [27, 31] have been applied to f waves extracted in the QRS interval. In fact, the envelope-based method analyzes an f wave signal resulting from the concatenation of consecutive TQ intervals, thus making f wave extraction superfluous [31]. Since concatenation sometimes leads to jumps at the interval boundaries, peaks located near the boundaries are excluded from polynomial interpolation. Moreover, some TQ intervals are so short that only a partial f wave is available for amplitude measurement.

The repeatability of f wave amplitude was investigated on a data set of 20 clinically stable patients with AF, using the average of the four largest peak-to-peak amplitudes [27]. For each patient, 10 ECGs of 10-s length were recorded at regular

[2] It may be noted that this procedure is closely related to the "sifting" procedure, which is part of empirical mode decomposition [33], where the lower and upper envelopes of the local extrema are used to compute the intrinsic mode functions.

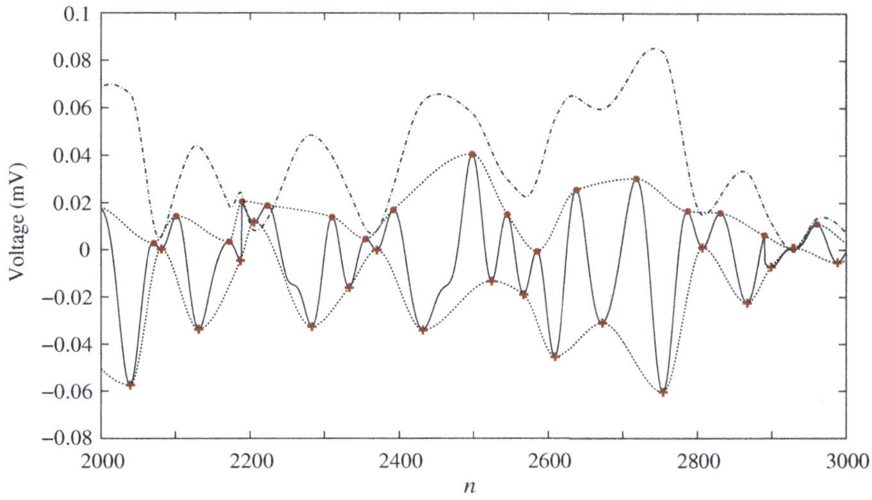

Fig. 6.1 Envelope-based measurement of f wave amplitude using polynomial interpolation: the f wave signal after 0.5–30 Hz bandpass filtering (solid line), the lower and upper envelopes $e_l(n)$ and $e_u(n)$ (dotted lines) obtained by connecting successive local minima and maxima (marked by "+" and "o," respectively), and the difference $e_u(n) - e_l(n)$ (dashed–dotted line) used to compute the amplitude in (6.1). (Reprinted from [31] with permission)

intervals over the course of 24 h. The results showed that the interpatient differences were substantial, with f wave amplitudes ranging from 60 to 350 μV (mean±standard deviation equal to 131±54 μV). On the other hand, the intrapatient differences were significantly smaller during the 24 h, ranging from 4 to 53 μV when determined over the 10 intrapatient ECGs, with an average standard deviation of 21 μV.

Assuming that f waves can be approximated by a sinusoid, the peak-to-peak amplitude can be compared to the RMS amplitude, since the former amplitude is approximately 2.8 times the latter amplitude. Using this approximation, a qualitative comparison can be made between the results reported in [27] and the histogram of f wave RMS amplitude displayed in Fig. 3.4b. The results are in fairly good agreement with each other, since, following multiplication of 2.8, the f wave amplitudes in Fig. 3.4b range from 35 to 340 μV (117±48 μV), to be compared with 60 to 350 μV (131±54 μV).

6.3 Atrial Fibrillatory Rate and Beyond

Atrial fibrillatory rate, being the other fundamental f wave characteristic, has received considerable clinical attention during the last two decades [3]. A spectral approach is commonly used to estimate the AFR, since estimation based on the occurrence times of the f waves is compounded by the difficulty to define a consistent fiducial point. Another reason is that the signal-to-noise ratio (SNR) may be poor. In contrast, when invasively recorded signals are subject to analysis, the AFR is often determined from the occurrence times of the local activations, with complex wavefront morphologies and low SNR as the factors which have the most influence on the accuracy of AFR estimation. Considerable research effort has been spent on developing techniques for estimation of local activation times [34–40] as an alternative to using spectral analysis [41–43].

Spectral analysis of the extracted f wave signal plays an important role not only in AFR estimation, but also in the characterization of f wave morphology. When changes in the spectral content of the f wave signal are of interest to investigate, whether spontaneous or due to intervention, time–frequency analysis is better suited for quantifying such changes.

In the engineering oriented literature, the term *dominant atrial frequency* (DAF) is usually substituted for AFR, where "dominant" refers to the largest spectral peak. In the clinical literature, the term *dominant atrial cycle length* (DACL) is sometimes substituted for AFR. Atrial fibrillatory rate, DAF, and DACL convey the same information, though they are expressed in units of fibrillations per minute (fpm), Hertz, and milliseconds, respectively. Since the DAF estimate is used to determine both AFR and DACL, DAF is the preferred terminology in the following.

6.3.1 Dominant Atrial Frequency

The position of the largest peak in the power spectrum of the extracted f wave signal defines the DAF, denoted ω_0. Nonparametric spectral estimation is typically employed, which, in most cases, is synonymous to Welch's method, where the signal is divided into shorter, overlapping segments, followed by windowing of each segment [44].[3] The power spectrum is obtained by averaging the power spectra (periodograms) of the segments. Each segment is padded with zeros so that the position of the spectral peak can be determined more accurately; however, zero padding does not improve spectral resolution in the sense that two closely spaced spectral peaks are better resolved when the original signal is padded with zeros. A signal length of a few seconds is needed to produce an acceptable variance of the power spectrum. If better spectral resolution is needed, longer segments need to be analyzed. For example, a 10-s segment yields, at best, a frequency resolution of 0.1 Hz depending on the window chosen.

Figure 6.2 displays the power spectra computed from extracted f wave signals in leads V_1, V_2, and V_3. The largest spectral peak occurs at approximately the same position in all three leads, where the f waves of V_1 have the largest amplitude. In this example, the position of the next largest peak in V_1 and V_2 is not harmonically related to the position of the largest peak; the next largest peak is likely the expression of a time-varying DAF, discussed below.

[3] Since the amplitude spectrum is analyzed in some studies, obtained as the square root of the power spectrum, caution should be exercised when comparing amplitude-related results.

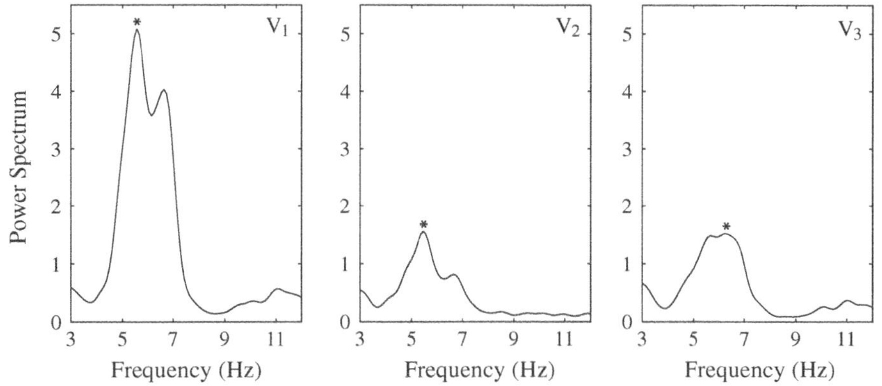

Fig. 6.2 Power spectra of extracted f wave signals in leads V_1, V_2, and V_3. The dominant peak is marked with "*"

Since f waves are mostly characterized by frequencies up to 25 Hz, a sampling rate much lower than that required for f wave extraction can be used. Thus, the original ECG sampling rate can be decimated to 50 Hz without loss of clinical information. Although sampling rate decimation is not a critical operation when performing nonparametric spectral analysis, it is critical when performing parametric spectral analysis based on autoregressive modeling due to the risk of producing spectra with spurious peaks for too high a sampling rate [45].

Instead of performing spectral analysis of the extracted f wave signal, the analysis may be confined to the samples of successive TQ intervals [46]. In such cases, a technique must be employed which can handle unevenly sampled signals. Using iterative singular spectrum analysis (SSA), cf. Sect. 5.7, an atrial subspace is first determined from several, consecutive TQ intervals, after which the f wave signal of the QRST intervals is estimated by projecting the QRST samples on the atrial subspace. The resulting signal, composed of interpolated samples in the QRST intervals and observed samples in the TQ intervals, is then subject to spectral analysis using, for example, Welch's method.

Using simulated f wave signals, all with 1-min duration and a 7-Hz DAF, iterative SSA was used to estimate the DAF [46]. The results showed that the estimation error rarely exceeded 1.0 Hz at heart rates up to 130–140 beats per minute (bpm) and relatively low SNRs. Recalling general results on the variance of frequency estimators [47, Chap. 3], the spectral estimation error is lower at higher frequencies, but higher at lower frequencies. Thus, not surprisingly, the best performing scenario for the iterative SSA is one with a slower heart rate, i.e., the TQ intervals are longer, and a higher DAF. The SSA-based technique was developed for estimating the DAF, whereas information on other harmonics, needed to compute some of the spectral parameters described below, is not captured.

Lomb's periodogram is another technique for estimating the power spectrum of an unevenly sampled signal [30, 48]. This periodogram is determined by minimizing the squared error between the observed samples and a sinusoidal model signal composed of different frequencies. The accuracy of DAF estimates obtained from Lomb's periodogram is similar to that of estimates obtained from iterative SSA, although the latter method tended to produce lower errors at lower heart rates [46].

As a rule, spectral analysis of multi-lead ECGs is performed on a lead-by-lead basis, resulting in a set of parameters characterizing the spatial distribution of spectral information. Another, less common approach is provided by the spectral envelope method [49] which combines spectral information of the different leads into a single power spectrum, where periodic components are emphasized and noise is suppressed [50, 51].

6.3.2 Spectral Parameters

The parameter *spectral organization* (SO) describes the harmonic structure of the f wave signal [52, 53]. A more organized signal, manifested by a harmonic spectrum with a dominant spectral peak, is hypothesized to reflect fewer wavelets circulating within the atria. Conversely, a less organized signal, manifested by "more frequency components added to the atrial signal," is hypothesized to reflect more wavelets. Spectral organization is defined by

$$P_{\mathrm{SO}} = \frac{\displaystyle\sum_{k=1}^{K} \int_{-\Delta\omega}^{\Delta\omega} S_x(\hat{\omega}_{k-1} + \omega) d\omega}{\displaystyle\int_{\omega_{\min}}^{\omega_{\max}} S_x(\omega) d\omega}, \tag{6.2}$$

where $S_x(\omega)$ is the power spectrum of $x(n)$, and $\omega_0, \ldots, \omega_{K-1}$ denote the positions of the K harmonics, i.e., the k-th harmonic is associated with ω_{k-1}. Four harmonics were analyzed in [52, 53], whereas two harmonics were analyzed in [51, 54]. The integration limits $\Delta\omega$ and $\omega_{\min}$ were set to 0.5 Hz and 2.5 Hz, respectively, and $\omega_{\max}$ was set to $((K+1)\hat{\omega}_0 - \Delta\omega)$. Since the actual positions of the second and higher harmonics often differ slightly from the expected positions at $k\hat{\omega}_0$, $k = 2, \ldots, K$, $\hat{\omega}_k$ is determined by a grid search restricted to an interval centered around $k\hat{\omega}_0$. A time-varying version of P_{SO} has been proposed in [54], involving an adaptive algorithm for tracking of the harmonics, see Sect. 6.4.1.

Another approach to characterizing the harmonic structure is based on the spectral line model, where the decay of the amplitude of the harmonics constitutes the crucial parameter [55]. The model is defined by the magnitude a_0 of the dominant spectral peak at ω_0, the exponential decay γ, referred to as the *harmonic decay* (HD), and the harmonic frequencies $\omega_0, \ldots, \omega_{K-1}$,

$$S_{\mathrm{HD}}(\omega) = a_0 e^{-\gamma k} \delta(\omega - \omega_k), \quad k = 0, \ldots, K-1, \tag{6.3}$$

where a_0 and γ are unknown parameters, whereas $\omega_0, \ldots, \omega_{K-1}$ may be determined as described above. By taking the logarithm of $S_{\text{HD}}(\omega)$, the estimation of a_0 and γ is transformed into a problem of fitting a line to $\ln S_x(\omega)$. Using the least squares (LS) method, joint minimization of the cost function

$$J(\ln a_0, \gamma) = \sum_{k=0}^{K-1} \left(\ln S_x(\hat{\omega}_k) - (\ln a_0 - \gamma k)\right)^2 \tag{6.4}$$

with respect to a_0 and γ yields the following two estimators:

$$\hat{a}_0 = \exp\left[\frac{2(2K-1)}{K(K+1)}\sum_{k=0}^{K-1} \ln S_x(\hat{\omega}_k) - \frac{6}{K(K+1)}\sum_{k=0}^{K-1} k \ln S_x(\hat{\omega}_k)\right], \tag{6.5}$$

$$\hat{\gamma} = -\frac{6}{K(K+1)}\sum_{k=0}^{K-1} \ln S_x(\hat{\omega}_k) + \frac{12}{K(K^2-1)}\sum_{k=0}^{K-1} k \ln S_x(\hat{\omega}_k), \tag{6.6}$$

where exponentiation is used to transform back to the original model parameters in (6.3). A wide range of f wave morphologies can be represented by the spectral line model, spanning from sawtooth-like waves, observed at an early stage of AF, to sinusoidal-like waves, observed in permanent AF, illustrated in Fig. 6.3. Since a slower AFR is usually associated with sawtooth-like waves, i.e., characterized by several harmonics, and a faster AFR with more sinusoidal-like waves, i.e., characterized by the fundamental frequency, it is plausible to assume that ω_0 and γ are positively correlated as AF progresses [55].

The logarithm of the *spectral power ratio* (SPR), defined by the harmonics positioned at $\hat{\omega}_0$ and $\hat{\omega}_1$, is yet another parameter for harmonic characterization [56],

$$P_{\text{SPR}} = \ln\left(\frac{S_x(\hat{\omega}_0)}{S_x(\hat{\omega}_1)}\right). \tag{6.7}$$

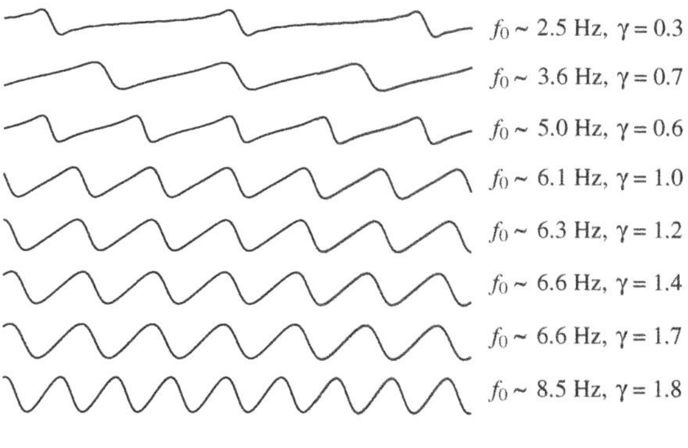

Fig. 6.3 Simulated f waves with different morphologies, obtained by varying the parameters f_0 and γ of the spectral line model in (6.3), assuming that $\omega_k = k2\pi f_0$

A large value of P_{SPR} reflects a spectrum with less pronounced harmonic structure, and vice versa.

The spectral parameters P_{SO}, γ, and P_{SPR} require that at least two harmonics are present. Using *spectral entropy* (SE), less emphasis is put on the harmonic structure of $S_x(\omega)$ and more on the complexity of the f wave signal [50, 51]. The spectral entropy of a narrowband signal is lower than that of a broadband signal. Since the entropy definition involves a probability mass function with unit area, the spectrum needs to be converted into such a function by normalizing each frequency component $S_x(\omega_l)$ with the sum of all L components,

$$\overline{S}_x(\omega_l) = \frac{S_x(\omega_l)}{\sum_{i=1}^{L} S_x(\omega_i)}, \quad l = 1, \ldots, L, \tag{6.8}$$

where ω_1 and ω_L denote the lower and upper frequency limits, respectively, and $\omega_2, \ldots, \omega_{L-1}$ are equidistantly spaced frequencies between ω_1 and ω_L; thus, ω_l does not denote a harmonic frequency in (6.8). The SE is defined by [57]

$$I_{SE} = -\sum_{l=1}^{L} \overline{S}_x(\omega_l) \log_2 \overline{S}_x(\omega_l). \tag{6.9}$$

The *spectral width* of the largest peak is yet another parameter which has been investigated in a few studies [56, 58, 59]. However, this measurement is influenced by the spectral leakage effect, manifested by the power of a sinusoid leaking into adjacent frequencies within a bandwidth of approximately $4\pi/N$, where N is the length of $x(n)$ [44]. Moreover, the temporal variation often observed in DAF has profound influence on the spectral width. Together, these two factors explain why the spectral width has had very limited significance in clinical studies.

6.3.3 Time–Frequency Analysis

Power spectral analysis reflects the average signal behavior of the analyzed interval, and the position of the largest spectral peak represents the main carrier of clinically significant information. In case of bi- or multimodal spectral peaks, the presence of joint frequencies is not necessarily reflected, but just as well that the DAF varies within the analyzed interval. Using time–frequency analysis in patients with permanent AF [60], the variation in the DAF was found to be as large as 2.5 Hz during just a few seconds, suggesting that temporal variation in the DAF is a characteristic of the underlying, complex electrical activation patterns in the atria. Another reason to pursue time–frequency analysis is the wish to characterize changes in the DAF due to intervention, e.g., drug administration and tilt table testing.

A plethora of techniques have been developed for time–frequency analysis, of which the simplest, and the most common, is the *short-term Fourier transform* (STFT), being a linear, nonparametric method. The STFT results from modifying the one-dimensional discrete-time Fourier transform to include a sliding time window $w(n)$ which extracts a segment from $x(n)$ for analysis, resulting in a two-dimensional function $X(n, \omega)$ defined by

$$X(n, \omega) = \sum_{l=-\infty}^{\infty} x(l)w(l - n)e^{-j\omega l}. \tag{6.10}$$

The length of $w(n)$ determines the resolution in time and frequency: a short window yields good time resolution but poor frequency resolution, and vice versa. By analogy with the computation of the periodogram, the spectrogram is obtained by computing the squared magnitude of the STFT,

$$S_x(n, \omega) = |X(n, \omega)|^2. \tag{6.11}$$

In certain clinical applications, it may be desirable to track changes in the DAF as small as 0.1 Hz, thus calling for a segment length of at least 10 s. On the other hand, the DAF may change so rapidly over time that a time resolution of 10 s is insufficient. These conflicting demands have proven difficult to achieve with the STFT, and, therefore, depending on the AF application at hand [59–61], other techniques for time–frequency analysis with better resolution in both time and frequency have been investigated.

The Wigner–Ville distribution (WVD) is a well-known quadratic, nonparametric transform offering better resolution than the STFT [30, 62, 63]. Unfortunately, the quadratic structure also means the introduction of cross-terms in the time–frequency domain, arising between different signal components as well as between signal and noise components. Although the influence of cross-terms can be reduced by including a kernel function, the practical use of the WVD is still limited when multicomponent signals are encountered. Since the tracking of changes in the DAF is an important aspect of time–frequency analysis, the *cross Wigner–Ville distribution* (XWVD) is an attractive choice as it integrates the estimation of a varying frequency with the computation of the WVD [64]. The XWVD is initiated by the frequency series $\hat{\omega}_{0,0}(n)$, determined from the STFT, where the two indices denote harmonic number and iteration number. The XWVD is computed between $x(n)$ and a sinusoid defined by $\hat{\omega}_{0,0}(n)$, from which an improved $\hat{\omega}_{0,1}(n)$ can be estimated using peak detection of the XWVD. Based on $\hat{\omega}_{0,1}(n)$, a new XWVD is computed, and so on, until the frequency series no longer changes from iteration to iteration.

Using the XWVD, spontaneous temporal variation can be uncovered in the DAF, illustrated in Fig. 6.4 where the XWVD of a 1-min extracted f wave signal is analyzed, obtained from a patient with permanent AF. The presence of such variation most likely explains why the dominant peak of the power spectrum is broad or bimodal as is the case in Fig. 6.2 [60].

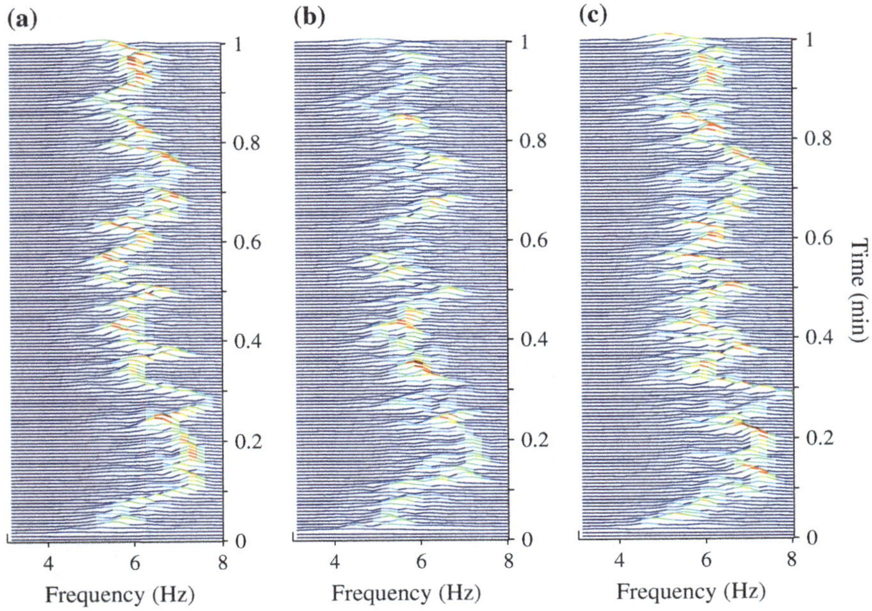

Fig. 6.4 The cross Wigner–Ville distribution of a 1-min f wave signal obtained from a patient with permanent AF, using a 2.5-s Hanning window. The distribution is displayed for leads **a** V_1, **b** V_2, and **c** V_3. The DAF is centered around 6 Hz in all three leads, with considerable variation ranging from about 5–7 Hz

The *spectral profile method* [55] was developed to address the limitation that the DAF is the focus of the XWVD, while other harmonics are ignored. The spectral profile results from averaging of frequency-aligned spectra of successive signal segments. By using a logarithmic frequency scale, rather than the conventional linear scale, spectra with different harmonic frequencies can be properly aligned and averaged. The resulting spectral profile exhibits a more distinct harmonic pattern than the spectra of separate segments, and, therefore, lends itself better to f wave characterization. In this method, the time–frequency distribution is similar to that produced by the STFT, except that a nonuniform, discrete-time Fourier transform is employed. The spectrum of each segment is aligned to the spectral profile by finding the frequency shift that minimizes the weighted LS error, after which the spectral profile is updated with the aligned spectrum.

In this method, each spectrum $\mathbf{q}_p$ of the time–frequency distribution is obtained by computing the nonuniform, discrete-time Fourier transform of $\mathbf{x}_p$,

$$\mathbf{q}_p = \mathbf{F}\mathbf{W}\mathbf{x}_p, \tag{6.12}$$

where the column vector $\mathbf{x}_p$ contains the N samples of the p-th signal segment; the computation is either made in overlapping or nonoverlapping segments. The resulting column vector $\mathbf{q}_p$ contains L different frequencies, the $N \times N$ diagonal matrix $\mathbf{W}$ defines the window function $w(n)$ applied to $\mathbf{x}_p$, and the $L \times N$ transform matrix $\mathbf{F}$ is defined by L nonuniformly sampled frequencies,

$$\mathbf{F} = \left[e^{-j0\boldsymbol{\omega}} \; e^{-j1\boldsymbol{\omega}} \; e^{-j2\boldsymbol{\omega}} \; \cdots \; e^{-j(N-1)\boldsymbol{\omega}} \right], \tag{6.13}$$

where $\boldsymbol{\omega} = \left[\nu_0 \; \cdots \; \nu_{L-1} \right]^T$ is a column vector with logarithmically spaced frequencies ν_l, defined by

$$\nu_l = \nu_{\text{low}} \pi^{\frac{\eta l}{L}}, \quad l = 0, \ldots, L-1. \tag{6.14}$$

The two parameters ν_{low} and η determine together the frequency interval relevant for f wave characterization, and L determines the sampling rate of the logarithmic frequency scale. Using $\nu_{\text{low}} = 0.31$ and $\eta = 2$, together with a 50-Hz sampling rate of the extracted f wave signal, the nonuniform Fourier transform is computed for frequencies ranging from 2.5 Hz to about 25 Hz [55].

Thanks to the logarithmic frequency sampling in (6.14), two spectra with different harmonic structures can be aligned. For example, a spectrum with harmonic frequencies at 5 and 10 Hz can be aligned to another spectrum with harmonic frequencies at 6 and 12 Hz, since the number of samples between the two harmonics is the same for logarithmically sampled spectra. Using linear frequency sampling, these two spectra cannot be aligned since the number of samples between the harmonics differ.

The magnitude of the spectrum, i.e., $|\mathbf{q}_p|$, is assumed to be described by a frequency-shifted (θ_p) and amplitude-scaled (a_p) version of the $L \times 1$ spectral profile vector $\boldsymbol{\phi}_p$, given by $a_p \, \mathbf{J}_{\theta_p} \, \boldsymbol{\phi}_p$. The shift matrix $\mathbf{J}_{\theta_p}$, defined in (5.25), takes care of the frequency shifting needed when updating $\boldsymbol{\phi}_p$ with new information. The weighted LS error criterion

$$J(\theta_p, a_p) = (|\mathbf{q}_p| - a_p \, \mathbf{J}_{\theta_p} \, \boldsymbol{\phi}_p)^T \, \mathbf{D} \, (|\mathbf{q}_p| - a_p \, \mathbf{J}_{\theta_p} \, \boldsymbol{\phi}_p) \tag{6.15}$$

is employed to estimate the unknown parameters θ_p and a_p. The primary purpose of the diagonal weight matrix $\mathbf{D}$ is to correct for the oversampling at lower frequencies due to the logarithmic sampling. However, the weight matrix $\mathbf{D}$ also makes it possible to emphasize frequencies which may be of special interest. Minimization of $J(\theta_p, a_p)$ with respect to θ_p and a_p yields the following estimators:

$$\hat{\theta}_p = \arg\max_{\theta_p} \left(|\mathbf{q}_p^T| \mathbf{D}^{\frac{1}{2}} \mathbf{J}_{\theta_p} \mathbf{D}^{\frac{1}{2}} \boldsymbol{\phi}_p \right), \tag{6.16}$$

$$\hat{a}_p = |\mathbf{q}_p^T| \mathbf{D}^{\frac{1}{2}} \mathbf{J}_{\hat{\theta}_p} \mathbf{D}^{\frac{1}{2}} \boldsymbol{\phi}_p. \tag{6.17}$$

Design considerations on $\mathbf{D}$, as well as details on the minimization of $J(\theta_p, a_p)$, are described in [55].

Since the spectral profile $\boldsymbol{\phi}_p$ is not known a priori, it can be estimated using exponential averaging of $|\mathbf{q}_p|$ once shifted to the position of the first harmonic in the spectral profile,

$$\hat{\boldsymbol{\phi}}_{p+1} = (1-\alpha_p)\hat{\boldsymbol{\phi}}_p + \alpha_p \frac{\mathbf{J}_{-\hat{\theta}_p}|\breve{\mathbf{q}}_p|}{\|\mathbf{J}_{-\hat{\theta}_p}|\breve{\mathbf{q}}_p|\|}, \quad p \geq 0, \tag{6.18}$$

where α_p is set to a positive value ($0 < \alpha_p < 1$), unless $\mathbf{x}_p$ contains an ectopic beat, large QRS-related residuals, or judged to be unreliable for some other reason, when α_p is set to zero. The spectral profile $\hat{\boldsymbol{\phi}}_0$ is initialized by setting one frequency equal to one at a position where the DAF is likely to occur, whereas all other frequencies are set to a value close to zero. The notation $\breve{\mathbf{q}}_p$ signifies that $\mathbf{q}_p$ has been pre- and appended with a sufficient number of samples to allow for frequency shifting; these additional samples are also set to a value close to zero. Normalization by $\|\mathbf{J}_{-\hat{\theta}_p}|\breve{\mathbf{q}}_p|\|$ in (6.18) is necessary to ensure that the spectral profile allows for meaningful estimation of a_p in (6.17).

In the spectral profile, the first harmonic has a fixed position throughout the analysis of $\mathbf{x}_p$, and, therefore, the spectral profile needs to be properly shifted before it can be interpreted as a spectrum. In particular, the first harmonic of the p-th segment, denoted $\hat{\omega}_{0,p}$, is obtained as

$$\hat{\omega}_{0,p} = \hat{\omega}_{0,0} - \hat{\theta}_p. \tag{6.19}$$

It should be noted that $\hat{a}_p$ is a measure of f wave amplitude, thus providing yet another definition to those earlier described in Sect. 6.2. The amplitude estimate $\hat{a}_p$ may also be used as a normalization factor when evaluating the model error $J(\hat{\theta}_p, \hat{a}_p)$ in successive signal segments [55].

Figure 6.5 shows that the harmonics of the spectral profile are considerably less smeared than are those of the amplitude spectrum obtained by Welch's method. This property can be ascribed to the frequency shifting which is part of the update of the spectral profile in (6.18).

The STFT, the XWVD, and the spectral profile method provide various degrees of insight into the time-varying nature of the DAF, as well as the harmonic composition of the f wave signal. Although time–frequency analysis provides more information than power spectral analysis, its impact on clinical studies has been rather limited. One reason may be the lack of an hypothesis connecting a certain property of the time–frequency distribution to an electrophysiological mechanism. Another reason may be that parameters are largely lacking for characterizing properties which are intrinsic to the time–frequency distribution, one of the few exceptions being the parameter tailored to investigate whether controlled respiration, mediated through the autonomic nervous system, influences the DAF in patients with permanent AF [65]. In that study, the frequency components in the interval 0.15–0.40 Hz of the power spectrum of the DAF series were quantified, since these components are known to reflect modulation of vagal tone, primarily through respiration, and therefore related to parasympathetic activation [66].

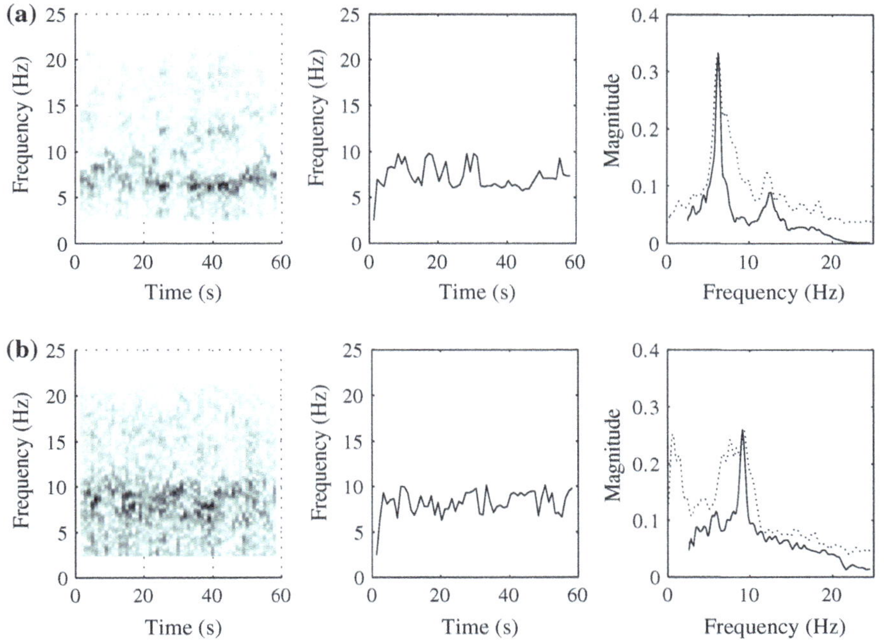

Fig. 6.5 Time–frequency analysis using the spectral profile method applied to a 60-s extracted f wave signal which either **a** contains a large second harmonic or **b** lacks a second harmonic. The time–frequency distribution, the DAF series, and the spectral profile (solid line) are displayed from left to right. The spectral profile obtained at the end of the 60-s interval is the one which is displayed. For comparison, the conventional amplitude spectrum (dotted line) is shown in the rightmost diagrams. In both **a** and **b**, the variation in the DAF is considerable

6.3.4 Frequency Tracking

When the time-varying characteristics of the harmonic components represent the main focus of investigation, time–frequency analysis may be replaced by *single frequency tracking* or *harmonic frequency tracking*, depending on whether one or more harmonic frequencies are of interest to analyze. Of the numerous techniques developed for single frequency tracking, the adaptive line enhancer is probably the most well-known [67, 68], composed of a time-varying bandpass filter $H(z; n)$ to enhance the harmonic component of the input signal $x(n)$, and an adaptive algorithm to estimate the instantaneous frequency $\omega_0(n)$ of the output signal $y(n)$. The resulting estimate $\hat{\omega}_0(n)$ is used to update the center frequency of the bandpass filter. Single frequency tracking can be extended to harmonic frequency tracking by assigning a time-varying bandpass filter and an adaptive algorithm to each of the harmonic components, resulting in a tracker with filter bank structure.

In the context of f wave characterization, single frequency tracking is part of a method developed for the purpose of selecting suitable patient candidates for restoration of sinus rhythm using catheter ablation [54]. The single frequency frequency tracker, belonging to the class of adaptive line enhancers, assumes that the input signal is modeled by [54, 69]

$$x(n) = A_0 e^{j\omega_0 n} + v(n), \tag{6.20}$$

where A_0 and ω_0 denote amplitude and fundamental frequency, respectively; the noise $v(n)$ is assumed to be white. Although the quantities in (6.20) are complex-valued, the model is still relevant to a real-valued signal since its complex-valued analytic representation can be used, defined by the observed, real-valued signal and its Hilbert transform, see Sect. 6.4.1.

A time-varying, first-order bandpass filter with complex-valued coefficients enhances the sinusoidal component in $x(n)$, defined by

$$H(z; n) = \frac{1 - \beta}{1 - \beta e^{j\omega(n)} z^{-1}}, \tag{6.21}$$

where $\omega(n)$ is the time-varying center frequency, and β $(0 \ll \beta < 1)$ defines the bandwidth. The filter $H(z; n)$ has unit gain and zero phase delay at $\omega(n)$, ensuring that the harmonic component is undistorted.

The center frequency $\omega(n)$ is estimated by an adaptive algorithm which, at each time instant n, tries to minimize the mean square error (MSE)

$$J(n) = E\left[|y(n) - e^{j\omega(n+1)} y(n-1))|^2\right], \tag{6.22}$$

where $y(n)$ denotes the output of $H(z; n)$. When $y(n) = e^{j\omega_0 n}$, $J(n)$ is minimized for $\omega(n) = \omega_0$, thus motivating the definition of $J(n)$ in (6.22). The MSE estimator of $\omega_0(n)$ is determined by differentiating $J(n)$ with respect to $\omega(n)$ and setting the result equal to zero, yielding

$$\hat{\omega}_0(n+1) = \arg(E\left[y(n) y^*(n-1)\right]). \tag{6.23}$$

Similar to the derivation of the well-known least mean square (LMS) algorithm [68], the expected value may be replaced by its instantaneous estimate at time n,

$$\hat{\omega}_0(n+1) \approx \arg(y(n) y^*(n-1)). \tag{6.24}$$

Since this estimator is sensitive to noise, exponential averaging is performed so that a smoothed estimate $Q(n)$ of the expected value in (6.23) is produced, while, at the same time, making sure that slow changes in $\omega(n)$ can be tracked. Hence, together with (6.21), the single frequency tracker is defined by the following two equations:

$$Q(n) = Q(n-1) + \alpha(y(n) y^*(n-1) - Q(n-1)), \tag{6.25}$$
$$\hat{\omega}_0(n+1) = \arg(Q(n)), \tag{6.26}$$

where $\hat{\omega}_0(n)$ is an estimate of the DAF and α $(0 < \alpha < 1)$ is a weight factor determining the speed of tracking. The estimate $\hat{\omega}_0(n)$ is inserted in $H(z; n)$ so that the next filtered sample can be computed, and so on. Single frequency tracking is illustrated in Fig. 6.6 for an extracted, bandpass filtered f wave signal, where the changes in the DAF are relatively small, oscillating at around 7 Hz, except for a marked increase to 9 Hz after 21 s due to an artifact; after a few seconds, however, $\hat{\omega}_0(n)$ returns to the earlier estimate.

The interest in harmonic analysis, which spurred the development of the spectral profile method, was also part of the motivation to extend the single frequency tracker to the handling of several harmonic frequencies. The starting point is the signal model with K harmonics [69],

$$x(n) = \sum_{k=1}^{K} A_k e^{jk\omega_0 n} + v(n). \tag{6.27}$$

This model implies that the tracker should have a filter bank structure, consisting of K bandpass filters $H_k(z; n)$, where each filter has its center frequency at an integer multiple of $\omega(n)$,

$$H_k(z; n) = \frac{1-\beta}{1-\beta e^{jk\omega(n)} z^{-1}}, \quad k = 1, \ldots, K, \tag{6.28}$$

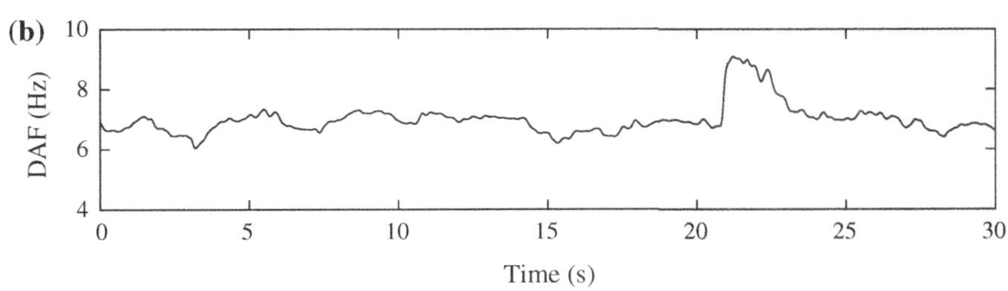

Fig. 6.6 **a** Extracted, bandpass filtered (4–12 Hz) f wave signal $x(n)$, and **b** related dominant atrial frequency (DAF), estimated using single frequency tracking ($\alpha = 0.05$, $\beta = 0.95$)

see Fig. 6.7. The adaptive algorithm used in single frequency tracking is also employed in harmonic frequency tracking, except that an estimate of the fundamental frequency is computed for each of the K harmonic components $y_k(n)$,

$$Q_k(n) = Q_k(n-1) + \alpha(y_k(n)y_k^*(n-1) - Q_k(n-1)), \tag{6.29}$$

$$\hat{\omega}_{0,k}(n+1) = \frac{\arg(Q_k(n))}{k}. \tag{6.30}$$

A global estimate of $\omega_0(n+1)$ is obtained as a linear combination of the different estimates $\hat{\omega}_{0,k}(n)$,

$$\hat{\omega}_0(n+1) = \sum_{k=1}^{K} w_k(n)\hat{\omega}_{0,k}(n+1). \tag{6.31}$$

The choice of the weights $w_k(n)$ is based on the same principle as that of weighted averaging, namely that $w_k(n)$ are inversely proportional to the noise variance, cf. (5.12). Since the noise variance is not defined for the harmonic model in (6.27), the minimum MSE error $J_{k,\min}(n)$ has been proposed as a surrogate measure of the noise variance [69]. Thus, before $\hat{\omega}_0(n+1)$ can be computed, $w_k(n)$ is computed using the following equations:

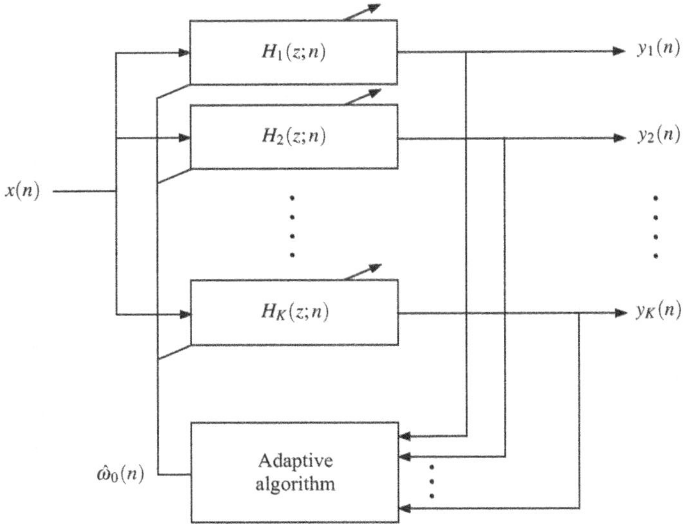

Fig. 6.7 Block diagram of the harmonic frequency tracker, composed of a filter bank with K bandpass filters $H_k(z;n)$ with harmonically coupled center frequencies and an adaptive algorithm for updating the center frequencies of the filters

$$\hat{J}_{k,\min}(n) = \hat{J}_{k,\min}(n-1) + \alpha(|y_k(n) - e^{jk\hat{\omega}_0(n)}y_k(n-1))|^2 - \hat{J}_{k,\min}(n-1)), \tag{6.32}$$

$$\hat{E}_k(n) = \hat{E}_k(n-1) + \alpha(|y_k(n)|^2 - \hat{E}_k(n-1)), \tag{6.33}$$

$$\hat{\sigma}^2_{\omega,k}(n) = \frac{\hat{J}_{k,\min}(n)}{\hat{E}_k(n)}, \tag{6.34}$$

$$w_k(n) = \frac{\dfrac{1}{\hat{\sigma}^2_{\omega,k}(n)}}{\displaystyle\sum_{i=1}^{K} \frac{1}{\hat{\sigma}^2_{\omega,i}(n)}}, \tag{6.35}$$

where $\hat{E}_k(n)$ is a smoothed estimate of the energy of $y_k(n)$ which is used to normalize $\hat{J}_{k,\min}(n)$ so that $w_k(n)$ reflects the local SNR. It should be noted that the second and higher harmonics are defined as integers of the fundamental frequency, although these frequencies are actually estimated by $\arg(Q_k(n))$ in (6.30). In contrast to time–frequency analysis, the harmonic frequency tracker produces harmonic signal components as a by-product, useful for various purposes such as the analysis of phase differences, which is the topic of the Sect. 6.4.

A precursor to single and harmonic frequency tracking is the DAF-controlled bandpass filter, designed to produce the first harmonic component, sometimes referred to as the *main atrial wave* [70, 71]. The DAF-controlled approach to bandpass filtering was introduced to reduce the effect of noise, being of critical importance to the computation of the sample entropy [70], but also as part of a method for characterizing f wave morphology [71]. While single and harmonic frequency tracking update the center frequency of the bandpass filter(s) on a sample-by-sample basis, the center frequency of the DAF-controlled filter is updated on a segment-by-segment basis, estimated in each segment from the power spectrum of the f wave signal.

6.4 f Wave Morphology and Regularity

Certain information on f wave morphology is provided by power spectral analysis and time–frequency analysis, for example, conveyed by the harmonic decay γ which reflects whether f waves have a sinusoid- or a sawtooth-looking morphology, cf. (6.3). However, the phase information is discarded in both these types of analysis, and, consequently, much of the morphologic information is discarded. By decomposing the extracted f wave signal into its harmonic components and comparing the respective phases, information on morphology can be retrieved (Sect. 6.4.1). Since phase analysis requires a relatively high SNR and relatively well-organized f waves, there is a need for robust approaches to morphologic characterization. One such approach considers the few largest eigenvalues of the correlation matrix of the f wave signal as a measure of regularity (Sect. 6.4.2). Another approach considers pairwise similarity of individual f waves, using a robust similarity measure (Sect. 6.4.3). These approaches have in common that they produce a parameter which characterizes the morphology of several, consecutive f waves, rather than the morphology of individual f waves. Hence, "f wave regularity" may be a more appropriate notion than "f wave morphology." Nonlinear techniques have also been considered for characterizing f wave regularity, including different measures of entropy (Sect. 6.4.4).

6.4.1 Phase Analysis

The classical approach to phase analysis of a lowpass signal $x(n)$ is based on its analytic representation, defined by

$$x_A(n) = x(n) + j\tilde{x}(n), \quad n = 0, \ldots, N-1, \tag{6.36}$$

where $\tilde{x}(n)$ denotes the Hilbert transform of $x(n)$. This transform shifts the phase of the positive frequency components by $-90°$ and the negative ones by $90°$ [72, 73]. Since the analytic signal $x_A(n)$ is complex-valued, it can alternatively be represented by its magnitude and phase,

$$x_A(n) = a(n)e^{j\psi(n)}, \tag{6.37}$$

where

$$a(n) = \sqrt{x^2(n) + \tilde{x}^2(n)}, \tag{6.38}$$

$$\psi(n) = \arctan\left(\frac{\tilde{x}(n)}{x(n)}\right). \tag{6.39}$$

Here, the function $\psi(n)$ defines the notion "phase" in a broad sense, without referring to sinusoidal phase. To interpret $\psi(n)$ as sinusoidal phase, the polar representation in (6.37) of a narrowband signal $y(n)$, obtained by bandpass filtering of $x(n)$ with center frequency ω_0, is considered:

$$\begin{aligned} y_A(n) &= y(n) + j\tilde{y}(n) = a(n)e^{j\psi(n)} \\ &= a(n)e^{j\phi(n)}e^{j\omega_0 n}. \end{aligned} \tag{6.40}$$

It is easily shown that the real-valued part of $y(n)$, i.e., the part with practical interest, can be expressed as

$$y(n) = a(n)\cos(\omega_0 n + \phi(n)). \tag{6.41}$$

Before computing the sinusoidal phase $\phi(n)$, $x(n)$ needs to be bandpass filtered to ensure that it is not a multi-component or broadband signal [74]. Even with the inclusion of bandpass filtering, $\phi(n)$ is still an instantaneous measurement which is vulnerable to noise.

Another approach to phase analysis is based on statistical modeling of the harmonic signal components [71]. As a first step, the observed signal $x(n)$ is decomposed into K different harmonic components $y_k(n)$, $k = 1, \ldots, K$, using a filter bank of linear, time-invariant bandpass filters. The center frequency of the filter $H_1(z)$, producing $y_1(n)$, is determined by the position of the largest spectral peak of $S_x(\omega)$, i.e., $\hat{\omega}_0$. Since the second and higher harmonic frequencies often differ slightly from their expected positions at $k\hat{\omega}_0$ due to changes in f wave morphology, the center frequencies of $H_2(z)$, $H_3(z)$, ... are determined by searching for the respective peaks in intervals centered around $k\hat{\omega}_0$, cf. the computation of P_{SO} in (6.2). Figure 6.8 illustrates a harmonic power spectrum and the passbands of the bandpass filter bank determined from the spectrum.

In a second step, the bandpass filtered signals $y_k(n)$ are subject to analysis in nonoverlapping segments with L samples,

$$y_{k,p}(n) = y_k(n - pL), \quad n = 0, \ldots, L - 1, \tag{6.42}$$

where p is the segment number. In [71], the lengths N and L were set to 10 and 0.5 s, respectively, where the latter setting implies that the analysis of f wave morphology is performed almost on a wave-to-wave basis (though the boundaries between f waves are not taken into consideration). In each segment, $y_{k,p}(n)$ is modeled by a sinusoid in Gaussian, white noise $v_{k,p}(n)$,

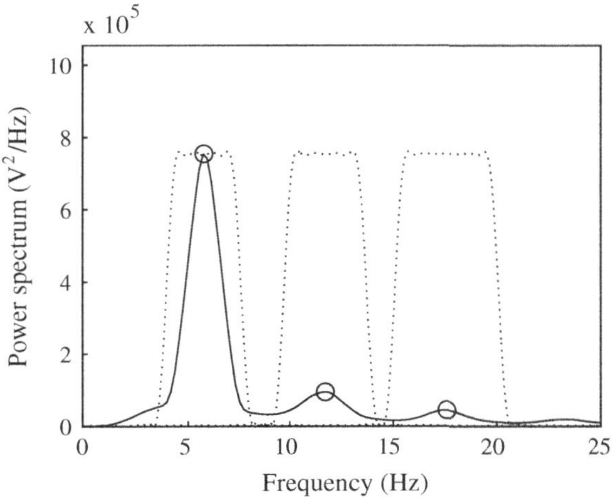

Fig. 6.8 Harmonic power spectrum and related passbands of the bandpass filters (dotted lines) for producing three harmonic components. The passbands are centered around the spectral peaks (marked with circles) and are increasingly wider at higher harmonic frequencies

$$y_{k,p}(n) = a_{k,p} \sin(\omega_{k,p}n + \phi_{k,p}) + v_{k,p}(n), \tag{6.43}$$

where $a_{k,p}$, $\omega_{k,p}$, and $\phi_{k,p}$ are unknown parameters. The maximum likelihood (ML) estimators of these three parameters are given by [47]

$$\hat{\omega}_{k,p} = \arg\max_{\omega} \left| \frac{1}{L} \sum_{n=0}^{L-1} y_{k,p}(n) e^{-j\omega n} \right|^2, \tag{6.44}$$

$$\hat{a}_{k,p} = \frac{2}{L} \left| \sum_{n=0}^{L-1} y_{k,p}(n) e^{-j\hat{\omega}_{k,p}n} \right|, \tag{6.45}$$

$$\hat{\phi}_{k,p} = \arctan \left(\frac{\sum_{n=0}^{L-1} y_{k,p}(n) \cos(\hat{\omega}_{k,p}n)}{\sum_{n=0}^{L-1} y_{k,p}(n) \sin(\hat{\omega}_{k,p}n)} \right). \tag{6.46}$$

Thus, $\hat{\omega}_{k,p}$ is determined by the position of the largest peak of the periodogram of $y_{k,p}(n)$, required before estimation of $a_{k,p}$ and $\phi_{k,p}$. The accuracy of sinusoidal modeling can be quantified by the MSE ε_p between $x_p(n)$ and its reconstructed, noise-free counterpart $\hat{x}_p(n)$,

$$\varepsilon_p = \frac{1}{L} \sum_{n=0}^{L-1} (x_p(n) - \hat{x}_p(n))^2, \tag{6.47}$$

where

$$\hat{x}_p(n) = \sum_{k=1}^{K} \hat{a}_{k,p} \sin(k\hat{\omega}_{0,p}n + \hat{\phi}_{k,p}). \tag{6.48}$$

Alternatively, $\hat{x}_p(n)$ may be obtained by replacing $k\hat{\omega}_{0,p}$ in (6.48) with $\hat{\omega}_{k,p}$ as suggested by the model in (6.43).

In a third step, the phase parameters characterizing f wave morphology are computed, defined by the differences between $\hat{\phi}_{2,p}$ and $\hat{\phi}_{1,p}$, $\hat{\phi}_{3,p}$ and $\hat{\phi}_{1,p}$, and so on. A straightforward comparison of two phase estimates is, however, not meaningful since the estimates relate to different frequencies, i.e., $\hat{\omega}_{k,p}$ and $\hat{\omega}_{1,p}$, and therefore not comparable. To solve this problem, $\hat{\phi}_{k,p}$ is converted to the same scale as $\hat{\phi}_{1,p}$ by division with k. Moreover, since the k-th harmonic completes about k periods when the first harmonic completes one period, the k-th harmonic is periodic by 2π in its own scale, and approximately periodic by $2\pi/k$ in the scale of $\hat{\phi}_{1,p}$. Therefore, the phase difference $\hat{\theta}_{k,p}$ is computed using the following expression:

$$\hat{\theta}_{k,p} = \frac{\hat{\phi}_{k,p}}{k} - \hat{\phi}_{1,p} \pm \frac{l \cdot 2\pi}{k}, \quad k = 2, \ldots, K, \tag{6.49}$$

where $\hat{\phi}_{1,p}$ is adjusted with an integer multiple l of $2\pi/k$ to become unique within the interval $[-\pi/k, \pi/k]$.

Characterization of f wave morphology using phase information is illustrated in Fig. 6.9, where f waves are positioned according to $\hat{\theta}_{2,p}$. The phase difference $\hat{\theta}_{3,p}$ usually plays a much more subordinate role, since $\hat{a}_{3,p}$ is usually much smaller than $\hat{a}_{2,p}$, and thus $\hat{\theta}_{3,p}$ has much less influence on f wave morphology. It is noted that a change of $\hat{\theta}_{2,p}$ by $\frac{\pi}{4}$ results in reversed wave polarity. Moreover, Fig. 6.9 shows that f waves positioned at about $-\frac{\pi}{8}$ have a steeper upslope than downslope, whereas f waves positioned at the opposite position, i.e., about $\frac{3\pi}{8}$, have a downslope steeper than the upslope.

Clustering of f wave segments is an application where the phase differences $\theta_{k,p}$ have been explored, with the aim of determining a representative, reconstructed f wave signal better suited for morphologic characterization than the observed f wave signal itself, see Fig. 6.10 [71].

Considering that the spectral characteristics of the f waves can change over time, there is a risk that the harmonic frequencies wander outside the passbands of the time-invariant bandpass filters $H_1(z), H_2(z), \ldots, H_K(z)$—a risk that increases with increasing length of the signal segment used for designing the filter bank. When such a situation arises, the harmonic components $y_{k,p}(n)$ become less reliable, with repercussions on the reliability of $\hat{\theta}_{k,p}$. This problem can be addressed by adaptively tracking the harmonic frequencies, using, for example, the algorithm described in Sect. 6.3.4 [54]. With such tracking, the filter passbands are updated on a sample-to-sample basis, implying that the phase differences can be estimated on a sample-to-

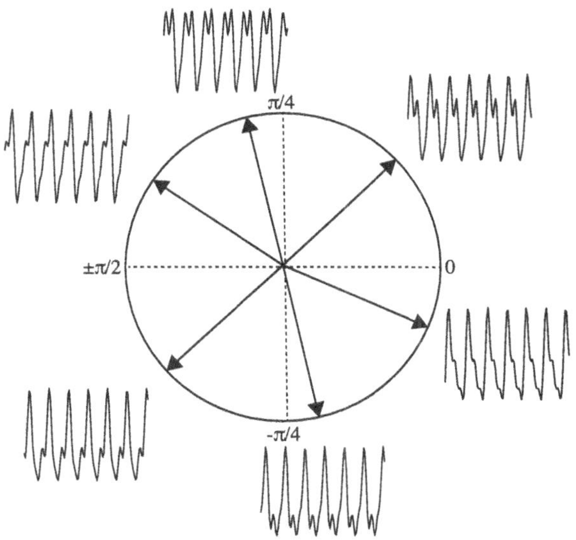

Fig. 6.9 Morphologic f wave characterization based on the phase difference $\hat{\theta}_{2,p}$, defined in (6.49) and confined to the interval $[-\frac{\pi}{2}, \frac{\pi}{2})$. The diagram is a variant of the well-known phasor diagram whose range is here adjusted to suit $\hat{\theta}_{2,p}$. The f waves are generated using the sawtooth model in (3.1)

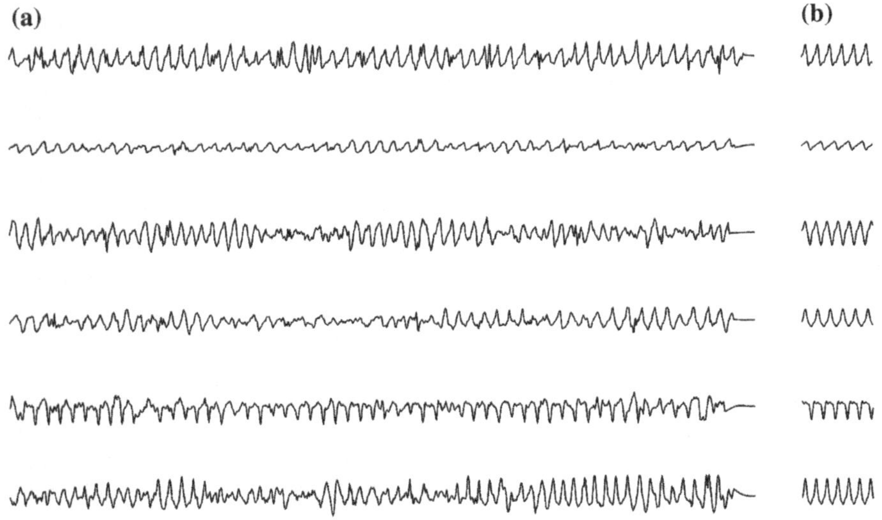

Fig. 6.10 **a** Extracted 10-s f wave signals obtained from six different patients with persistent AF, and **b** reconstructed f waves judged to be representative of the corresponding signals in (**a**). Nonoverlapping 0.5-s segments of $x(n)$ are clustered based on $\theta_{1,p}$ and $\theta_{2,p}$, after which the f waves belonging to the largest cluster are reconstructed; for details, see [71]

sample basis from the harmonic components $y_k(n)$. Thus, the segment-based estimate $\hat{\theta}_{k,p}$ is replaced by $\hat{\theta}_k(n)$.

Once $y_k(n)$ is available, the instantaneous phase $\phi_k(n)$ is computed by

$$\hat{\phi}_k(n) = \arctan\left(\frac{\tilde{y}_k(n)}{y_k(n)}\right), \quad k = 1, \ldots, K, \tag{6.50}$$

followed by computation of the instantaneous phase difference $\hat{\theta}_k(n)$. Since $\hat{\phi}_k(n)$ is vulnerable to noise, lowpass filtering of the phase difference $\hat{\theta}_k(n)$ has been suggested [54]. The filtering was accompanied by the hypothesis that a change in f wave morphology is reflected by a change in the slope of a straight line which, in a sliding window, is fitted to $\hat{\theta}_2(n)$; higher-order phase differences were not analyzed. Morphologic regularity was quantified by the variance of the resulting slopes: a variance close to zero indicated a strong coupling between the first and the second harmonics, and vice versa.

6.4.2 PCA-Based Characterization of Regularity

Since phase analysis is only suitable for f wave signals with a relatively high SNR, PCA-based approaches have been investigated which, to some extent, trade morphologic detail for robustness. In particular, the mapping of estimated parameters to wave morphology offered by phase analysis, cf. the signal model in (6.43), is traded for a more robust, data-driven characterization of f wave regularity where the link to a signal model is lost.

The starting point of PCA is the data matrix $\mathbf{X}$, formed by dividing the extracted f wave signal into M nonoverlapping segments containing N samples each,

$$\mathbf{X} = \left[\mathbf{x}_1\ \mathbf{x}_2\ \cdots\ \mathbf{x}_M\right], \tag{6.51}$$

where each column $\mathbf{x}_p$ has been centered. The signal segments, forming the columns in $\mathbf{X}$, have not been aligned relative to any fiducial point. Thus, the definition of $\mathbf{X}$ in (6.51) differs from the one in (5.100), where the columns have been aligned relative to the occurrence times of the QRS complexes. Time alignment was also involved in the study which first pursued PCA-based characterization of AF signals [75]; in that study, the occurrence times of the atrial activations in the intracardiac electrogram were used for alignment.

The principal components are associated with the variances given by the eigenvalues $\lambda_1, \ldots, \lambda_N$ of the $N \times N$ sample correlation matrix $\hat{\mathbf{R}}_x = \frac{1}{M}\mathbf{X}\mathbf{X}^T$, cf. Sect. (5.6.1). When f wave morphology is regular across the analyzed signal segments, only a few eigenvectors are required to represent the f waves. A measure of how well the K most significant eigenvectors represent, on average, the M signals in $\mathbf{X}$ is provided by the normalized, cumulative sum of the K largest eigenvalues [30, 75–77]:

$$R_K = \frac{\sum_{i=1}^{K} \lambda_i}{\sum_{i=1}^{N} \lambda_i}, \quad 0 < R_K \leq 1, \tag{6.52}$$

where λ_i are sorted in decreasing order $\lambda_1 > \lambda_2 > \cdots > \lambda_N$ and $K \ll N$.[4] Interestingly, R_5 has been used to quantify the overall quality of ECG signals in various types of arrhythmia [79], though not on extracted f wave signals.

Figure 6.11 illustrates R_3 for two different signals: one with regular f wave morphology, and another with more irregular morphology and higher noise level. The difference in signal characteristic is well-reflected by R_3. Since the f wave signals $\mathbf{x}_p$ are not aligned, the ensemble $\mathbf{X}$ is heterogenous, leading to much lower values of R_3 than what is often reported in studies on ECG analysis.

A minor variation on R_K as a measure of regularity is to determine the number of eigenvalues K needed to make R_K exceed a certain preset level, and then use that particular value of K as a measure of regularity [80]. Obviously, a smaller K indicates a more regular signal since fewer eigenvectors are, on average, required to reconstruct the analyzed signal.

---

[4] When the data matrix $\mathbf{X}$ is composed of overlapping segments, defined by a sliding window shifted with one sample at a time, R_K is known as the *fractional spectral radius* and used to quantify the stochastic complexity of a signal [57], see also [78].

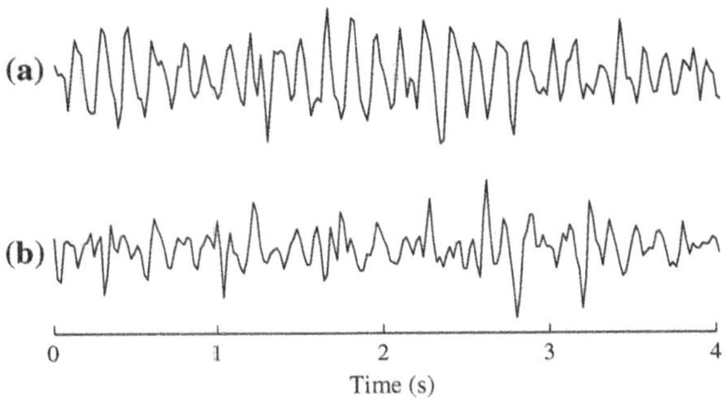

Fig. 6.11 **a** Regular and **b** irregular f wave signals characterized by $R_3 = 0.40$ and 0.25, respectively [77]

As a complement to R_K which characterizes the overall regularity of all signal segments in $\mathbf{X}$, the reconstruction error associated with $\mathbf{x}_p$, using the K most significant eigenvectors, may serve as a measure of regularity in individual segments. The reconstruction error of the p-th segment is defined by

$$\varepsilon_{p,K} = \frac{1}{N}(\mathbf{x}_p - \hat{\mathbf{s}}_{p,K})^T(\mathbf{x}_p - \hat{\mathbf{s}}_{p,K}), \tag{6.53}$$

where the reconstructed signal $\hat{\mathbf{s}}_{p,K}$ results from projecting $\mathbf{x}_p$ on the K most significant eigenvectors of $\hat{\mathbf{R}}_x$,

$$\hat{\mathbf{s}}_{p,K} = \boldsymbol{\Phi}_K \boldsymbol{\Phi}_K^T \mathbf{x}_p, \tag{6.54}$$

with

$$\boldsymbol{\Phi}_K = \left[\boldsymbol{\varphi}_1\ \boldsymbol{\varphi}_2 \cdots \boldsymbol{\varphi}_K\right]. \tag{6.55}$$

It is noted that the expected value of $\varepsilon_{p,K}$ is related to R_K through the following expression [30]:

$$E\left[\varepsilon_{p,K}\right] = \frac{1}{N}\sum_{i=K+1}^{N}\lambda_i = \frac{1}{N}(1 - R_K)\sum_{i=1}^{N}\lambda_i. \tag{6.56}$$

Early on in the history of automated ECG analysis, $\varepsilon_{p,K}$ was used to exclude noisy QRS complexes and artifacts from classifying QRS complexes in single-lead ECGs. However, a set of Gaussian functions were then used instead of the eigenvectors in (6.55) [81]. More recently, related to f wave characterization, the definition in (6.53) has been generalized so that it applies to multi-lead ECGs, with the aim of characterizing stationarity of atrial wavefront patterns during AF [80, 82], see Sect. 6.6.

6.4.3 Similarity-Based Characterization of Regularity

Morphologic similarity is a crucial feature when clustering QRS complexes [83–85], which may be quantified by the correlation between two QRS complexes once they have been properly aligned in time, cf. (5.159). The correlation-based approach requires that QRS detection and QRS delineation have been performed. This approach can be applied to f waves as well, but then requiring that detection and delineation of individual f waves have been performed [86]. Compared to QRS detection and QRS delineation, the conditions under which the corresponding f wave algorithms should operate are much more challenging since f waves wax and wane and sometimes completely disappear. Moreover, since there is no clinical consensus on what defines f wave onset and end, delineation performance cannot be evaluated on annotated databases. For the algorithms proposed in [86], the occurrence time and onset of each f wave are determined using mathematical morphology operators [87–89]. It should be noted that only f wave onset needs to be determined since f wave end is identical to the onset of the subsequent f wave.

The main idea behind the correlation-based approach is to first assess morphologic similarity for all pairwise combinations of the M different f waves $x_i(n)$, $i = 1, \ldots, M$, contained in the analyzed segment. The resulting correlation coefficients are then merged into one single parameter describing morphologic regularity. Since Pearson's correlation coefficient suffers from the disadvantages of being invariant to changes in amplitude and vulnerable to impulsive noise, the *signed correlation coefficient* (SCC) has been proposed, avoiding these disadvantages by coarse quantization of the observed signal $x_i(n)$ into three parts ("trichotomization") [90]:

$$x_{t,i}(n) = \begin{cases} 1, & x_i(n) \in S_p, \\ 0, & x_i(n) \in S_z, \\ -1, & x_i(n) \in S_n. \end{cases} \tag{6.57}$$

The signal space is spanned by the positive subspace S_p, the zero subspace S_z, and the negative subspace S_n, which are mutually disjunct. Each subspace is defined by a set of signal-dependent thresholds which can be fixed or variable over time. Before trichotomization, $x_i(n)$ is normalized by its maximum amplitude or some other suitable signal feature.

The products computed in Pearson's correlation coefficient are replaced by signed products of the two trichotomized signals $x_{t,i}(n)$ and $x_{t,j}(n)$, denoted $\otimes$ and defined by

$$x_{t,i}(n) \otimes x_{t,j}(n) = \begin{cases} 1, & x_{t,i}(n) = x_{t,j}(n), \\ -1, & x_{t,i}(n) = -x_{t,j}(n) \text{ and } x_{t,i}(n) \neq 0, \\ 0, & \text{otherwise,} \end{cases} \tag{6.58}$$

where $i, j = 1, \ldots, M$. Hence, the SCC is given by

$$P_{\mathrm{SCC},i,j} = \frac{\sum_{n=0}^{N-1} x_{t,i}(n) \otimes x_{t,j}(n)}{\sqrt{\sum_{n=0}^{N-1} x_{t,i}(n) \otimes x_{t,i}(n)} \sqrt{\sum_{n=0}^{N-1} x_{t,j}(n) \otimes x_{t,j}(n)}}$$
$$= \frac{1}{N} \sum_{n=0}^{N-1} x_{t,i}(n) \otimes x_{t,j}(n). \tag{6.59}$$

Similar to Pearson's correlation coefficient, the signed correlation coefficient is limited to $-1 \leq P_{\mathrm{SCC},i,j} \leq 1$, where 1 and -1 correspond to identical morphology but with equal or opposite polarity, respectively. Due to trichotomization, the product of the square root terms in the denominator of (6.59) equals N. Since the length of $x_{t,i}(n)$ typically varies from f wave to f wave, the shortest signal of $x_{t,i}(n)$ and $x_{t,j}(n)$ determines N; the length is determined after alignment.

In a simplified version of the SCC, the trichotomization in (6.57) is omitted, i.e., $x_{t,i}(n) \equiv x_i(n)$, and the signed product is redefined so that dichotomization is performed on the difference between $x_{t,i}(n)$ and $x_{t,j}(n)$ [86],

$$x_{t,i}(n) \otimes x_{t,j}(n) = \begin{cases} 1, & |x_{t,i}(n) - x_{t,j}(n)| \leq \eta, \\ -1, & |x_{t,i}(n) - x_{t,j}(n)| > \eta. \end{cases} \tag{6.60}$$

The threshold η can be taken as a percentage of the combined peak-to-peak amplitudes of $x_{t,i}(n)$ and $x_{t,j}(n)$.

Based on $P_{\mathrm{SCC},i,j}$, $i, j = 1, \ldots, M$, whether determined using dicho- or trichotomization, morphologic regularity can be quantified by the following function [86]:

$$\kappa(r) = \frac{2}{M(M-1)} \sum_{i=1}^{M} \sum_{j=i+1}^{M} \exp\left[-\frac{(P_{\mathrm{SCC},i,j} - 1)^2}{r^2}\right], \tag{6.61}$$

where $0 \leq \kappa(r) \leq 1$. The function $\kappa(r)$ reaches its maximum when all f waves have identical morphology, i.e., $P_{\mathrm{SCC},i,j} = 1$ for all combinations of i and j. The parameter r ($r > 0$) can be viewed as a threshold determining whether pairs of f waves are similar, i.e., fewer pairs are similar when r is set to a value close to zero. While exponents other than two in (6.61) have been investigated, this choice has been found to yield good overall performance [86, 91]. Thus, the three parameters M, η, and r need to be set in the correlation-based approach to characterizing f wave regularity.

Figure 6.12 illustrates the use of $\kappa(r)$ for an extracted f wave signal exhibiting substantial variation in both amplitude and morphology. It is noted that $\kappa(r)$ approaches zero in intervals with waning f waves, but is close to one in intervals with waxing f waves.

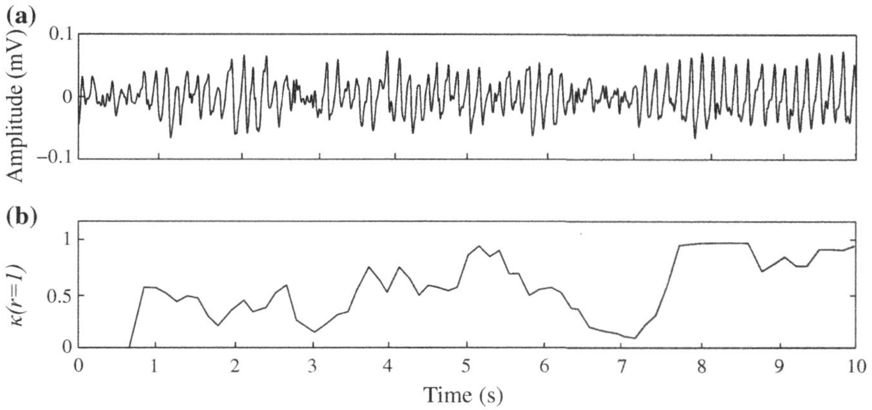

Fig. 6.12 **a** Extracted f wave signal and **b** related regularity function $\kappa(r = 1)$ computed in a sliding window, using $M = 5$ and $\eta = 0.15$

6.4.4 Entropy-Based Characterization of Regularity

Entropy measures provide information on nonlinear characteristics of a signal which is complementary to the information provided by linear transformation methods such as spectral analysis and PCA. The signal characteristic quantified by entropy is usually referred to as complexity, with regularity, predictability, repeatability, and self-similarity as alternative descriptions. For f wave signals, entropy may also be viewed as a measure of "AF organization" [92]—a term originating from electrogram-based analysis where the aim is to quantify the organization of local activity as well as the spatial organization (coordination) between different regions of the atria [93]. However, a widely accepted definition of "AF organization" is unfortunately missing.

A large number of entropy measures have been proposed, most of them resulting from different approaches to estimation [94, 95]. Shannon entropy I_{ShEn} [96], approximate entropy I_{ApEn} [97], sample entropy I_{SampEn} [98], spectral entropy I_{SE} [99], wavelet entropy [100], conditional entropy [101], and fuzzy entropy [91] have all been investigated in the realm of AF, either to characterize RR interval irregularity in AF detection and AF management (Chaps. 4 and 7, respectively) or f wave regularity, i.e., the topic of this section.

In an early study, I_{SampEn} was used to predict the termination of AF episodes in ambulatory ECG recordings [78]. The results showed that I_{SampEn} could not distinguish terminating from nonterminating AF, probably due to the often poor signal quality which precluded reliable computation of I_{SampEn}. In a later study, it was shown that both I_{SampEn} and I_{ApEn} are sensitive to the presence of spike artifacts [102], i.e., QRS-related residuals, which would lead to improper characterization of f wave regularity. Thus, the accuracy of I_{SampEn} depends on the prevailing signal quality.

A means to reduce the influence of noise is to bandpass filter the extracted f wave signal, implemented either by reconstructing the signal from the wavelet coefficients of the scale containing the DAF [103, 104], or using the output of a bandpass filter whose center frequency is defined by the DAF [70], i.e., the approach employed in phase analysis, cf. Sect. 6.4.1. Interestingly, when computing I_{SampEn} from a DAF-controlled bandpass filtered signal with a 3-Hz bandwidth [70], termination of paroxysmal AF could be predicted in the database previously analyzed in [78] without success. This result demonstrates that AF termination is associated with a change in f wave regularity which becomes increasingly more regular just before termination. It also demonstrates that entropy-based prediction calls for bandpass filtering of the f wave signal.

The idea to use a DAF-controlled bandpass filter was later expanded into a DAF-controlled filter bank, composed of harmonically-related bandpass filters, cf. Sect. 6.4.1, thus making it possible to compute I_{SampEn} for each harmonic component [105]. Since I_{SampEn} does not in itself convey any information on the strength of a harmonic component, a measure of strength is needed to judge the significance of the harmonics. In [105], strength was quantified by the relative energy of the second and the third harmonic components.

Before computation of I_{SampEn}, three parameters need to be set: the length m of the two subsequences to be compared, the similarity tolerance r, and the number of samples N, cf. the definition in (4.12). With respect to m and r, an early recommendation was to use $m = 1$ or 2 together with $0.1 \leq r/\sigma_x \leq 0.2$, where σ_x denotes the standard deviation of the analyzed signal [106–108]. This recommendation, which was based on biomedical signals with relatively slow dynamics, was later found to be less appropriate for signals with fast dynamics [109], thus motivating an investigation of how to choose optimal values of m and r in applications where f wave characterization is required. Using I_{SampEn} to predict termination of paroxysmal AF and outcome of electrical cardioversion in persistent AF, the choice of m and r was found to have significant influence on prediction performance [110]. In particular, when optimizing the performance of a predictor or classifier, the results suggested that a wider range of values of m and r should be considered than suggested by the early recommendation.

The sampling rate of the f wave signal influences the computation of I_{SampEn}, since the probability that two subsequences are identical, i.e., the maximum norm in (4.11) is below r, becomes increasingly higher as the sampling rate becomes increasingly faster, i.e., the sample-to-sample changes become increasingly smaller. To mitigate the problem that oversampling can produce misleading values of I_{SampEn}, a lag of L samples may be introduced between successive samples in the two subsequences for comparison, where L is related to the degree of oversampling [111]. When counting the number of similar subsequences in (4.12), only those which are L samples apart are considered. The lag may be determined from the properties of the autocorrelation function of the analyzed signal, e.g., its first zero-crossing [111]. Using simulated signals, the lag-based definition of I_{SampEn} was found to produce consistent results at different sampling rates, while the original definition did not.

A straightforward approach to choosing the sampling rate is to rely on knowledge from spectral analysis of the f wave signal, suggesting that frequencies up to about 25 Hz are relevant and thus a sampling rate of at least 50 Hz should be used. However, higher frequencies may still be relevant to the computation of I_{SampEn}, therefore motivating the use of a sampling rate higher than 50 Hz. Yet another approach is to choose the sampling rate which offers the best performance, for example, when the aim is to predict AF termination or to predict the outcome of electrical cardioversion [110]; for these two prediction problems, the best-performing sampling rate was found to be as high as 250 Hz.

The number of samples N should be chosen large enough so that the dynamics of several f waves is captured, where at least one second of the f wave signal is used to compute I_{SampEn} [110]. While the choice of N is related to the sampling rate, there seems to be general consensus that N should not be less than 200–250 samples, irrespective of sampling rate, to provide reasonably accurate estimates of I_{SampEn} [110, 112, 113].

6.5 Signal Quality Control

Several indices have been proposed for assessing the overall quality of ECG signals, e.g., the relative power of baseline variation, signal kurtosis, and the ratio of the number of beats detected by two different QRS detectors where one detector is tuned to be more sensitive to noise than the other [79, 114]. Unfortunately, these indices do not provide information on whether f wave characterization can be reliably performed. Therefore, a few methods have been developed for assessment of the signal quality, operating either in the time domain (Sect. 6.5.1) or the frequency domain (Sect. 6.5.2). Segments are discarded if the signal quality index (SQI) fulfills certain criteria. A completely different approach to dealing with poor signal quality is to postprocess the series of DAF estimates resulting from time–frequency analysis of the f wave signal [115].

6.5.1 Time Domain Analysis

Model-based assessment of signal quality explores basic information of the f wave signal, such as the variational patterns of amplitude and repetition rate. The harmonic model in (6.27), but with phase also included, is useful for such assessment [116]. Building on the observation that the variation in the DAF is restricted in short signal segments, a model signal can be reconstructed accounting for local variation in frequency and amplitude. The SQI is defined by the error between the observed signal and the reconstructed model signal.

The f-waves are modeled by a complex signal defined by the sum of K harmonically related, complex exponentials with fundamental frequency ω_0, corrupted by additive, white, complex Gaussian noise $v(n)$,

$$x(n) = \sum_{k=1}^{K} A_k e^{j(k\omega_0 n + \phi_k)} + v(n), \quad n = 0, \ldots, N-1, \tag{6.62}$$

where A_k and ϕ_k denote the amplitude and phase, respectively, of the k-th harmonic. The parameters $A_1, \phi_1, \ldots, A_K, \phi_K$, contained in the $2K \times 1$ vector

$$\boldsymbol{\theta} = \left[A_1 \; \phi_1 \; \cdots \; A_K \; \phi_K\right]^T, \tag{6.63}$$

and ω_0 are assumed to be deterministic, but unknown. In matrix format, the model in (6.62) is given by

$$\mathbf{x} = \mathbf{Z}(\omega_0)\mathbf{a}(\boldsymbol{\theta}) + \mathbf{v}, \tag{6.64}$$

where $\mathbf{a}(\boldsymbol{\theta})$ is a $K \times 1$ vector,

$$\mathbf{a}(\boldsymbol{\theta}) = \left[A_1 e^{j\phi_1} \; \cdots \; A_K e^{j\phi_K}\right]^T. \tag{6.65}$$

and $\mathbf{Z}(\omega_0)$ is an $N \times K$ Vandermonde matrix containing the frequency information,

$$\mathbf{Z}(\omega_0) = \begin{bmatrix} 1 & 1 & \cdots & 1 \\ e^{j\omega_0 1} & e^{j2\omega_0 1} & \cdots & e^{jK\omega_0 1} \\ \vdots & \vdots & \ddots & \vdots \\ e^{j\omega_0(N-1)} & e^{j2\omega_0(N-1)} & \cdots & e^{jK\omega_0(N-1)} \end{bmatrix}. \tag{6.66}$$

Unfortunately, joint ML estimation of $\mathbf{a}(\boldsymbol{\theta})$ and ω_0, defined by [116],

$$[\hat{\omega}_0, \hat{\boldsymbol{\theta}}] = \arg\min_{\omega_0, \boldsymbol{\theta}} \|\mathbf{x} - \mathbf{Z}(\omega_0)\mathbf{a}(\boldsymbol{\theta})\|^2, \tag{6.67}$$

does not result in closed-form expressions of the estimators $\hat{\omega}_0$ and $\hat{\boldsymbol{\theta}}$. Therefore, a suboptimal, two-step approach is considered in which $\mathbf{a}(\boldsymbol{\theta})$ is first determined by LS estimation, followed by insertion of the resulting $\hat{\mathbf{a}}(\boldsymbol{\theta})$ into the ML estimator of ω_0. For a given ω_0, the LS estimator is given by [117], see also (5.53):

$$\hat{\mathbf{a}}(\boldsymbol{\theta}) = (\mathbf{Z}(\omega_0)^H \mathbf{Z}(\omega_0))^{-1} \mathbf{Z}(\omega_0)^H \mathbf{x}. \tag{6.68}$$

Inserting $\hat{\mathbf{a}}(\boldsymbol{\theta})$ in (6.67), the ML estimator of ω_0 is defined by

$$\hat{\omega}_0 = \arg \min_{\omega_{0,\min} \leq \omega_0 \leq \omega_{0,\max}} \|\mathbf{x} - \mathbf{Z}(\omega_0)(\mathbf{Z}(\omega_0)^H \mathbf{Z}(\omega_0))^{-1} \mathbf{Z}(\omega_0)^H \mathbf{x}\|^2, \tag{6.69}$$

where minimization is performed using a grid search over the frequency interval $[\omega_{0,\min}, \omega_{0,\max}]$ in which the DAF is likely to be found. The estimate $\hat{\omega}_0$ represents a *global frequency estimate* as it is based on the segment with N samples, having a length of several seconds.

Variation in the DAF is allowed by dividing $\mathbf{x}$ into P overlapping subsegments $\mathbf{x}_p$, $p = 1, \dots, P$. Each subsegment contains L samples, with L chosen so that the subsegment contains at least one f-wave. For each subsegment, a *local frequency estimate* $\hat{\omega}_{0,p}$ is determined, using

$$\hat{\omega}_{0,p} = \arg \min_{|\omega_{0,p}-\hat{\omega}_0| \le \Delta\omega_0} \|\mathbf{x}_p - \mathbf{Z}_L(\omega_{0,p})(\mathbf{Z}_L(\omega_{0,p})^H \mathbf{Z}_L(\omega_{0,p}))^{-1} \mathbf{Z}_L(\omega_{0,p})^H \mathbf{x}_p\|^2, \tag{6.70}$$

where $\mathbf{Z}_L(\omega_{0,p})$ consists of the first L rows of $\mathbf{Z}(\omega_{0,p})$ and $\Delta\omega_0$ is the maximum deviation from $\hat{\omega}_0$ in any of the P subsegments. This implies that $\hat{\omega}_{0,p}$ accounts for short-time variation as long as it does not deviate more than $\Delta\omega_0$ from $\hat{\omega}_0$.

Reconstruction in terms of the signal part in (6.62) has the disadvantage of yielding a fixed amplitude and a fixed phase within the analyzed N-sample segment, thus motivating the use of a basis vector approach which can produce a signal with time-varying amplitude. The local DAF estimates $\hat{\omega}_{0,p}$ are used to create constant-amplitude basis vectors $\mathbf{b}_k, k = 1, \dots, K$, describing the phase variation of the signal. The vector $\hat{\mathbf{a}}_p(\boldsymbol{\theta}_p)$, containing local amplitude and phase information, is obtained using the LS estimator in (6.68), but with $\mathbf{Z}_L(\hat{\omega}_{0,p})$ replacing $\mathbf{Z}(\omega_0)$ and $\mathbf{x}_p$ replacing $\mathbf{x}$. The vector $\mathbf{y}_{k,p}$ is then computed from $\phi_{k,p}$, i.e., the phase of the k-th element of $\hat{\mathbf{a}}_p(\boldsymbol{\theta}_p)$, and $\hat{\omega}_{0,p}$,

$$\mathbf{y}_{k,p} = \begin{bmatrix} k\hat{\omega}_{0,p}0 + \phi_{k,p} \\ \vdots \\ k\hat{\omega}_{0,p}(L-1) + \phi_{k,p} \end{bmatrix}, \quad p = 1, \dots, P. \tag{6.71}$$

Since the related phase vectors $\mathbf{y}_{k,p}$ are overlapping, the overlapping parts are averaged to produce a global $N \times 1$ phase vector $\mathbf{y}_k$ which then is used to construct the constant-amplitude basis vector $\mathbf{b}_k$,

$$\mathbf{b}_k = \cos(\mathbf{y}_k), \tag{6.72}$$

capturing the phase variation in $\mathbf{x}$.

The time-varying amplitude of the reconstructed signal is described by the $N \times 1$ vectors $\boldsymbol{\alpha}_k$,

$$\boldsymbol{\alpha}_k = \left[\alpha_k(0)\ \alpha_k(1) \cdots \alpha_k(N-1)\right]^T, \tag{6.73}$$

whose maximum sample-to-sample variation in $\alpha_k(n)$ is limited by $\Delta\alpha_k$,

$$|\alpha_k(n) - \alpha_k(n-1)| \le \Delta\alpha_k. \tag{6.74}$$

The tolerance $\Delta\alpha_k$ should be chosen so that the variation in f-wave amplitude is captured, but not the variation due to noise. The model signal $\hat{\mathbf{s}}$ is obtained by summing the elementwise product of the basis vectors and the amplitude estimates of the harmonic components,

$$\hat{s}(n) = \sum_{k=1}^{K} \hat{\alpha}_k(n) b_k(n), \tag{6.75}$$

where $b_k(n)$ denotes the n-th element of $\mathbf{b}_k$. The amplitude estimator $\hat{\boldsymbol{\alpha}}_k$ is obtained by minimizing the following expression:

$$\hat{\boldsymbol{\alpha}}_k = \arg\min_{\boldsymbol{\alpha}_k} \sum_{n=0}^{N-1} \|\alpha_k(n) b_k(n) - \text{Re}\,[x(n)]\,\|^2, \tag{6.76}$$

which, along with the $N-1$ constraints in (6.74), defines a convex optimization problem which is solved numerically; the notion "Re" denotes the real part.

The SQI is defined by the normalized RMS of the model error $\hat{\mathbf{e}} = \mathbf{x} - \hat{\mathbf{s}}$,

$$S = 1 - \frac{\sigma_{\hat{e}}}{\sigma_x}, \tag{6.77}$$

where $\sigma_{\hat{e}}$ and σ_x denote the RMS of $\hat{\mathbf{e}}$ and $\mathbf{x}$, respectively. For any reasonable estimate of $\hat{\mathbf{s}}$, S is restricted to the interval [0, 1], where 0 indicates poor signal quality and 1 indicates perfect modeling of $\mathbf{x}$. A fixed threshold η_S can be used to indicate whether the f-waves in the analyzed segment have sufficient quality for characterization, see Fig. 6.13.

6.5.2 Frequency Domain Analysis

A disadvantage with the spectral profile method is its lack of control of what goes into the update of the spectral profile: the spectrum of a segment with large QRS-related residuals is just as influential as is the spectrum of a segment with noise-free f waves. Although the spectral profile can have a slow adaptation rate which limits the sensitivity to occasional noisy segments, several consecutive noisy segments will cause the spectral profile to lose its structure and, accordingly, the DAF estimates can no longer be trusted. Once the spectral profile has lost its structure, the recovery time may become unacceptably long, even if subsequent segments are associated with a harmonic structure. This limitation can be remedied by adopting a spectral modeling approach in which the spectrum of each segment is checked before entering the update of the spectral profile [118]. A harmonic spectrum is modeled as a sum of Gaussian functions (cf. (5.68)),

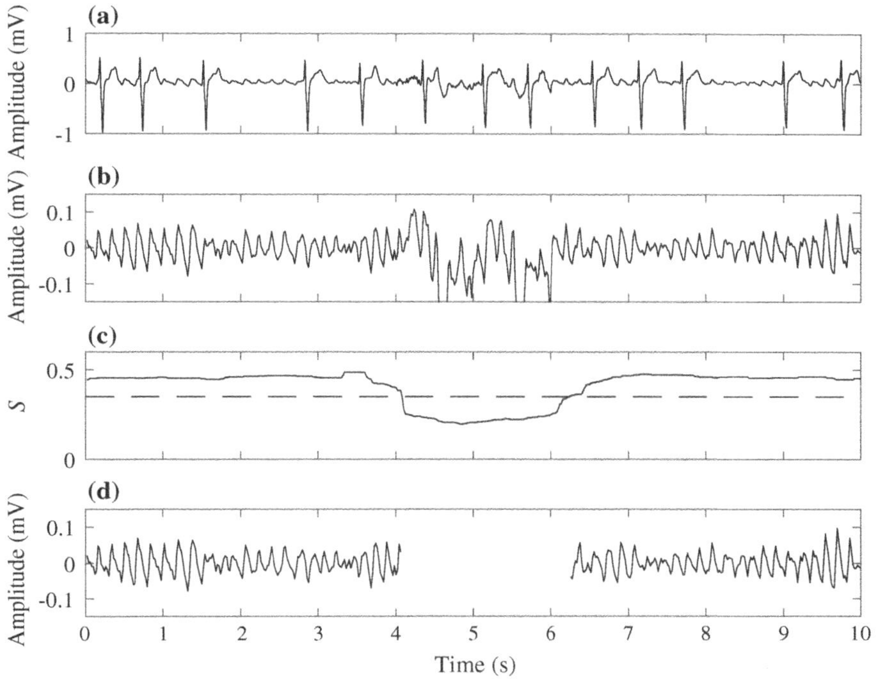

Fig. 6.13 Illustration of signal quality assessment. **a** ECG signal obtained from a patient with AF, **b** extracted f wave signal containing a noisy episode, **c** signal quality index S (solid line) and threshold η_S defining acceptable signal quality (dashed line), and **d** extracted f wave signal where the low-quality segment has been removed based on the information in **c**. The segment lengths N and L were set to 5 and 0.5 s, respectively

$$S_x(\omega, \boldsymbol{\theta}_p) = \sum_{k=1}^{K} A_{k,p} \exp\left[-\frac{(\omega - k\omega_{0,p} - \Delta_{k,p})^2}{2\sigma_{k,p}^2}\right], \tag{6.78}$$

where K is the number of Gaussians, $A_{k,p}$ is the spectral magnitude, $\sigma_{k,p}$ is the width, and $\Delta_{k,p}$ is the frequency jitter associated with the second and higher frequencies $k\omega_{0,p}$, $k = 2, \ldots, K$; thus, $\Delta_{1,p} = 0$. The model parameter vector $\boldsymbol{\theta}_p$, containing

$$\boldsymbol{\theta}_p = \left[A_{1,p} \cdots A_{K,p}\ \sigma_{1,p} \cdots \sigma_{K,p}\ \Delta_{2,p} \cdots \Delta_{K,p}\ \omega_{0,p}\right]^T, \tag{6.79}$$

is estimated by minimizing the following weighted LS error criterion with respect to $\boldsymbol{\theta}_p$:

$$J(\boldsymbol{\theta}_p) = (|\mathbf{q}_p| - \mathbf{s}(\boldsymbol{\theta}_p))^T \mathbf{D}\mathbf{E}_p(|\mathbf{q}_p| - \mathbf{s}(\boldsymbol{\theta}_p)), \tag{6.80}$$

where $\mathbf{q}_p$ is the nonuniform, windowed Fourier transform of the analyzed signal segment, defined in (6.12). The vector $\mathbf{s}(\boldsymbol{\theta}_p)$ is obtained by sampling the Gaussian model in (6.78) at the logarithmic frequencies ν_l defined in (6.14), yielding

$$\mathbf{s}(\boldsymbol{\theta}_p) = \left[S_x(\nu_0, \boldsymbol{\theta}_p) \cdots S_x(\nu_{L-1}, \boldsymbol{\theta}_p)\right]^T . \tag{6.81}$$

The matrices $\mathbf{D}$ and $\mathbf{E}_p$ are both diagonal, but handle different aspects of spectral weighting. Identical to the spectral profile method, $\mathbf{D}$ corrects for the oversampling at lower frequencies due to the logarithmic sampling. The matrix $\mathbf{E}_p$, on the other hand, is designed so that the frequency intervals in $|\mathbf{q}_p|$ with harmonic components are weighted with one, whereas the remaining intervals are weighted with a value close to zero; thus, this matrix is segment-dependent, while $\mathbf{D}$ is not. Details on the design of the matrices $\mathbf{D}$ and $\mathbf{E}_p$, as well as the multidimensional optimization procedure associated with $J(\boldsymbol{\theta}_p)$, can be found in [118].

A set of parameters characterizing the harmonic pattern is introduced to decide whether $\mathbf{q}_p$ should be excluded from the spectral profile update, i.e., whether or not α_p in (6.18) should be set to zero. The following three parameters, of which the first two relate to the model in (6.78), are used to exclude spectra which do not exhibit a harmonic structure [118]:

1. The minimized error $J(\hat{\boldsymbol{\theta}}_p)$, quantifying the similarity between $\mathbf{q}_p$ and the model spectrum $\mathbf{s}(\hat{\boldsymbol{\theta}}_p)$.
2. The width $\hat{\sigma}_{1,p}$, characterizing the spectral peak of the first harmonic.
3. The ratio of the maximum magnitude between the first and the second harmonics and the magnitude of the first harmonic, picking up the occurrence of spurious peaks between the first and the second harmonic.

For poor-quality signals, Fig. 6.14a, b present the spectral profile when computed without and with application of the exclusion criteria. It is obvious that the dominant peak becomes much more distinct when noisy segments are excluded from the update of the spectral profile. For good-quality signals, the spectral profile remains essentially unchanged after application of the exclusion criteria, see Fig. 6.14c, d.

6.6 Spatial Characterization

Most parameters proposed for f wave characterization are defined with reference to single-lead analysis, and extended to multi-lead ECG analysis by simply computing the parameters on a lead-by-lead basis. This approach has the disadvantage of ignoring intrinsic spatial information resulting from joint analysis of available leads. The vectorcardiographic f wave loops, defined by the orthogonal leads X, Y, and Z, provide basic spatial information (Sect. 6.6.1), whereas body surface potential mapping (BSPM) can provide much more comprehensive spatial information on AF activation patterns (Sect. 6.6.2). For example, the regions which are responsible for AF maintenance may be localized from such maps, with potential implications on AF treatment since regional information may contribute to improve the planning of

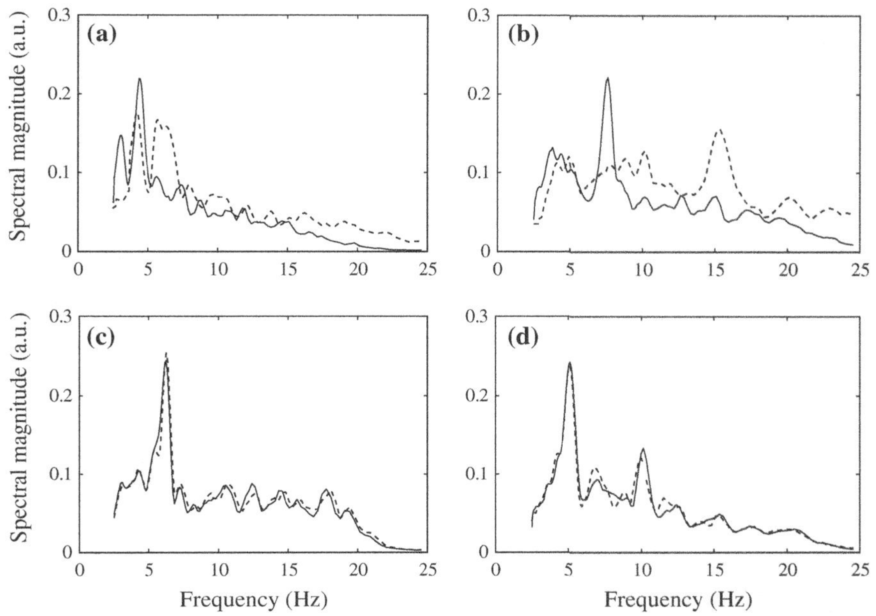

Fig. 6.14 Spectral profiles **a** before (dashed line) and **b** after application of exclusion criteria (solid line), obtained from extracted f wave signals containing large-amplitude QRS residuals. Spectral profiles **c** before and **d** after application of exclusion criteria, obtained from f wave signals with good quality

an ablation procedure [119]. From an engineering viewpoint, spatial characterization of body surface maps is still in its infancy, leaving much room for the development of robust, tailored signal processing algorithms. So far, most types of spatial analysis are extended versions of single-lead analysis, e.g., estimation of the DAF and phase analysis.

Body surface potential mapping is also the starting point for reconstruction of the potentials on the epicardial surface of the heart—a technique known as *ECG imaging* (ECGI). From the time sequence of epicardial potentials, electrograms can be constructed at different locations on the epicardium. Since ECG imaging involves several advanced aspects which are far outside the scope of this book, such as techniques for solving the inverse problem and imaging techniques to obtain subject-specific information on the geometries of the heart and the torso surfaces (based on computer tomography or magnetic resonance imaging), the interested reader is referred to the literature in this area [20, 120–124].

6.6.1 Vectorcardiogram Loop Analysis

The precursor to vectorcardiogram (VCG) loop analysis of f waves was a study which investigated the characteristics of loops in atrial flutter [125]. Since the reentry circuit of isthmus-dependent atrial flutter is known to contribute significantly to the VCG, it was hypothesized that flutter loops would be mostly contained in a two-dimensional plane whose orientation is approximately parallel to the reentry circuit. To corroborate this hypothesis, the planarity of each flutter loop was determined, as well as the orientation of the plane, described by the azimuth and elevation angles relative to the frontal plane. By analyzing the VCG synthesized from the 12-lead ECG,[5] recorded in patients before undergoing catheter ablation of atrial flutter, it was shown that flutter loops were mainly planar and had orientations concentrated to a narrow region of azimuth and elevation angles, likely corresponding anatomically with the expected flutter circuit. Atrial flutter waves in intervals without ventricular activity were analyzed on a wave-by-wave basis, i.e., each flutter wave was delineated manually.

This study laid the foundation for a number of studies investigating f wave loops [128–130], see also [131]. In contrast to flutter waves, f waves are less organized, and, therefore, spatial f wave analysis is more difficult to pursue. Spatial analysis can either be based on individual f waves in TQ intervals [128, 129] or an extracted signal containing multiple f waves. The latter case is preferable when low-amplitude f waves and noise, in combination with short TQ intervals, are to be analyzed [130]. Moreover, in the latter case, there is no need to delineate individual f waves, but a segment of the extracted signal can be analyzed. The data matrix is formed by the three orthogonal leads X, Y, and Z,

$$\mathbf{X} = \left[\mathbf{x}_{\mathrm{X}}\ \mathbf{x}_{\mathrm{Y}}\ \mathbf{x}_{\mathrm{Z}}\right]^T . \tag{6.82}$$

where each column vector, i.e., lead, contains N samples. Segment lengths of 1-s and 60-s were analyzed in [130].

The orientation of the plane-of-best-fit is defined as the two-dimensional projection of the loop producing the minimum MSE with respect to the original loop. The plane is determined from eigenanalysis of the sample correlation matrix of the data in $\mathbf{X}$, resulting in the three eigenvectors $\boldsymbol{\varphi}_1$, $\boldsymbol{\varphi}_2$, and $\boldsymbol{\varphi}_3$ associated with the eigenvalues $\lambda_1 \geq \lambda_2 \geq \lambda_3$. The eigenvector $\boldsymbol{\varphi}_1$ defines the principal axis, i.e., the axis with the largest correlation among the data, $\boldsymbol{\varphi}_2$ spans the plane-of-best-fit together with the principal axis, and $\boldsymbol{\varphi}_3 = \left[\varphi_{3,\mathrm{X}}, \varphi_{3,\mathrm{Y}}, \varphi_{3,\mathrm{Z}}\right]^T$ is the perpendicular axis which defines the azimuth and elevation angles of the plane-of-best-fit:

[5] The orthogonal leads X, Y, and Z can be synthesized from the 12-lead ECG using, for example, the inverse Dower matrix [126, 127], see also page 64.

$$\phi_{AZ} = \arctan\left(\frac{\varphi_{3,Z}}{\varphi_{3,X}}\right), \tag{6.83}$$

$$\phi_{EL} = \left|\arctan\left(\frac{\varphi_{3,Y}}{\sqrt{\varphi_{3,X}^2 + \varphi_{3,Z}^2}}\right)\right|, \tag{6.84}$$

where $-90° < \phi_{AZ} < 90°$ and $0 < \phi_{EL} < 90°$. Loop planarity is defined as [132]

$$\psi_{PL} = \frac{\lambda_3}{\lambda_1 + \lambda_2 + \lambda_3}, \tag{6.85}$$

which is close to zero when when the loop is essentially planar. Thus, the characterization of a segment containing several f waves embraces the three parameters ϕ_{AZ}, ϕ_{EL}, and ψ_{PL} [130].

Although the results from VCG loop analysis have had few implications on AF treatment, they have still provided certain qualitative information. Notably, varying degrees of organization have been observed, where the more organized cases have their plane-of-best-fit near the sagittal plane [128]. Moreover, a relatively weak coupling between loop morphology and the DAF was observed, suggesting that both these parameters may have a place in AF classification [130]. Analysis of the pseudo-VCG, defined by the leads V_5, aVF, and V_1, suggests that changes in loop morphology may be used to predict conversion from AF to atrial tachycardia, information which in turn may be used to establish when the therapy is on an effective path [133].

6.6.2 Body Surface Potential Mapping

Noninvasive, spatiotemporal analysis of electrical activation patterns may be performed on a body surface map constructed from a large number of leads which are placed on the anterior and posterior thorax. In the context of AF, such analysis was first considered in [134], with the overall aim of establishing whether single wavefronts as well as multiple simultaneous wavefronts, previously observed in intracardiac maps [135–137], could also be observed in body surface maps. Of the 56 leads, recorded during four minutes, 40 were arranged in matrix format on the anterior thorax and 16 on the posterior thorax. The traditional approaches to cardiac mapping of invasive data, i.e., *isopotential mapping* and *isochronal mapping*, were adopted for visualizing and analyzing cardiac activation [134]. The isopotential map displays the voltage for different electrode positions on the body surface at a given time instant, with contour lines connecting points of equal voltage. The isochronal map displays contour lines which connect points of equal activation time, often accompanied by one or several arrows to indicate the major propagation path. While it is straightforward to construct a isopotential map from the samples of the multi-lead f wave signal, the isochronal map requires that the activation time is determined for each electrode position. Each isochronal contour line is identified from the isopotential map as the line for which the voltage is equal to zero; to have a single representation of each activation wavefront, instead of having both forward and backward movement of the wavefront (i.e., atrial de- and repolarization), only points with a positive slope should be used for identification of the contour line. To improve spatial resolution, interpolation can be applied to the isopotential map, which in turn implies improved resolution of the isochronal map.

The information conveyed by noninvasive isochronal maps has been assessed qualitatively by classifying maps into the following three types [134], originally developed for electrogram-based analysis [138]: Type I (single wavefront), Type II (single wavefront with wave breakages and splitting), or Type III (multiple simultaneous wavefronts or none at all). On a data set consisting of 14 patients with persistent AF, all three types were represented, leading the authors to conclude that isochronal mapping has the potential to characterize activation patterns in AF. However, no comparison was made to invasively recorded activation maps. Figure 6.15 illustrates isopotential and isochronal maps, in both cases determined from a subinterval of an f wave.

Accurate identification of isochronal contour lines calls for high-quality signals, which in BSPM analysis implies the use of bandpass filtering to reduce the influence of baseline wander (particularly critical when finding the time for zero voltage) and myoelectrical noise. So far, TQ-based f wave analysis has been performed instead of f wave extraction to avoid the risk of analyzing QRS-related residuals [14, 80, 119, 134, 139]. Even when these precautions are taken, f wave amplitude may be so low that accurate determination of the activation times is not possible, especially for leads positioned far away from the atria. Since an isochronal map displays only one activation, variation in f wave amplitude and morphology may call for multiple maps, rendering the interpretation more complex [140].

Noninvasive *isofrequency mapping* in AF means the construction of a map displaying the spatial distribution of the DAF ("DAF map"), where the DAF is estimated in each lead using any of the techniques described in Sect. 6.3.1. Since the DAF map does not require the determination of activation times, its computation is much more straightforward. An important application of the DAF map is the identification of high-frequency sources which play an important role in the maintenance of AF [119]. Knowledge on the location of such sources are expected to improve the planning and outcome of ablation—an expectation supported by results obtained from invasive DAF maps showing that ablation guided by the identification of high-frequency sources increases the likelihood for long-term maintenance of sinus rhythm [141]. A comparison of the locations of the highest frequency source in the surface and invasive DAF maps, where the latter map served as the reference, demonstrated statistically significant correlation [119]. The agreement between these two types of DAF map is illustrated in Fig. 6.16, where the highest frequency source has similar location in both types of map.

Phase mapping is a tool particularly well-suited for characterizing temporal changes in spatial activation patterns in cardiac fibrillation, notably rotor activ-

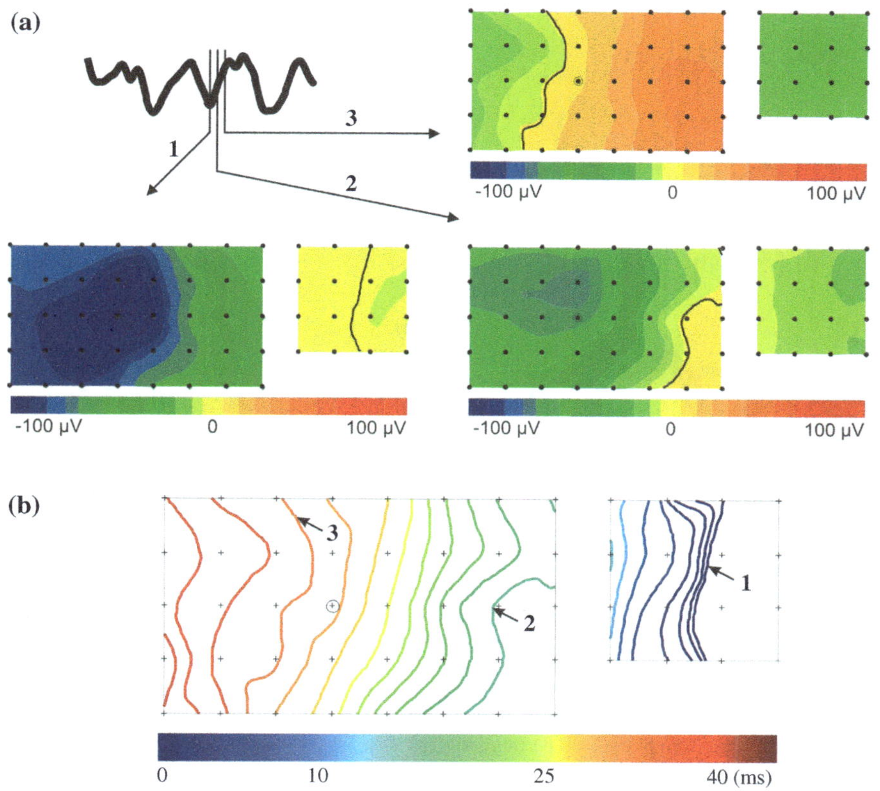

Fig. 6.15 **a** Isopotential maps obtained at three time instants of an f wave, using a 56-lead system for body surface potential mapping. Each isopotential map is composed of two submaps: one based on the anterior leads and another, smaller based on the posterior leads. The solid, black line in each map connects the points with zero voltage. **b** Isochronal map of the f wave in (**a**), where contour lines are drawn every 2 ms. Note that the three zero-voltage lines in (**a**) are also part of the isochronal map, indicated by the numbers 1, 2, and 3. (Reprinted from [134] with permission)

ity [142]. The term "rotor" refers to an activation wavefront circulating in an organized fashion around a center of rotation ("phase singularity point"). The engine in phase mapping is the Hilbert-based instantaneous phase computation, defined in (6.39), performed at regular time intervals in all the available leads to produce a time sequence of phase maps ("phase movie"). From this movie, the presence of a phase singularity point is identified as the site where the curved activation wavefront and wavetail of the rotor meet each other, i.e., a point where the phase of the rotating waves progresses through a complete cycle from $-\pi$ to π [142, 143], see also [144, 145].

Identification of phase singularities is important since they pinpoint where the tissue is capable of supporting rotors which drive AF. Hence, such points repre-

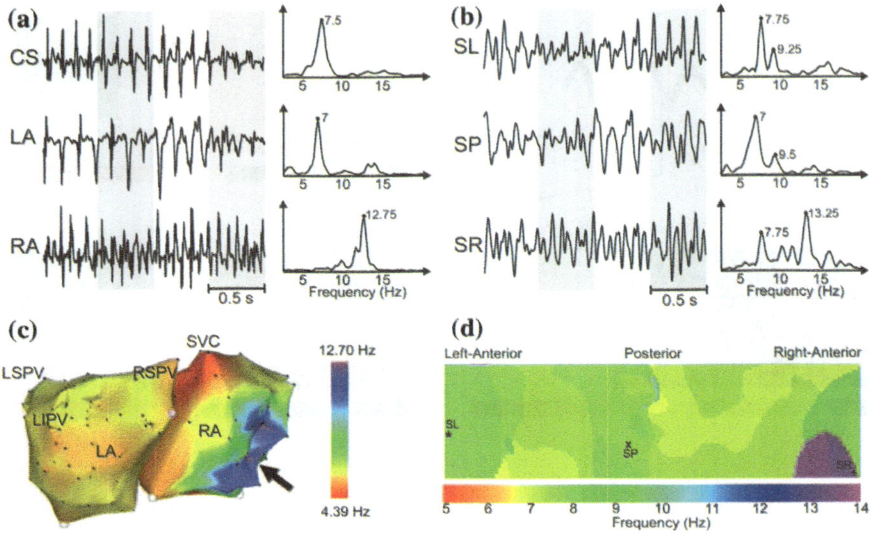

Fig. 6.16 **a** Electrograms recorded at different atrial sites and related power spectra, with the dominant atrial frequency (DAF) indicated, and **b** surface ECG leads and related power spectra. **c** Invasive DAF map obtained by electroanatomical mapping. The arrow points to the right atrial (RA) region with highest DAF. **d** Noninvasive DAF map with superimposed locations of the electrodes used in (**b**). The following acronyms are used: coronary sinus (CS), left atrial (LA), left inferior pulmonary vein (LIPV), left superior pulmonary vein (LSPV), right superior pulmonary vein (RSPV), surface left (SL), surface posterior (SP), surface right (SR), and superior vena cava (SVC). (Reprinted from [119] with permission)

sent potential targets for ablation. The significance of rotor-guided ablation has been studied in patients with persistent AF, mostly with promising results [146–148], although poor efficacy has also been reported [149]. In these studies, the instantaneous phase map was computed from intracardiac electrograms.

As noted in Sect. 6.4.1, stable, one-dimensional phase analysis requires that the f wave signal is bandpass filtered before phase computation—an operation which is equally needed in phase mapping. It has been demonstrated that bandpass filtering, with center frequency defined by the highest DAF of all available ECG leads, provides more accurate identification of phase singularity points than when bandpass filtering is omitted [14], see also [150]. By performing bandpass filtering, rotors were found to be more long-lasting, thereby facilitating the study of rotor characteristics such as trajectory, stability, and life span, and promoting atrial sites as potential targets for ablation.

The isopotential, isochronal, isofrequency, and phase maps have in common that they provide a basis for identification of features with electrophysiological interpretation. An overall approach to noninvasive BSPM analysis, disregarding map-specific features, is based on PCA of the temporal sequence of isopotential maps, proposed for quantifying *spatial complexity* of atrial wavefronts [80], see also [51]. In this approach, spatial complexity is linked to dimensionality reduction: a map which can be approximated by a few eigenvectors is considered less complex (more organized) than a map which requires several eigenvectors. The starting point for analysis is the $L \times N$ data matrix

$$\mathbf{X} = \left[\mathbf{x}(0)\ \mathbf{x}(1)\ \cdots\ \mathbf{x}(N-1)\right] \tag{6.86}$$

whose columns $\mathbf{x}(n)$ contain L leads at time n,

$$\mathbf{x}(n) = \begin{bmatrix} x_1(n) \\ x_2(n) \\ \vdots \\ x_L(n) \end{bmatrix}, \quad n = 0, \ldots, N-1, \tag{6.87}$$

where N is the number of samples subject to analysis. Each column $\mathbf{x}(n)$ contains a spatial map, and thus $\mathbf{X}$ contains the entire temporal sequence of maps. Each row of $\mathbf{X}$, i.e., $x_l(0), \ldots, x_l(N-1)$, contains the samples of successive, concatenated TQ intervals of the l-th lead.[6] The onset and end of each TQ interval is determined either by the intervals related to the occurrence times of the surrounding QRS complexes [80], or delineation of T wave end and QRS onset [139]. As already noted on page 155, the presence of f waves makes delineation challenging, especially when using a delineation algorithm not designed for, nor evaluated on, ECG signals in AF [139, 153].

The normalized cumulative sum R_K of the K largest eigenvalues λ_i, defined in (6.52), obtained from the sample correlation matrix of $\mathbf{X}$, cf. (5.106), provides a statistical measure of how well $\mathbf{X}$ is approximated by $\tilde{\mathbf{X}}$, obtained as a truncated series expansion of separable matrices resulting from SVD of $\mathbf{X}$,

$$\tilde{\mathbf{X}} = \sum_{k=1}^{K} \sigma_k \mathbf{u}_k \mathbf{v}_k^T, \tag{6.88}$$

where σ_k are the ordered singular values and $\mathbf{u}_k$ and $\mathbf{v}_k$ are the associated left and right singular vectors, respectively. Thus, for a fixed K, $\mathbf{X}$ is considered less complex when R_K is close to one, and vice versa; K was set to 3 in [80, 139]. Alternatively, K can be set to that value which makes R_K exceed 0.95 [80], and thus $K_{0.95}$ replaces R_K as the main information carrier; a larger $K_{0.95}$ implies higher spatial complexity. To smooth out the influence of temporal variation, R_3 and $K_{0.95}$ were computed in six consecutive 10-s segments and averaged.

---

[6] Principal component analysis of 180-lead isopotential maps, recorded in sinus rhythm, was pursued already in 1964, but then motivated by the completely different question "What is the minimum number of leads which can contain all of the electrocardiographic information available on the body surface?" [151], see also [152]. In those studies, the data matrix was defined by one single isopotential map, while $\mathbf{X}$ in (6.86) contains N maps. Thus, the former approach is purely spatial, while the approach in [80] may be labelled "spatiotemporal".

It should be emphasized that the approximation in (6.88) is identical to the one earlier encountered in (6.54). This is realized by forming a data matrix with the reconstructed signals $\hat{\mathbf{s}}_{p,K}$, i.e., $\tilde{\mathbf{X}} = \left[\hat{\mathbf{s}}_{1,K} \cdots \hat{\mathbf{s}}_{P,K}\right]$, so that (6.54) can be expressed as $\tilde{\mathbf{X}} = \boldsymbol{\Phi}_K \boldsymbol{\Phi}_K^T \mathbf{X}$. Since $\boldsymbol{\Phi} = \mathbf{U}$ and $\mathbf{X} = \mathbf{U}\boldsymbol{\Sigma}\mathbf{V}^T$, cf. page 184, then

$$\tilde{\mathbf{X}} = \mathbf{U}_K \mathbf{U}_K^T \mathbf{X} = \mathbf{U}_K \boldsymbol{\Sigma}_K \mathbf{V}_K^T = \sum_{k=1}^{K} \sigma_k \mathbf{u}_k \mathbf{v}_k^T. \tag{6.89}$$

For overall characterization of spatial complexity, the number of leads is not as critical as it is for the maps which offer an electrophysiological interpretation. Using PCA, this aspect was investigated by computing a complexity measure closely related to R_K for a 64-lead map, as well as for 32- and 10-lead maps, where the latter two maps were subsets of the 64-lead map. In particular, the 10-lead map was chosen such that it closely approximated the standard 12-lead ECG [139]. The results demonstrated that similar information can be derived from all three maps, suggesting that the standard 12-lead ECG is actually useful for determining spatial complexity.

6.7 f Wave Characterization in Clinical Applications

This section provides a brief overview of popular clinical applications, where f wave characteristics are explored with the goal of monitoring, detecting, or predicting changes in the atrial activity, either due to procedural intervention or spontaneous in origin. These applications, having emerged during the last decade, call for advances in methodological development as well as for further clinical studies to better establish the significance of f wave characteristics.

Whether monitoring, detection, or prediction is of interest, a single-parameter approach is usually pursued first, involving measurements from the lead with the most prominent f waves. The natural extension of this approach is to consider multi-lead measurements of a single parameter. In decision-oriented applications, for example, the prediction of catheter ablation outcome, a multi-parameter approach is likely to achieve better performance than a single-parameter approach. However, the more parameters involved in the decision-making, the larger needs the data set to be to adequately characterize performance.

6.7.1 *Monitoring of Drug Response*

The use of antiarrhythmic drugs is one of several approaches to long-term AF management which aims at restoring and maintaining sinus rhythm, an approach known as "rhythm-control therapy," cf. Sect. 1.8.3. Since antiarrhythmic drugs are moderately effective and may have serious side effects including life-threatening ventricular arrhythmias, it is important to develop ECG-based tests for quantifying the feasibility and dosage of a selected drug by monitoring various f wave characteristics. Such tests may also prove useful for drug development as they avoid the complexity of invasive electrophysiological testing, and offer a valuable complement to pharmacokinetic studies.

Dominant atrial frequency has been extensively studied for a great number of antiarrhythmic drugs designed to increase refractoriness and/or delay conduction of the atrial myocardium [154]. Most studies report on a substantial decrease in the DAF in patients responding to the drug [155–160]. This is a desirable result since a lower DAF usually means a more favorable outcome of rhythm-control therapy as it may lead to conversion to sinus rhythm. A decrease in the DAF is illustrated in Fig. 6.17 for an antiarrhythmic drug administered at several occasions during a time span of almost three days; the largest decrease in the DAF took place during the first day.

For a drug under development, administered to patients with persistent AF, the short-term dynamics of the DAF was studied using the spectral profile method [161]. The results showed that the "baseline" DAF, i.e., the DAF determined just before the time of the first drug administration, was not predictive of conversion to sinus rhythm. On the other hand, the decrease in the DAF was significantly more rapid in patients converting to sinus rhythm than in those not converting. A similar rapid decrease was observed in the harmonic decay and the standard deviation of the DAF, computed in 1-min intervals, suggesting that drug treatment increases AF organization, as reflected by more pronounced harmonics, and stabilizes the DAF.

So far, entropy and other nonlinear measures have not been considered for non-invasive monitoring and evaluation of drug response.

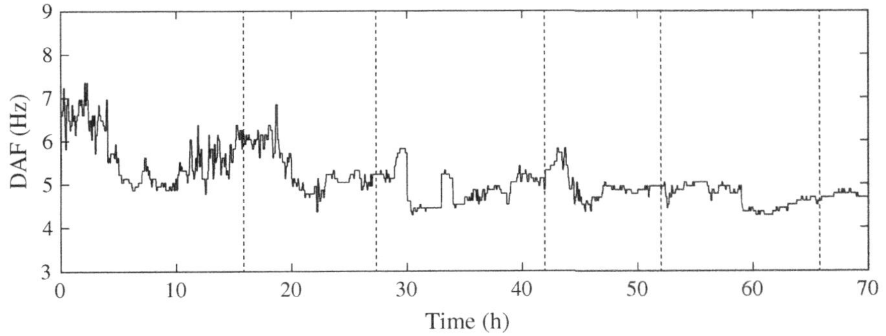

Fig. 6.17 Response of the dominant atrial frequency (DAF) to an antiarrhythmic drug (flecainide). The drug was administered at the onset of the recording and repeated after 16, 27, 42, 52, and 66 h (indicated by dashed lines)

6.7.2 Prediction of Catheter Ablation Outcome

Outcome prediction performed before catheter ablation can prevent unnecessary procedural risk in patients with low chance of successful AF termination [162]. Conversely, outcome prediction can be useful for selecting patients who require more aggressive ablation techniques than what is offered by catheter ablation. The significance of preoperative outcome prediction applies particularly to patients with persistent AF, since catheter ablation in patients with paroxysmal AF is associated with better success rate. The time span of prediction may differ from study to study: short-term prediction concerns successful AF termination in direct connection with catheter ablation, i.e., intraprocedural outcome [31, 163], while long-term prediction concerns maintenance of sinus rhythm a few months or longer following catheter ablation [19, 56, 164, 165]. Short-term prediction usually represents a simpler task than long-term prediction and is therefore associated with better performance—an observation which should be kept in mind when comparing the results of different studies on outcome prediction.

The significance of f wave amplitude in prediction of catheter ablation outcome has been investigated in patients with persistent AF [19, 31, 56, 163]. Clinical studies have shown that patients with lower f wave amplitude are less likely to benefit from catheter ablation [4, 5].[7] The lower amplitude may be related to a more disorganized (complex) form of AF, characterized by several activation wavefronts propagating in different directions which lead to wavefront collisions and a lower f wave amplitude.

Outcome prediction can be restricted to analyzing only the lead with the most prominent f waves, typically lead V_1, [56], or all available leads so that lead-dependent measurements can be produced [19, 31, 163]. In [19, 56], both addressing long-term prediction and applying traditional amplitude measures, i.e., peak-to-peak amplitude and spectral power $|\hat{S}_x(\hat{\omega}_0)|^2$, to the preoperative ECG, no statistically significant difference was found in f wave amplitude between terminating and non-terminating AF. Thus, these two studies, using automated amplitude measurements, do not support the results of the above-mentioned clinical studies [4, 5] which showed that a lower f wave amplitude is predictive of AF recurrence.

Alternatively, amplitude measurements can be derived from a PCA-based rank-one approximation of the data matrix containing the preoperative 12-lead ECG [31], cf. (6.88) with $K = 1$. The main reason for performing PCA-based dimensionality reduction is to retain the main f wave characteristics, while at the same time making amplitude measurements less sensitive to noise due to, for example, loosely attached electrodes. The envelope-based definition of f wave amplitude, illustrated in Fig. 6.1, is applied to the rank-one approximated data matrix. Using this approach in short-term prediction, f wave amplitude was found to differ significantly between terminating and nonterminating AF.

---

[7] It is somewhat remarkable that outcome prediction was based on manual f wave amplitude measurements in recent studies [4, 5], although algorithms for amplitude measurements have been available for many years.

Invasive studies have shown that a low DAF is predictive of long-term catheter ablation outcome in patients with persistent AF [166–168]. Similar results have been reported in noninvasive studies, where the DAF also differed significantly between terminating and nonterminating AF, either in lead V_1 [164] or in leads I, aVR, and V_5 [163]; however, no such difference was reported in [56]. Out of several spectral parameters, including the DAF, the position of the second harmonic, the harmonic decay, the spectral concentration, and the spectral power, it was only the harmonic decay that differed significantly between the two groups [56]. The results suggested that patients with more organized AF, reflected by more harmonics, are less likely to relapse to AF following catheter ablation.

Sample entropy could not predict AF termination, irrespective of whether DAF-controlled bandpass filtering was performed [163] or not [31]. Neither could spectral entropy predict AF termination [163].

While the results reported from single-parameter prediction may not be particularly striking, it has been noted that the performance of ECG-derived parameters to predict AF termination and long-term success of catheter ablation in patients with persistent AF is at least as good as that achieved by clinical parameters [163].

6.7.3 Prediction of Cardioversion Outcome

Electrical cardioversion is a well-established, noninvasive procedure with which AF is converted to sinus rhythm by delivering a high energy electrical shock, usually by placing two electrodes on the chest [169], cf. page 14. The shock is synchronized with the QRS complex to avoid delivery during ventricular repolarization, i.e., the T wave, which can induce ventricular fibrillation. Electrical cardioversion is usually accompanied by administration of an antiarrhythmic drug to increase the likelihood of conversion.

Unfortunately, as many as 35% of patients with persistent AF who undergo cardioversion relapse to AF, most of them within two weeks [170]. Consequently, in the same way as prediction of catheter ablation outcome can provide better selection of patients who will maintain sinus rhythm after ablation, prediction of cardioversion outcome can provide better selection of patients. From an engineering perspective, however, there is little difference between the problems of predicting catheter ablation and cardioversion outcome.

Early studies on ECG-based predictors in patients with persistent AF suggest that a lower DAF may be used as a long-term predictor of maintenance of sinus rhythm [7, 171]. Subsequent studies demonstrated the significance of a lower DAF for maintenance, especially when prediction was performed in AF of short duration [172] or when prediction was based on the DAF computed after an unsuccessful shock [173]. However, one study found the harmonic decay, being faster in patients relapsing to AF than in patients maintaining sinus rhythm, to be a more powerful predictor than the DAF, although the DAF was also a statistically significant predictor [174]. In all these studies, the spectral parameters were determined from the extracted f wave signal.

Rather than focusing on the dominant spectral peak, some studies have proposed predictive parameters for quantifying the spectral content of certain scales of the wavelet transform. Using the original ECG, rather than an extracted f wave signal or TQ intervals, the wavelet entropy was proposed as a predictor, computed from the scales containing 20–30 Hz components [175]. Using instead the extracted f wave signal, the sample entropy [103] and the central tendency [176], i.e., a measure describing the degree of signal variability, were computed from the scale containing the DAF and used as independent predictors. The results of these three studies showed that wavelet-based parameters may be used to predict maintenance of sinus rhythm following cardioversion.

6.7.4 Prediction of Spontaneous AF Termination

The question whether it is possible to predict spontaneous termination of an AF episode was highlighted to the engineering community in the PhysioNet/Computing in Cardiology Challenge in 2004 [177, 178]. As a result, several subsequent papers addressed this question using the AF Termination Database (AFTDB) which was made available for this challenge. Prediction of spontaneous termination relates to the hypothesis that subtle changes in f wave characteristics precede AF termination. With successful prediction, the parameters employed to characterize the f wave signal may help to explain why AF is terminating in certain individuals, but not in others. Such information may not only lead to more effective therapy, but also to avoidance of ineffective therapeutic intervention and reduced patient risk.

Early experimental studies, analyzing intracardiac electrograms, showed that prolongation of the DACL is a significant determinant of spontaneously terminating AF episodes in many patients [179, 180]. This result has been shown to carry over to the analysis of the surface ECG, where spontaneous termination is also preceded by a decrease in the DAF [59, 78, 178, 181]. The time course of the decrease before termination differs from study to study, where periods of about 5 to 10 minutes have been reported. In one study, a decrease in the DAF was only observed in patients who converted to sinus rhythm during morning hours, but not in those who converted in the afternoon or evening—results suggesting that electrophysiological mechanisms of termination may be different depending on the time of day [181]. For studies using the AFTDB, the decrease occurred immediately before spontaneous termination [78, 178].

Using parameters derived from the spectral profile, spontaneous termination in AFTDB was best predicted by a low DAF, a slow harmonic decay, and a stable DAF, while f wave amplitude, defined by (6.17), sample entropy, and spectral entropy could not discriminate between terminating and nonterminating AF [78]. Using parameters derived from the STFT, the DAF was, together with the average heart rate, the best-performing predictors [59], while f wave amplitude, defined by $|\hat{S}_x(\hat{\omega}_0)|^2$, and spectral width did not contribute to better prediction.

Introducing DAF-controlled bandpass filtering of the extracted f wave signal, a decrease in sample entropy was observed before termination [70]. Interestingly, the prediction performance achieved using sample entropy was identical to that achieved using the DAF [78], thus emphasizing the importance of prefiltering to reduce the sensitivity of sample entropy to noise. Similar prediction performance was achieved when the sample entropy was computed from a filtered f wave signal, obtained by reconstructing the signal from the wavelet coefficients of the scale containing the DAF [103, 104]. Wavelet decomposition was later considered for prediction of spontaneous AF termination [182], but then accompanied by computation of the wavelet entropy, defined by the Shannon entropy of the relative energies of the different scales, cf. page 207. However, other nonlinear parameters than entropy have been found to offer better prediction performance on AFTDB; for details, see [92].

6.7.5 Detection and Characterization of Circadian Variation

It is well-known that heart rate and blood pressure increase during daytime and decrease during night-time in healthy subjects. However, many other bodily functions also exhibit circadian variation. Information on circadian rhythms can help to establish proper timing of drug administration so that the effect of a drug can be maximized (chronotherapy) [183, 184]. The attenuation or absence of circadian variation may be indicative of certain risk conditions.

Circadian variation is driven by various external factors, e.g., sleep–wake routine, meal consumption, emotional state, and intrinsic activity of the autonomic nervous system. The latter type of activity is well-studied in the literature, with results demonstrating that sympathetic tone dominates during daytime activity, while vagal tone dominates during night-time sleep.

Detection and characterization of circadian variation usually involve a sinusoidal model which is fitted to the observed data using LS techniques [184, 185]. In this approach, the offset, commonly referred to as the "midline estimating statistic of rhythm" (MESOR), the amplitude, and the phase of the sinusoid, whose period is 24 h, constitute the model parameters. Detection can be based on a comparison of the MSE associated with two different models, namely 1. the MSE between the observed data and the non-circadian model defined by the MESOR only, and 2. the MSE between the observed data and the sinusoidal model. The most relevant of these two models is determined using a statistical test, for example, a paired bootstrap hypothesis test [77].

With respect to f wave characteristics and circadianity, the DAF was the first parameter to be investigated, determined every sixth hour from 24-h ambulatory recordings in patients with persistent AF [186]. A significant decrease in the DAF was observed at night, and an increase during the morning hours, reaching its maximum during the afternoon hours. To a large extent, these results were reproduced in subsequent studies on patients with persistent or permanent AF, although circadian variation was not detected in all patients [187, 188]. It has been pointed out that the short-term variation often observed in the DAF, uncovered by time–frequency analysis, may exceed the circadian variation, with implications on the accuracy of detecting circadian variation [188].

These studies share the limitation of a short recording duration, ranging from 15 to 24 h [186–188]. Hence, less than one sinusoidal period was available for parameter estimation, implying a large variance of the resulting estimates. To address this limitation, the DAF was studied on 7-day recordings in patients with persistent AF [77]. The results showed that the circadian variation detected in a 7-day recording was not always detected in all seven 24-h periods of the same recording, thus casting doubt on the validity of the conclusions made in [186–188]. In addition to the DAF, the eigenvalue-based parameter R_3, defined in (6.52), and the sample entropy were also studied. These parameters exhibited circadian variation although not always in the same patient.

References

1. A. Bollmann, D. Husser, L.T. Mainardi, F. Lombardi, P. Langley, A. Murray, J.J. Rieta, J. Millet, S.B. Olsson, M. Stridh, L. Sörnmo, Analysis of surface electrocardiograms in atrial fibrillation: techniques, research, and clinical applications. Europace **8**, 911–926 (2006)
2. T.A.R. Lankveld, S. Zeemering, H.J.G.M. Crijns, U. Schotten, The ECG as a tool to determine atrial fibrillation complexity. Heart **100**, 1077–1084 (2014)
3. P.G. Platonov, V.D.A. Corino, M. Seifert, F. Holmqvist, L. Sörnmo, Atrial fibrillatory rate in the clinical context: natural course and prediction of intervention outcome. Europace **16**, iv110–iv119 (2014)
4. Z. Cheng, H. Deng, K. Cheng, T. Chen, P. Gao, M. Yu, Q. Fang, The amplitude of fibrillatory waves on leads aVF and V1 predicting the recurrence of persistent atrial fibrillation patients who underwent catheter ablation. Ann. Noninvasive Electrocardiol. **18**, 352–358 (2013)
5. I. Nault, N. Lellouche, S. Matsuo, S. Knecht, M. Wright, K.T. Lim, F. Sacher, P. Platonov, A. Deplagne, P. Bordachar, N. Derval, M.D. O'Neill, G.J. Klein, M. Hocini, P. Jaïs, J. Clémenty, M. Haïssaguerre, Clinical value of fibrillatory wave amplitude on surface ECG in patients with persistent atrial fibrillation. J. Interv. Card. Electrophysiol. **26**, 11–19 (2009)
6. A. Bollmann, A. Tveit, D. Husser, M. Stridh, L. Sörnmo, P. Smith, S.B. Olsson, Fibrillatory rate response to candesartan in persistent atrial fibrillation. Europace **10**, 1138–1144 (2008)
7. A. Bollmann, D. Husser, R. Steinert, M. Stridh, L. Sörnmo, S.B. Olsson, D. Polywka, J. Molling, C. Geller, H.U. Klein, Echo- and electrocardiographic predictors for atrial fibrillation recurrence following cardioversion. J. Cardiovasc. Electrophysiol. **14**, 162–165 (2003)
8. D. Husser, M. Stridh, L. Sörnmo, P. Platanov, S.B. Olsson, A. Bollmann, Analysis of the surface electrocardiogram for monitoring and predicting antiarrhythmic drug effects in atrial fibrillation. Cardiovasc. Drugs Therapy **18**, 377–386 (2004)
9. M. Aunes-Jansson, N. Edvardsson, M. Stridh, L. Sörnmo, L. Frison, A. Berggren, Decrease of the atrial fibrillatory rate, increased organization of the atrial rhythm and termination of atrial fibrillation by AZD7009. J. Electrocardiol. **46**, 29–35 (2013)
10. S. Nattel, Atrial electrophysiological remodeling caused by rapid atrial activation: underlying mechanisms and clinical relevance to atrial fibrillation. Cardiovasc. Res. **42**, 298–308 (1999)
11. S. Petrutiu, A. Sahakian, S. Swiryn, Time domain analysis of atrial fibrillation, in *Understanding Atrial Fibrillation: The Signal Processing Contribution* ed. by L.T. Mainardi, L. Sörnmo, S. Cerutti (Morgan & Claypool, San Francisco, 2008), pp. 33–51 (Chap. 2)

12. R. Alcaraz, J.J. Rieta, A review on sample entropy applications for the non-invasive analysis of atrial fibrillation electrocardiograms. Biomed. Signal Process. Control **5**, 1–14 (2010)
13. R. Alcaraz, J.J. Rieta, Applications of nonlinear methods to atrial fibrillation, in *Complexity and Nonlinearity in Cardiovascular Signals*, ed. by R. Barbieri, E.P. Scilingo, G. Valenza (Springer, 2017), pp. 387–426
14. M. Rodrigo, M. de la Salud Guillem, A.M. Climent, J. Pedrón-Torrecilla, A. Liberos, J. Millet, F. Fernández-Avilés, F. Atienza, O. Berenfeld, Body surface localization of left and right atrial high-frequency rotors in atrial fibrillation patients: a clinical-computational study. Heart Rhythm **9**, 1584–1591 (2014)
15. V. Jacquemet, A. van Oosterom, J.-M. Vesin, L. Kappenberger, Analysis of electrocardiograms during atrial fibrillation: a biophysical approach. IEEE Med. Biol. Eng. Mag. **25**, 79–88 (2006)
16. V. Jacquemet, A. van Oosterom, Modeling atrial fibrillation: From myocardial cells to ECG, in *Understanding Atrial Fibrillation: The Signal Processing Contribution*, ed. by L.T. Mainardi, L. Sörnmo, S. Cerutti (Morgan & Claypool, San Francisco, 2008) Chap. 7, pp. 151–173
17. O.V. Aslanidi, M.A. Colman, J. Stott, H. Dobrzynski, M.R. Boyett, A.V. Holden, H. Zhang, 3D virtual human atria: a computational platform for studying clinical atrial fibrillation. Prog. Biophys. Mol. Biol. **107**, 156–168 (2011)
18. M. Meo, V. Zarzoso, O. Meste, D.G. Latcu, N. Saoudi, Catheter ablation outcome prediction in persistent atrial fibrillation using weighted principal component analysis. Biomed. Signal Process. Control **8**, 958–968 (2013)
19. L.Y. Di Marco, D. Raine, J.P. Bourke, P. Langley, Recurring patterns of atrial fibrillation in surface ECG predict restoration of sinus rhythm by catheter ablation. Comput. Biol. Med. **54**, 172–179 (2014)
20. R. Dubois, A.J. Shah, M. Hocini, A. Denis, N. Derval, H. Cochet, F. Sacher, L. Bear, J. Duchateau, P. Jaïs, M. Haïssaguerre, Non-invasive cardiac mapping in clinical practice: application to the ablation of cardiac arrhythmias. J. Electrocardiol. **48**, 966–974 (2015)
21. A. Shah, M. Hocini, M. Haïssaguerre, P. Jaïs, Non-invasive mapping of cardiac arrhythmias. Curr. Cardiol. Rep. **17**, 1–11 (2015)
22. M. Thurmann, J. Janney, The diagnostic importance of fibrillatory wave size. Circulation **25**, 991–994 (1962)
23. M.R. Culler, J.A. Boone, P.C. Gazes, Fibrillatory wave size as a clue to etiological diagnosis. Am. Heart J. **66**, 435–436 (1963)
24. H. Åberg, Atrial fibrillation II. A study of fibrillatory wave size on the regular scalar electrocardiogram. Acta Med. Scand. **185**, 381–385 (1969)
25. M.H. Aysha, A.S. Hassan, Diagnostic importance of fibrillatory wave amplitude: a clue to echocardiographic left atrial size and etiology of atrial fibrillation. J. Electrocardiol. **21**, 247–251 (1988)
26. J.L. Blackshear, R.E. Safford, L.A. Pearce on behalf of the stroke prevention in atrial fibrillation investigators, F-amplitude, left atrial appendage velocity, and thromboembolic risk in nonrheumatic atrial fibrillation. Clin. Cardiol. **19**, 309–313 (1996)
27. Q. Xi, A.V. Sahakian, J. Ng, S. Swiryn, Atrial fibrillatory wave characteristics on surface electrogram: ECG to ECG repeatability over twenty-four hours in clinically stable patients. J. Cardiovasc. Electrophysiol. **15**, 911–917 (2004)
28. Q. Xi, A.V. Sahakian, T.G. Frohlich, J. Ng, S. Swiryn, Relationship between pattern of occurrence of atrial fibrillation and surface electrocardiographic fibrillatory wave characteristics. Heart Rhythm **1**, 656–663 (2004)
29. R. Alcaraz, F. Hornero, J.J. Rieta, Noninvasive time and frequency predictors of long-standing atrial fibrillation early recurrence after electrical cardioversion. Pacing Clin. Electrophysiol. **34**, 1241–1250 (2011)
30. L. Sörnmo, P. Laguna, *Bioelectrical Signal Processing in Cardiac and Neurological Applications* (Elsevier (Academic Press), Amsterdam, 2005)
31. M. Meo, V. Zarzoso, O. Meste, D.G. Latcu, N. Saoudi, Spatial variability of the 12-lead surface ECG as a tool for noninvasive prediction of catheter ablation outcome in persistent atrial fibrillation. IEEE Trans. Biomed. Eng. **60**, 20–27 (2013)

32. V. Zarzoso, D.G. Latcub, A.R. Hidalgo-Muñoz, M. Meo, O. Meste, I. Popescu, N. Saoudi, Non-invasive prediction of catheter ablation outcome in persistent atrial fibrillation by fibrillatory wave amplitude computation in multiple electrocardiogram leads. Arch. Cardiovasc. Dis. **109**, 679–688 (2016)
33. N.E. Huang, Z. Shen, S.R. Long, M.C. Wu, H.H. Shi, Q. Zheng, N.-C. Yen, C.C. Tung, H.H. Liu, The empirical mode decomposition and the Hilbert spectrum for nonlinear and non-stationary time series analysis. Proc. R. Soc. Lond. A **454**, 903–995 (1998)
34. D.W. Botteron, J.M. Smith, A technique for measurement of the extent of spatial organization of atrial activation during atrial fibrillation in the intact human heart. IEEE Trans. Biomed. Eng. **42**, 579–586 (1995)
35. M. Holm, R. Johansson, S.B. Olsson, J. Brandt, C. Lührs, A new method for analysis of atrial activation during chronic atrial fibrillation in man. IEEE Trans. Biomed. Eng. **43**, 198–210 (1996)
36. L. Faes, G. Nollo, R. Antolini, F. Gaita, F. Ravelli, A method for quantifying atrial fibrillation organization based on wave morphology similarity. IEEE Trans. Biomed. Eng. **49**, 1504–1513 (2002)
37. R.P.M. Houben, N.M.S. de Groot, M.A. Allessie, Analysis of fractionated atrial fibrillation electrograms by wavelet decomposition. IEEE Trans. Biomed. Eng. **57**, 1388–1398 (2010)
38. J. Ng, V. Sehgal, J.K. Ng, D. Gordon, J.J. Goldberger, Iterative method to detect atrial activations and measure cycle length from electrograms during atrial fibrillation. IEEE Trans. Biomed. Eng. **61**, 273–278 (2014)
39. M. El Haddad, R.P.M. Houben, R. Stroobandt, F. Van Heuverswyn, R. Tavernier, M. Duytschaever, Novel algorithmic methods in mapping of atrial and ventricular tachycardia. Circ. Arrhythm. Electrophysiol. **7**, 463–472 (2014)
40. C.D. Cantwell, C.H. Roney, F.S. Ng, J.H. Siggers, S.J. Sherwin, N.S. Peters, Techniques for automated local activation time annotation and conduction velocity estimation in cardiac mapping. Comput. Biol. Med. **65**, 229–242 (2015)
41. K.M. Ropella, A.V. Sahakian, J.M. Baerman, S. Swiryn, Effects of procainamide on intra-atrial electrograms during atrial fibrillation: implications for detection algorithms. Circulation **77**, 1047–1054 (1988)
42. S. Lazar, S. Dixit, F.E. Marchlinski, D.J. Callans, E.P. Gerstenfeld, Presence of left-to-right atrial frequency gradient in paroxysmal but not persistent atrial fibrillation in humans. Circulation **110**, 3181–3186 (2004)
43. Y.J. Lin, C.T. Tai, T. Kao, H.W. Tso, S. Higa, H.M. Tsao, S.L. Chang, M.H. Hsieh, S.A. Chen, Frequency analysis in different types of paroxysmal atrial fibrillation. J. Am. Coll. Cardiol. **47**, 1401–1407 (2006)
44. M. Hayes, *Statistical Digital Signal Processing and Modeling* (Wiley, New York, 1996)
45. R. Sassi, L.T. Mainardi, P. Maison-Blanche, S. Cerruti, Estimation of spectral parameters of residual ECG signals during atrial fibrillation using autoregressive models. Folia Cardiologica **12**, 108–110 (2005)
46. R. Sassi, V.D.A. Corino, L.T. Mainardi, Analysis of surface atrial signals: time series with missing data? Ann. Biomed. Eng. **37**, 2082–2092 (2009)
47. S.M. Kay, *Fundamentals of Statistical Signal Processing. Estimation Theory* (Prentice-Hall, New Jersey, 1993)
48. N.R. Lomb, Least-squares frequency analysis of unequally spaced data. Astrophys. Space Sci. **39**, 447–462 (1976)
49. D.S. Stoffer, D.E. Tyler, D.A. Wendt, The spectral envelope and its applications. Stat. Sci. **15**, 224–253 (2000)
50. L. Uldry, J. Van Zaen, Y. Prudat, L. Kappenberger, J.-M. Vesin, Measures of spatiotemporal organization differentiate persistent from long-standing atrial fibrillation. Europace **14**, 1125–1131 (2012)
51. S. Zeemering, T.A.R. Lankveld, P. Bonizzi, I. Limantoro, S.C.A.M. Bekkers, H.J.G.M. Crijns, U. Schotten, The electrocardiogram as a predictor of successful pharmacological cardioversion and progression of atrial fibrillation. Europace, eux234 (2017)

52. T.H. Everett, L.-C. Kok, R.H. Vaughn, J.R. Moorman, D.E. Haines, Frequency domain algorithm for quantifying atrial fbrillation organization to increase defibrillation efficiency. IEEE Trans. Biomed. Eng. **48**, 969–978 (2001)
53. T.H. Everett, J.R. Moorman, L.-C. Kok, J.G. Akar, D.E. Haines, Assessment of global atrial fibrillation organization to optimize timing of atrial defibrillation. Circulation **103**, 2857–2861 (2001)
54. A. Buttu, E. Pruvot, J. Van Zaen, A. Viso, A. Forclaz, P. Pascale, S.M. Narayan, J. Vesin, Adaptive frequency tracking of the baseline ECG identifies the site of atrial fibrillation termination by catheter ablation. Biomed. Signal Process. Control **8**, 969–980 (2013)
55. M. Stridh, L. Sörnmo, C.J. Meurling, S.B. Olsson, Sequential characterization of atrial tachyarrhythmias based on ECG time–frequency analysis. IEEE Trans. Biomed. Eng. **51**, 100–114 (2004)
56. R. Alcaraz, F. Hornero, J.J. Rieta, Electrocardiographic spectral features for long-term outcome prognosis of atrial fibrillation catheter ablation. Ann. Biomed. Eng. **44**, 3307–3318 (2016)
57. I.A. Rezek, S.J. Roberts, Stochastic complexity measures for physiological signal analysis. IEEE Trans. Biomed. Eng. **45**, 1186–1191 (1998)
58. S. Pehrson, M. Holm, C. Meurling, M. Ingemansson, B. Smideberg, L. Sörnmo, S.B. Olsson, Non-invasive assessment of magnitude and dispersion of atrial cycle length during chronic atrial fibrillation in man. Eur. Heart J. **19**, 1836–1844 (1998)
59. F. Chiarugi, M. Varanini, F. Cantini, F. Conforti, G. Vrouchos, Noninvasive ECG as a tool for predicting termination of paroxysmal atrial fibrillation. IEEE Trans. Biomed. Eng. **54**, 1399–1406 (2007)
60. M. Stridh, L. Sörnmo, C.J. Meurling, S.B. Olsson, Characterization of atrial fibrillation using the surface ECG: time-dependent spectral properties. IEEE Trans. Biomed. Eng. **48**, 19–27 (2001)
61. C. Vayá, J.J. Rieta, Time and frequency series combination for non-invasive regularity analysis of atrial fibrillation. Med. Biol. Eng. Comput. **47**, 687–696 (2009)
62. S. Qian, D. Chen, *Joint Time–Frequency Analysis Methods and Applications* (Prentice-Hall, New Jersey, 1996)
63. L. Cohen, *Time–Frequency Analysis* (Prentice-Hall, New Jersey, 1995)
64. B. Boashash, Estimating and interpreting the instantaneous frequency of a signal–Part 2: algorithms and applications. Proc. IEEE **80**, 540–568 (1992)
65. M. Stridh, L. Sörnmo, C.J. Meurling, S.B. Olsson, Detection of autonomic modulation in permanent atrial fibrillation. Med. Biol. Eng. Comput. **41**, 625–629 (2003)
66. M. Malik, Standard measurements of heart rate variability, in *Dynamic Electrocardiography*, ed. by M. Malik, A.J. Camm (Wiley, New York, 2004), pp. 13–21 (Chap. 2)
67. B. Widrow, S.D. Stearns, *Adaptive Signal Proccessing* (Prentice-Hall, New Jersey, 1985)
68. S. Haykin, *Adaptive Filter Theory*, 5th edn. (Pearson, New Jersey, 2014)
69. J. Van Zaen, L. Uldry, C. Duchêne, Y. Prudat, R.A. Meuli, M.M. Murray, J.-M. Vesin, Adaptive tracking of EEG oscillations. J. Neurosci. Meth. **186**, 97–106 (2010)
70. R. Alcaraz, J.J. Rieta, Sample entropy of the main atrial wave predicts spontaneous termination of paroxysmal atrial fibrillation. Med. Eng. Phys. **31**, 917–922 (2009)
71. M. Stridh, D. Husser, A. Bollmann, L. Sörnmo, Waveform characterization of atrial fibrillation using phase information. IEEE Trans. Biomed. Eng. **56**, 1081–1089 (2009)
72. A.V. Oppenheim, R.W. Schafer, J.R. Buck, *Discrete-Time Signal Processing*, 2nd edn. (Prentice-Hall, New Jersey, 1999)
73. B. Picinbono, On instantaneous amplitude and phase of signals. IEEE Trans. Signal Process. **45**, 552–560 (1997)
74. M. Chavez, M. Besserve, C. Adam, J. Martinerie, Towards a proper estimation of phase synchronization from time series. J. Neurosci. Meth. **154**, 149–160 (2006)
75. L. Faes, G. Nollo, M. Kirchner, E. Olivetti, F. Gaita, R. Riccardi, R. Antolini, Principal component analysis and cluster analysis for measuring the local organisation of human atrial fibrillation. Med. Biol. Eng. Comput. **39**, 656–663 (2001)

76. S.G. Priori, D.W. Mortara, C. Napolitano, L. Diehl, V. Paganini, F. Cantù, G. Cantù, P.J. Schwartz, Evaluation of the spatial aspects of T-wave complexity in the long-QT syndrome. Circulation **96**, 3006–3012 (1997)
77. R. Goya-Esteban, F. Sandberg, Ó. Barquero-Pérez, A. García Alberola, L. Sörnmo, J.L. Rojo-Álvarez, Long-term characterization of persistent atrial fibrillation: wave morphology, frequency, and irregularity analysis. Med. Biol. Eng. Comput. **52**, 1053–1060 (2014)
78. F. Nilsson, M. Stridh, A. Bollmann, L. Sörnmo, Predicting spontaneous termination of atrial fibrillation using the surface ECG. Med. Eng. Phys. **26**, 802–808 (2006)
79. J. Behar, J. Oster, Q. Li, G.D. Clifford, ECG signal quality during arrhythmia and its application to false alarm reduction. IEEE Trans. Biomed. Eng. **60**, 1660–1666 (2013)
80. P. Bonizzi, M. de la Salud Guillem, A.M. Climent, J. Millet, V. Zarzoso, F. Castells, O. Meste, Noninvasive assessment of the complexity and stationarity of the atrial wavefront patterns during atrial fibrillation. IEEE Trans. Biomed. Eng. **57**, 2147–2157 (2010)
81. M.E. Nygårds, J. Hulting, An automated system for ECG monitoring. Comput. Biomed. Res. **12**, 181–202 (1979)
82. P. Bonizzi, S. Zeemering, J.M.H. Karel, L.Y. Di Marco, L. Uldry, J. Van Zaen, J.-M. Vesin, U. Schotten, Systematic comparison of non-invasive measures for the assessment of atrial fibrillation complexity: a step forward towards standardization of atrial fibrillation electrogram analysis. Europace **17**, 318–325 (2015)
83. M. Lagerholm, C. Peterson, G. Braccini, L. Edenbrandt, L. Sörnmo, Clustering ECG complexes using Hermite functions and self-organizing maps. IEEE Trans. Biomed. Eng. **47**, 838–848 (2000)
84. Z. Syed, J. Guttag, C. Stultz, Clustering and symbolic analysis of cardiovascular signals: Discovery and visualization of medically relevant patterns in long-term data using limited prior knowledge. J. Adv. Signal Process., 1–16 (2007)
85. J.H. Abawajy, A.V. Kelarev, M. Chowdhury, Multistage approach for clustering and classification of ECG data. Comput. Meth. Prog. Biomed. **112**, 720–730 (2013)
86. R. Alcaraz, F. Hornero, A. Martínez, J.J. Rieta, Short-time regularity assessment of fibrillatory waves from the surface ECG in atrial fibrillation. Physiol. Meas. **33**, 969–984 (2012)
87. P. Sun, Q.H. Wu, A.M. Weindling, A. Finkelstein, K. Ibrahim, An improved morphological approach to background normalization of ECG signals. IEEE Trans. Biomed. Eng. **50**, 117–121 (2003)
88. Y. Sun, K. Chan, S. Krishnan, Characteristic wave detection in ECG signal using morphological transform. BMC Cardiovasc. Disord. **5**, 28 (2005)
89. F. Zhang, L. Yong, QRS detection based on multiscale mathematical morphology for wearable ECG devices in body area networks. IEEE Trans. Biomed. Circuits Syst. **3**, 220–228 (2009)
90. J. Lian, G. Garner, D. Muessig, V. Lang, A simple method to quantify the morphological similarity between signals. Signal Process. **90**, 684–688 (2010)
91. W. Chen, J. Zhuang, W. Yu, Z. Wang, Measuring complexity using FuzzyEn, ApEn, and SampEn. Med. Eng. Phys. **31**, 61–68 (2009)
92. M. Julián, R. Alcaraz, J.J. Rieta, Comparative assessment of nonlinear metrics to quantify organization-related events in surface electrocardiograms of atrial fibrillation. Comput. Biol. Med. **48**, 66–76 (2014)
93. F. Ravelli, M. Masè, Computational mapping in atrial fibrillation: how the integration of signal-derived maps may guide the localization of critical sources. Europace **16**, 714–723 (2014)
94. Z. Liang, Y. Wang, X. Sun, D. Li, L.J. Voss, J.W. Sleigh, S. Hagihira, X. Li, EEG entropy measures in anesthesia. Front. Comput. Neurosci. **9**, 16 (2015)
95. W. Xiong, L. Faes, P.C. Ivanov, Entropy measures, entropy estimators, and their performance in quantifying complex dynamics: effects of artifacts, nonstationarity, and long-range correlations. Phys. Rev. E **95**, 062114 (2017)
96. C.E. Shannon, A mathematical theory of communication. Bell Sys. Tech. J. **27**, 379–423 (1948)

97. S.M. Pincus, Approximate entropy as a measure of system complexity. Proc. Natl. Acad. Sci. USA **88**, 2297–2301 (1991)
98. S.J. Richman, J.R. Moorman, Physiological time-series analysis using approximate entropy and sample entropy. Am. J. Physiol. **278**, H2039–H2049 (2000)
99. T. Inouye, K. Shinosaki, H. Sakamoto, S. Toi, S. Ukai, A. Iyama, Y. Katsuda, M. Hirano, Quantification of EEG irregularity by use of the entropy of the power spectrum. Electroencephal. Clin. Neurophysiol. **79**, 204–210 (1991)
100. O.A. Rosso, S. Blanco, J. Yordanova, V. Kolev, A. Figliola, M. Schürmann, E. Başar, Wavelet entropy: a new tool for analysis of short duration brain electrical signals. J. Neurosci. Meth. **105**, 65–75 (2001)
101. A. Porta, G. Baselli, D. Liberati, N. Montano, C. Cogliati, T. Gnecchi-Ruscone, A. Malliani, S. Cerutti, Measuring regularity by means of a corrected conditional entropy in sympathetic outflow. Biol. Cybern. **78**, 71–78 (1998)
102. A. Molina-Picó, D. Cuesta-Frau, M. Aboy, C. Crespo, P. Miró-Martínez, S. Oltra-Crespo, Comparative study of approximate entropy and sample entropy robustness to spikes. Artif. Intell. Med. **53**, 97–106 (2011)
103. R. Alcaraz, J.J. Rieta, A non-invasive method to predict electrical cardioversion outcome of persistent atrial fibrillation. Med. Biol. Eng. Comput. **46**, 625–635 (2008)
104. R. Alcaraz, J.J. Rieta, Wavelet bidomain sample entropy analysis to predict spontaneous termination of atrial fibrillation. Physiol. Meas. **29**, 65–80 (2008)
105. R. Alcaraz, F. Sandberg, L. Sörnmo, J.J. Rieta, Classification of paroxysmal and persistent atrial fibrillation in ambulatory ECG recordings. IEEE Trans. Biomed. Eng. **58**, 1441–1449 (2011)
106. D.T. Kaplan, M.I. Furman, S.M. Pincus, S.M. Ryan, L.A. Lipsitz, A.L. Goldberger, Aging and the complexity of cardiovascular dynamics. Biophys. J. **59**, 945–949 (1991)
107. S.M. Pincus, D.L. Keefe, Quantification of hormone pulsatility via an approximate entropy algorithm. Am. J. Physiol. **262**, E741–E754 (1992)
108. M.S. Pincus, A.L. Goldberger, Physiological time-series analysis: what does regularity quantify? Am. J. Physiol. (Heart Circ. Physiol.)**266**, H1643–H1656 (1994)
109. S. Lu, X. Chen, J.K. Kanters, I.C. Solomon, K.H. Chon, Automatic selection of the threshold value r for approximate entropy. IEEE Trans. Biomed. Eng. **55**, 1966–1972 (2008)
110. R. Alcaraz, D. Abásolo, R. Hornero, J.J. Rieta, Optimal parameters study for sample entropy-based atrial fibrillation organization analysis. Comput. Meth. Progr. Biomed. **99**, 124–132 (2010)
111. F. Liao, Y.-K. Jan, Using modified sample entropy to characterize aging-associated microvascular dysfunction. Front. Physiol. **7**, 126 (2016)
112. X. Chen, I.C. Solomon, K.H. Chon, Comparison of the use of approximate entropy and sample entropy: applications to neural respiratory signal. in *Proceedings of IEEE Conference on Engineering in Medicine and Biology (EMBS)*, vol. 27 (2005), pp. 4212–4215
113. J.M. Yentes, N. Hunt, K.K. Schmid, J.P. Kaipust, D. McGrath, N. Stergiou, The appropriate use of approximate entropy and sample entropy with short data sets. Ann. Biomed. Eng. **41**, 349–365 (2013)
114. G.D. Clifford, J. Behar, Q. Li, I. Rezek, Signal quality indices and data fusion for determining clinical acceptability of electrocardiograms. Physiol Meas. **33**, 1419–1433 (2012)
115. F. Sandberg, M. Stridh, L. Sörnmo, Robust time–frequency analysis of atrial fibrillation using hidden Markov models. IEEE Trans. Biomed. Eng. **55**, 502–511 (2008)
116. M. Henriksson, A. Petrėnas, V. Marozas, F. Sandberg, L. Sörnmo, Model-based assessment of f-wave signal quality in patients with atrial fibrillation. IEEE Trans. Biomed. Eng. (2018) (accepted)
117. P. Stoica, H. Li, J. Li, Amplitude estimation of sinusoidal signals: survey, new results, and an application. IEEE Trans. Signal Process. **48**, 338–352 (2000)
118. V.D.A. Corino, L.T. Mainardi, M. Stridh, L. Sörmno, Improved time–frequency analysis of atrial fibrillation signals using spectral modelling. IEEE Trans. Biomed. Eng. **56**, 2723–2730 (2008)

119. M.S. Guillem, A.M. Climent, J. Millet, Á. Arenal, F. Fernández-Avilés, J. Jalife, F. Atienza, O. Berenfeld, Noninvasive localization of maximal frequency sites of atrial fibrillation by body surface potential mapping. Circ. Arrhythm. Electrophysiol. **6**, 294–301 (2013)
120. Y. Rudy, J.E. Burnes, Noninvasive electrocardiographic imaging. Ann. Noninvasive Electrocardiol. **4**, 340–359 (1999)
121. P.S. Cuculich, Y. Wang, B.D. Lindsay, M.N. Faddis, R. Schuessler, R.J. Damiano Jr., L. Li, Y. Rudy, Noninvasive characterization of epicardial activation in humans with diverse atrial fibrillation patterns. Circulation **122**, 1364–1372 (2010)
122. M. Haïssaguerre, M. Hocini, A. Denis, A.J. Shah, Y. Komatsu, S. Yamashita, M. Daly, S. Amraoui, S. Zellerhoff, M.Q. Picat, A. Quotb, L. Jesel, H. Lim, S. Ploux, P. Bordachar, G. Attuel, V. Meillet, P. Ritter, N. Derval, F. Sacher, O. Bernus, H. Cochet, P. Jaïs, R. Dubois, Driver domains in persistent atrial fibrillation. Circulation **130**, 530–538 (2014)
123. Y. Rudy, B.D. Lindsay, Electrocardiographic imaging of heart rhythm disorders: from bench to bedside. Card. Electrophysiol. Clin. **7**, 17–35 (2015)
124. Z. Zhou, Q. Jin, L.Y. Chen, L. Yu, L. Wu, B. He, Noninvasive imaging of high-frequency drivers and reconstruction of global dominant frequency maps in patients with paroxysmal and persistent atrial fibrillation. IEEE Trans. Biomed. Eng. **63**, 1333–1340 (2016)
125. J. Ng, A.V. Sahakian, W.G. Fisher, S. Swiryn, Atrial flutter loops derived from the surface ECG: does the plane of the loop correspond anatomically to the macroreentrant circuit? J. Electrocardiol. **36**, S181–186 (2003)
126. G.E. Dower, H.B. Machado, J.A. Osborne, On deriving the electrocardiogram from vectorcardiographic leads. Clin. Cardiol. **3**, 87–95 (1980)
127. L. Edenbrandt, O. Pahlm, Vectorcardiogram synthesized from a 12-lead ECG: superiority of the inverse Dower matrix. J. Electrocardiol. **21**, 361–367 (1988)
128. J. Ng, A.V. Sahakian, W.G. Fisher, S. Swiryn, Surface ECG vector characteristics of organized and disorganized atrial activity during atrial fibrillation. J. Electrocardiol. **37**, 91–97 (2004)
129. S.R. Dibs, J. Ng, R. Arora, R.S. Passman, A.H. Kadish, J.J. Goldberger, Spatiotemporal characterization of atrial activation in persistent human atrial fibrillation: multisite electrogram analysis and surface electrocardiographic correlations–A pilot study. Heart Rhythm **5**, 686–693 (2008)
130. U. Richter, D. Husser, A. Bollmann, M. Stridh, L. Sörnmo, Spatial characteristics of atrial fibrillation electrocardiograms. J. Electrocardiol. **41**, 165–172 (2008)
131. S. Petrutiu, J. Ng, G.M. Nijm, H. Al-Angari, S. Swiryn, A.V. Sahakian, Atrial fibrillation and waveform characterization: a time domain perspective in the surface ECG. IEEE Eng. Med. Biol. Mag. **25**, 24–30 (2006)
132. F. Badilini, J. Fayn, P. Maison-Blanche, A. Leenhardt, M.C. Forlini, I. Denjoy, P. Coumel, P. Rubel, Quantitative aspects of ventricular repolarization: relationship between three-dimensional T wave loop morphology and scalar QT dispersion. Ann. Noninvasive Electrocardiol. **2**, 146–157 (1997)
133. T. Baykaner, R. Trikha, J.A.B. Zaman, D.E. Krummen, P.J. Wang, S.M. Narayan, Electrocardiographic spatial loops indicate organization of atrial fibrillation minutes before ablation-related transitions to atrial tachycardia. J. Electrocardiol. **50**, 307–315 (2017)
134. M.S. Guillem, A.M. Climent, F. Castells, D. Husser, J. Millet, A. Arya, C. Piorkowski, A. Bollmann, Noninvasive mapping of human atrial fibrillation. J. Cardiovasc. Electrophysiol. **20**, 507–513 (2009)
135. M. Mansour, R. Mandapati, O. Berenfeld, J. Chen, F. Samie, J. Jalife, Left-to-right gradient of atrial frequencies during acute atrial fibrillation in the isolated sheep heart. Circulation **103**, 2631–2636 (2001)
136. J. Jalife, O. Berenfeld, M. Mansour, Mother rotors and fibrillatory conduction: a mechanism of atrial fibrillation. Cardiovasc. Res. **54**, 204–216 (2002)
137. J. Jalife, Rotors and spiral waves in atrial fibrillation. J. Cardiovasc. Electrophysiol. **14**, 776–780 (2003)
138. K.T. Konings, C.J. Kirchhof, J.R. Smeets, H.J. Wellens, O.C. Penn, M.A. Allessie, High-density mapping of electrically induced atrial fibrillation in humans. Circulation **89**, 1665–1680 (1994)

139. L.Y. Di Marco, J.P. Bourke, P. Langley, Spatial complexity and spectral distribution variability of atrial activity in surface ECG recordings of atrial fibrillation. Med. Biol. Eng. Comput. **50**, 439–446 (2012)
140. J.M. Rogers, P.V. Bayly, Quantitative analysis of complex rhythms, in *Quantitative Cardiac Electrophysiology*, ed. by D. Rosenbaum (CRC Press, New York, 2002), pp. 403–428 (Chap. 12)
141. F. Atienza, J. Almendral, J. Jalife, S. Zlochiver, R. Ploutz-Snyder, E.G. Torrecilla, A. Arenal, J. Kalifa, F. Fernández-Avilés, O. Berenfeld, Real-time dominant frequency mapping and ablation of dominant frequency sites in atrial fibrillation with left-to-right frequency gradients predicts long-term maintenance of sinus rhythm. Heart Rhythm **6**, 33–40 (2009)
142. K. Umapathy, K. Nair, S. Masse, S. Krishnan, J. Rogers, M.P. Nash, K. Nanthakumar, Phase mapping of cardiac fibrillation. Circ. Arrhythm. Electrophysiol. **3**, 105–114 (2010)
143. S.V. Pandit, J. Jalife, Rotors and the dynamics of cardiac fibrillation. Circ. Res. **112**, 849–862 (2013)
144. P. Kuklik, S. Zeemering, B. Maesen, J. Maessen, H.J. Crijns, S. Verheule, A.N. Ganesan, U. Schotten, Reconstruction of instantaneous phase of unipolar atrial contact electrogram using a concept of sinusoidal recomposition and Hilbert transform. IEEE Trans. Biomed. Eng. **62**, 296–302 (2015)
145. R. Vijayakumar, S.K. Vasireddi, P.S. Cuculich, M.N. Faddis, Y. Rudy, Methodology considerations in phase mapping of human cardiac arrhythmias. Circ. Arrhythm. Electrophysiol. **9**, 1–11 (2016)
146. J.M. Miller, R.C. Kowal, V. Swarup, J.P. Daubert, E.G. Daoud, J.D. Day, K.A. Ellenbogen, J.D. Hummel, T. Baykaner, D.E. Krummen, S.M. Narayan, V.Y. Reddy, K. Shivkumar, J.S. Steinberg, K.R. Wheelan, Initial independent outcomes from focal impulse and rotor modulation ablation for atrial fibrillation: multicenter FIRM Registry. J. Cardiovasc. Electrophysiol. **25**, 921–929 (2014)
147. S.M. Narayan, T. Baykaner, P. Clopton, A. Schricker, G.G. Lalani, D.E. Krummen, K. Shivkumar, J.M. Miller, Ablation of rotor and focal sources reduces late recurrence of atrial fibrillation compared with trigger ablation alone. J. Am. Coll. Cardiol. **63**, 1761–1768 (2014)
148. D. Calvo, J. Rubín, D. Pérez, C. Morís, Ablation of rotor domains effectively modulates dynamics of human long-standing persistent atrial fibrillation. Circ. Arrhythm. Electrophysiol. **10** (2017)
149. E. Buch, M. Share, R. Tung, P. Benharash, P. Sharma, J. Koneru, R. Mandapati, K.A. Ellenbogen, K. Shivkumar, Long-term clinical outcomes of focal impulse and rotor modulation for treatment of atrial fibrillation: a multicenter experience. Heart Rhythm **13**, 636–641 (2016)
150. M. Rodrigo, A.M. Climent, A. Liberos, F. Fernández-Avilés, O. Berenfeld, F. Atienza, M.S. Guillem, Technical considerations on phase mapping for identification of atrial reentrant activity in direct- and inverse-computed electrograms. Circ. Arrhythm. Electrophysiol. **11** (2018)
151. L.G. Horan, N.C. Flowers, D.A. Brody, Principal factor waveforms of the thoracic QRS-complex. Circ. Res. **14**, 131–145 (1964)
152. A.M. Scher, A.C. Young, W.M. Meredith, Factor analysis of the electrocardiogram test of electrocardiographic theory: normal hearts. Circ. Res. **8**, 519–526 (1960)
153. L.Y. Di Marco, L. Chiari, A wavelet-based ECG delineation algorithm for 32-bit integer online processing. BioMed. Eng. Online **10**, 23 (2011)
154. M.C. Wijffels, R. Dorland, F. Mast, M.A. Allessie, Widening of the excitable gap during pharmacological cardioversion of atrial fibrillation in the goat: effects of cibenzoline, hydroquinidine, flecainide, and d-sotalol. Circulation **102**, 260–267 (2000)
155. M. Holm, S. Pehrsson, M. Ingemansson, L. Sörnmo, R. Johansson, L. Sandhall, M. Sunemark, B. Smideberg, C. Olsson, S.B. Olsson, Non-invasive assessment of atrial refractoriness during atrial fibrillation in man–Introducing, validating, and illustrating a new ECG method. Cardiovasc. Res. **38**, 69–81 (1998)
156. A. Bollmann, N. Kanuru, K. McTeague, P. Walter, D.B. DeLurgio, J. Langberg, Frequency analysis of human atrial fibrillation using the surface electrocardiogram and its response to ibutilide. Am. J. Cardiol. **81**, 1439–1445 (1998)

157. A. Fujiki, T. Tsuneda, M. Sugao, K. Mizumaki, H. Inoue, Usefulness and safety of bepridil in converting persistent atrial fibrillation to sinus rhythm. Am. J. Cardiol. **92**, 472–475 (2003)
158. D. Raine, P. Langley, A. Murray, A. Dunuwille, J.P. Bourke, Surface atrial frequency analysis in patients with atrial fibrillation: a tool for evaluating the effects of intervention. J. Cardiovasc. Electrophysiol. **15**, 1021–1026 (2004)
159. D. Husser, M. Stridh, L. Sörnmo, C. Geller, H.U. Klein, S.B. Olsson, A. Bollmann, Time-frequency analysis of the surface electrocardiogram for monitoring antiarrhythmic drug effects in atrial fibrillation. Am. J. Cardiol. **95**, 526–528 (2005)
160. V.P. Raygor, J. Ng, J.J. Goldberger, Surface ECG f wave analysis of dofetilide drug effect in the atrium. J. Cardiovasc. Electrophysiol. **26**, 644–648 (2015)
161. M. Aunes, K. Egstrup, L. Frison, A. Berggren, M. Stridh, L. Sörnmo, N. Edvardsson, Rapid slowing of the atrial fibrillatory rate after administration of AZD7009 predicts conversion of atrial fibrillation. J. Electrocardiol. **47**, 316–323 (2014)
162. E.P. Gerstenfeld, S. Duggirala, Atrial fibrillation ablation: indications, emerging techniques, and follow-up. Prog. Cardiovasc. Dis. **58**, 202–212 (2015)
163. T. Lankveld, S. Zeemering, D. Scherr, P. Kuklik, B.A. Hoffmann, S. Willems, B. Pieske, M. Haïssaguerre, P. Jaïs, H.J. Crijns, U. Schotten, Atrial fibrillation complexity parameters derived from surface ECGs predict procedural outcome and long-term follow-up of stepwise catheter ablation for atrial fibrillation. Circ. Arrhythm. Electrophysiol. **9**, e003354 (2016)
164. S. Matsuo, N. Lellouche, M. Wright, M. Bevilacqua, S. Knecht, I. Nault, K.T. Lim, L. Arantes, M.D. O'Neill, P.G. Platonov, J. Carlson, F. Sacher, M. Hocini, P. Jaïs, M. Haïssaguerre, Clinical predictors of termination and clinical outcome of catheter ablation for persistent atrial fibrillation. J. Am. Coll. Cardiol. **54**, 788–795 (2009)
165. K. Yoshida, A. Chugh, E. Good, T. Crawford, J. Myles, S. Veerareddy, S. Billakanty, W.S. Wong, M. Ebinger, F. Pelosi, K. Jongnarangsin, F. Bogun, F. Morady, H. Oral, A critical decrease in dominant frequency and clinical outcome after catheter ablation of persistent atrial fibrillation. Heart Rhythm **7**, 295–302 (2010)
166. I. Drewitz, S. Willems, T.V. Salukhe, D. Steven, B.A. Hoffmann, H. Servatius, K. Bock, M. Aydin, K. Wegscheider, T. Meinertz, T. Rostock, Atrial fibrillation cycle length is a sole independent predictor of a substrate for consecutive arrhythmias in patients with persistent atrial fibrillation. Circ. Arrhythm. Electrophysiol. **3**, 351–360 (2010)
167. E.K. Heist, F. Chalhoub, C. Barrett, S. Danik, J.N. Ruskin, M. Mansour, Predictors of atrial fibrillation termination and clinical success of catheter ablation of persistent atrial fibrillation. Am. J. Cardiol. **110**, 545–551 (2012)
168. L.Y. Di Marco, D. Raine, J.P. Bourke, P. Langley, Characteristics of atrial fibrillation cycle length predict restoration of sinus rhythm by catheter ablation. Heart Rhythm **10**, 1303–1310 (2013)
169. P. Kirchhof, S. Benussi, D. Kotecha, A. Ahlsson, D. Atar, B. Casadei, M. Castella, H.C. Diener, H. Heidbuchel, J. Hendriks, G. Hindricks, A.S. Manolis, J. Oldgren, B.A. Popescu, U. Schotten, B. Van Putte, P. Vardas, S. Agewall, J. Camm, G. Baron Esquivias, W. Budts, S. Carerj, F. Casselman, A. Coca, R. De Caterina, S. Deftereos, D. Dobrev, J.M. Ferro, G. Filippatos, D. Fitzsimons, B. Gorenek, M. Guenoun, S.H. Hohnloser, P. Kolh, G.Y. Lip, A. Manolis, J. McMurray, P. Ponikowski, R. Rosenhek, F. Ruschitzka, I. Savelieva, S. Sharma, P. Suwalski, J.L. Tamargo, C.J. Taylor, I.C. Van Gelder, A.A. Voors, S. Windecker, J.L. Zamorano, K. Zeppenfeld, 2016 ESC guidelines for the management of atrial fibrillation developed in collaboration with EACTS. Eur. Heart J. **37**, 2893–2962 (2016)
170. N.P. Gall, F.D. Murgatroyd, Electrical cardioversion for AF–the state of the art. Pacing Clin. Electrophysiol. **30**, 554–567 (2007)
171. A. Bollmann, M. Mende, A. Neugebauer, D. Pfeiffer, Atrial fibrillatory frequency predicts atrial defibrillation threshold and early arrhythmia recurrence in patients undergoing internal cardioversion of persistent atrial fibrillation. Pacing Clin. Electrophysiol. **25**, 1179–1184 (2002)
172. F. Holmqvist, M. Stridh, J.E. Waktare, L. Sörnmo, S.B. Olsson, C.J. Meurling, Atrial fibrillatory rate and sinus rhythm maintenance in patients undergoing cardioversion of persistent atrial fibrillation. Eur. Heart J. **27**, 2201–2207 (2006)

173. R. Alcaraz, J.J. Rieta, F. Hornero, Analysis of atrial fibrillation recidivity under successive attempts of electrical cardioversion based on fibrillatory wave amplitude and dominant frequency. J. Med. Biol. Eng. **33**, 455–462 (2013)
174. F. Holmqvist, M. Stridh, J.E.P. Waktare, L. Sörnmo, S.B. Olsson, C.J. Meurling, Atrial fibrillation signal organization predicts sinus rhythm maintenance in patients undergoing cardioversion of atrial fibrillation. Europace **8**, 559-565 (2006)
175. J.N. Watson, P.S. Addison, N. Uchaipichat, A. Shah, N.R. Grubb, Wavelet transform analysis predicts outcome of DC cardioversion for atrial fibrillation patients. Comput. Biol. Med. **37**, 517–523 (2007)
176. R. Alcaraz, J.J. Rieta, Central tendency measure and wavelet transform combined in the non-invasive analysis of atrial fibrillation recordings. Biomed. Eng. Online **11**, 46 (2012)
177. G.B. Moody, Spontaneous termination of atrial fibrillation: a challenge from Physionet and Computers in Cardiology 2004, in *Proceedings of Computers in Cardiology*, vol. 31 (2004), pp. 101–104
178. S. Petrutiu, A.V. Sahakian, S. Swiryn, Abrupt changes in fibrillatory wave characteristics at the termination of paroxysmal atrial fibrillation in humans. Europace **9**, 466–470 (2007)
179. Y. Asano, J. Saito, K. Matsumoto, K. Kaneko, T. Yamamoto, U. Masatsugu, On the mechanism of termination and perpetuation of atrial fibrillation. Am. J. Cardiol. **69**, 1033–1038 (1992)
180. H. Sih, K. Ropella, S. Swiryn, E. Gerstenfeld, A.V. Sahakian, Observations from intra-atrial recordings on the termination of electrically induced atrial fibrillation in humans. Pacing Clin. Electrophysiol. **17**, 1231–1242 (1994)
181. A. Fujiki, M. Sakabe, K. Nishida, K. Mizumaki, H. Inoue, Role of fibrillation cycle length in spontaneous and drug-indcued termination of human atrial fibrillation–Spectral analysis of fibrillation waves from surface electrocardiogram. Circ. J. **67**, 391–395 (2003)
182. R. Alcaraz, J.J. Rieta, Application of wavelet entropy to predict atrial fibrillation progression from the surface ECG. Comput. Math. Meth. Med. **13**, 1–9 (2012)
183. R. Hermida, D. Ayala, F. Portaluppi, Circadian variation of blood pressure: the basis for the chronotherapy of hypertension. Adv. Drug Delivery Rev **59**, 904–922 (2007)
184. J.R. Fernández, R.C. Hermida, A. Mojón, Chronobiological analysis techniques: application to blood pressure. Phil. Trans. R. Soc. A **367**, 431–445 (2009)
185. C. Bingham, B. Arbogast, C.C. Guillaume, J.K. Lee, F. Halberg, Inferential statistical methods for estimating and comparing cosinor parameters. Chronobiologia **9**, 397–439 (1982)
186. A. Bollmann, K. Sonne, H. Esperer, I. Toepffer, H. Klein, Circadian variations in atrial fibrillatory frequency in persistent human atrial fibrillation. Pacing Clin. Electrophysiol. **23**, 1867–1871 (2000)
187. C.J. Meurling, J.E. Waktare, F. Holmqvist, A. Hedman, A.J. Camm, S.B. Olsson, M. Malik, Diurnal variations of the dominant cycle length of chronic atrial fibrillation. Am. J. Physiol. **280**, H401–H406 (2001)
188. F. Sandberg, A. Bollmann, D. Husser, M. Stridh, L. Sörnmo, Circadian variation in dominant atrial fibrillation frequency in persistent atrial fibrillation. Physiol. Meas. **31**, 531–542 (2010)

Chapter 7
Modeling and Analysis of Ventricular Response in Atrial Fibrillation

Valentina D. A. Corino, Frida Sandberg, Luca T. Mainardi and Leif Sörnmo

7.1 Introduction

The ventricular response in atrial fibrillation (AF) is highly irregular, mainly due to the atrial impulses arriving irregularly at the atrioventricular (AV) node. As a result, the RR intervals differ dramatically in length. Despite the irregularity, the ventricular response is not completely random, but exhibits weak correlation [1] and certain short- or long-term predictability [2]. Another characteristic is that the ventricular rate is often higher in AF than in normal sinus rhythm, a characteristic explored in AF detection, see Chap. 4. The RR interval series observed in normal sinus rhythm and AF differ with respect to both variability and irregularity, two aspects which are illustrated in Fig. 7.1.

The ventricular response plays a significant role in the management of patients with AF [3]. In fact, the control of ventricular rate effectively reduces the risk of complication and improves the quality of life. Therefore, the study of factors influencing ventricular rate and its dynamics is of great importance as it may lead to strengthened decision-making in AF management.

To characterize normal sinus rhythm, a wide range of parameters have been investigated, often categorized into dispersion parameters to characterize RR variability,

V. D. A. Corino · L. T. Mainardi
Department of Electronics, Information and Bioengineering, Politecnico di Milano, Milan, Italy
e-mail: valentina.corino@polimi.it

L. T. Mainardi
e-mail: luca.mainardi@polimi.it

F. Sandberg · L. Sörnmo (✉)
Department of Biomedical Engineering and Center for Integrative Electrocardiology, Lund University, Lund, Sweden
e-mail: leif.sornmo@bme.lth.se

F. Sandberg
e-mail: frida.sandberg@bme.lth.se

L. Sörnmo (ed.), *Atrial Fibrillation from an Engineering Perspective*,
Series in BioEngineering, https://doi.org/10.1007/978-3-319-68515-1_7

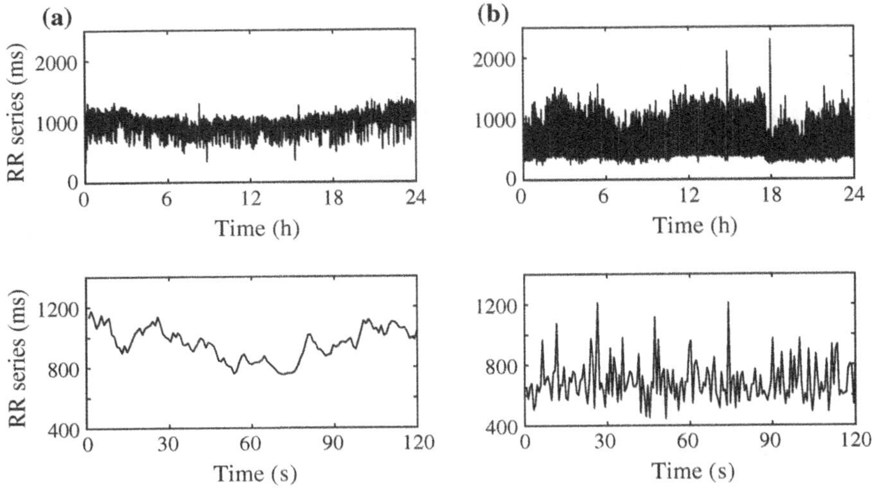

Fig. 7.1 24-h RR interval series in **a** normal sinus rhythm and **b** AF. The 24-h plots (top row) show that the dispersion is much larger in AF than in normal sinus rhythm. The zoomed-in segments (bottom row) demonstrate that the RR interval series in AF not only has larger dispersion, but it is also much more irregular

spectral parameters to characterize autonomic influence on the sinus node, and different types of entropy to characterize RR irregularity. To characterize AF, spectral parameters have received very limited interest since the RR interval spectrum is essentially flat, and lacks peaks which may carry physiological information [4]. On the other hand, dispersion parameters and entropy measures have conveyed clinically valuable information: for example, lower variability and/or irregularity of the RR interval series have been associated with poor outcome in patients with AF [5, 6]. Given the growing clinical interest to understand the characteristics of the RR intervals, an overview of results reported in clinical studies is provided in Sect. 7.2.

The analysis of ventricular response can be augmented with information on f waves so that the coupling between the atria and the ventricles, through the AV node, can be investigated. The AV node plays a particularly important role in AF by acting as a "filter" which blocks certain atrial impulses, with repercussions on ventricular activation. By developing methods for analyzing AV nodal properties, patient-specific information may be obtained which describes the effect of a certain antiarrhythmic drug. The properties can be studied by means of mathematical modeling, considered either for simulation of various scenarios or estimation of model parameters. In the latter case, the observed signal is acquired either invasively or from the surface ECG. Signal processing techniques are usually required to separate the atrial from the ventricular activity before parameter estimation can be performed, see Chap. 5.

Section 7.3 provides a brief overview of methods for heuristic assessment of the AV node. Section 7.4 describes a method for analyzing the relationship between atrial input and ventricular response during AF. Sections 7.5 and 7.6 review several AV node models for simulation and parameter estimation, respectively. The chapter concludes with a comparison of AV node models in Sect. 7.7.

7.2 RR Interval Analysis

Classical dispersion parameters such as the coefficient of variation and the root mean square of successive differences (RMSSD), defined in (4.2) and (4.3), respectively, have been found useful for characterizing the variability of RR intervals in AF [2, 7].[1] However, variability parameters provide an incomplete characterization of RR intervals, since they cannot characterize irregularity, i.e., the degree of unpredictability. Therefore, variability parameters have been complemented with different entropy measures to characterize irregularity, including approximate entropy I_{ApEn}, sample entropy I_{SampEn}, and Shannon entropy I_{ShEn} (see Sect. 4.2.1 for definitions), as well as conditional entropy [8]. Figure 7.2 illustrates the difference between variability

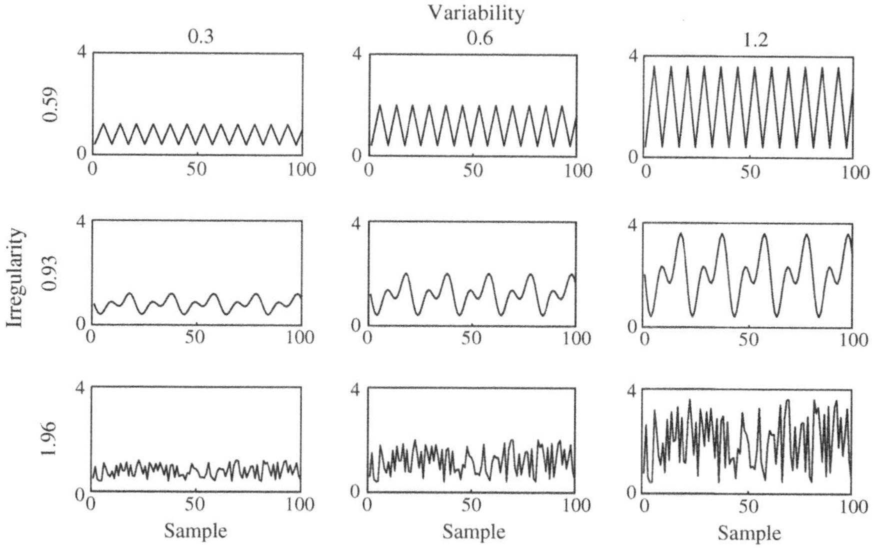

Fig. 7.2 Illustration of the difference between variability, quantified by the standard deviation, and irregularity, quantified by the sample entropy. Each row shows a time series with identical irregularity (given by the numbers to the left of the diagrams), but increasing variability from left to right, whereas each column shows series with identical variability (given by the numbers on above the diagrams), but increasing irregularity from top to bottom. The units of the horizontal and vertical axes are arbitrary

[1]The reason for not using the term "ventricular response" in this section is that it implies, at least in this book, that an atrial input is also part of the analysis.

and irregularity for different time series with identical variability but different irregularity, and vice versa.

Two early studies analyzed the RR interval series in patients with AF, demonstrating that reduced variability is associated with worse outcome [5, 6]. Using the 24-h ambulatory ECG, reduced RR interval irregularity was found to have independent prognostic value for cardiac mortality during long-term follow-up in patients with permanent AF [9].

The association between RR intervals and long-term clinical outcome has been evaluated in a population of ambulatory patients with mild-to-moderate heart failure and AF at baseline. Patients with symptomatic heart failure were enrolled in a multicenter study on sudden death [10]. Both I_{ApEn} and I_{ShEn} were found to be significantly lower in nonsurvivors than in survivors for all subgroups of death (total mortality, sudden death, and heart failure death). Patients with a lower I_{ApEn} had significantly lower survival: Kaplan–Meier analysis [11] showed that a lower I_{ApEn} was associated with a nearly fourfold higher total mortality (40% vs. 12%) and more than six times higher mortality due to progression of heart failure (19% vs. 3%) and sudden death (18 vs. 3%). The criterion $I_{\mathrm{ApEn}} < 1.68$, where 1.68 is the lower tertile of the data set, was found to be a significant predictor of all types of mortality after adjustment for significant clinical covariates [12], leading up to the main finding that lower irregularity is associated with worse outcome in AF patients.

Results in the literature suggest that irregularity parameters may be used as risk indicators. Thus, it is of interest to investigate to what extent irregularity is affected by commonly used rate-control drugs. The effect of the selective A1-receptor agonist tecadenoson, alone as well as in combination with the beta blocker esmolol, was assessed in a small group of AF patients [13]. Tecadenoson was found to reduce heart rate and increase variability, but did not have any effect on irregularity. Beta blockade with intravenous esmolol further increased variability and decreased heart rate. In another study [14], no significant differences in RR irregularity, as quantified by I_{ApEn}, were observed in patients with AF and congestive heart failure when treated with beta blockers, digoxin, or amiodarone. The effect of rate-control drugs on RR variability/irregularity was investigated in 60 patients with permanent AF, involving the drugs diltiazem, verapamil (both calcium channel blockers), metoprolol, and carvedilol (beta blocker) [15]. Variability was assessed by well-known parameters such as the standard deviation and the RMSSD, whereas irregularity was assessed by I_{ApEn}, I_{ShEn}, and a measure based on conditional entropy [16]. A significantly lower heart rate was obtained for all investigated drugs, reaching its lowest rate for the calcium channel blockers. Moreover, all drugs were found to increase RR variability significantly relative to the baseline recording, whereas only the beta blockers increased RR irregularity significantly.

Using the data set in [15], circadian variation was investigated by means of five ambulatory recordings per patient, obtained at baseline as well as during the four different drug regimens [17]. Variability and irregularity parameters were computed in nonoverlapping, 20-min segments. Circadianity was assessed using cosinor analysis of the resulting series, characterized by the 24-h mean and the excursion over the mean described by the amplitude of the cosine fitted to the data [18]. Heart rate and variability parameters, including the standard deviation and the RMSSD, exhibited significant circadian variation in most patients, whereas circadian variation in I_{ApEn} and I_{SampEn} was detected in only a few patients. When circadian variation was detected in I_{ApEn} at baseline, the patients had more severe symptoms. All drugs decreased the rhythm-adjusted mean of the heart rate and increased the rhythm-adjusted mean of variability parameters (the rhythm-adjusted mean is also referred to as "midline estimating statistic of rhythm", MESOR [19]). Only carvedilol and metoprolol decreased the normalized amplitude over the 24 h of the irregularity parameters and heart rate. The results suggested that circadian variation can be observed in most patients using variability parameters, but only in a few patients using irregularity parameters.

The above-mentioned clinical studies are limited by small patient groups. Therefore, further studies are needed to better assess whether variability and irregularity parameters are predictive of patient status.

7.3 Heuristic Assessment of the Atrioventricular Node

Rate-control drugs act on atrial and/or AV nodal properties to lower the ventricular rate. During drug development, electrophysiological effects of antiarrhythmic drugs are usually assessed invasively in sinus rhythm. Since an atrial pacing protocol cannot be applied in patients with AF, the electrophysiological drug effects on the AV node are still not completely understood. When optimizing drug therapy, noninvasive assessment of AV nodal electrophysiology may help to select optimal therapy. During the early clinical phases of drug development, noninvasive characterization of the AV node may facilitate data collection from large patient cohorts and favor patient-tailored therapy. Estimation of the AV nodal refractory period using the surface ECG has been attempted in several studies, employing different heuristic approaches [20–24].

Heuristic assessment of AV nodal electrophysiology has relied on simple approaches to characterizing the RR intervals. Noninvasive estimation of the functional refractory period of the AV node during AF has been attempted by simply selecting the shortest RR interval or the 5-th percentile of the RR interval series [20, 23, 25]. In dogs, it was demonstrated that the shortest RR interval correlated statistically with the functional refractory period, determined using a pacing protocol. Therefore, the shortest RR interval was used as a surrogate measurement of the functional refractory period [20].

Using the Poincaré plot, where each RR interval is plotted against the preceding RR interval, an estimate of the functional refractory period can be obtained as well. The value of the lower envelope, determined as a regression line, at 1 s ("1 s intercept") and the degree of scatter above the lower envelope have been proposed as surrogate measurements of AV nodal refractoriness and concealed AV conduction, i.e., the effect of blocked impulses on the conduction of subsequent impulses, respectively. The circadian variation of the 1-s intercept of the lower envelope was investigated in 120 patients who underwent 24-h ambulatory monitoring at baseline [24]. During an observation period of 33 ± 16 months, there were 25 deaths, including 13 cardiac and 8 stroke deaths. All patients showed significant circadian rhythms in the lower envelope, however, patients dying subsequently from cardiac causes, but not from fatal stroke, had less pronounced circadian rhythm, with amplitudes which were less than 55% of those in surviving patients. It was suggested that blunted circadian rhythm of AV conduction represents an independent risk of cardiac death in patients with permanent AF.

The presence of clusters in the histogram-based Poincaré plot, based on the RR intervals derived from the 24-h ambulatory ECG, has been suggested as a marker of higher AF organization to predict the outcome of electrical cardioversion [26]. A cluster was considered to be present when a peak in the histogram plot could be identified visually. Later, the histogram-based Poincaré plot served as the basis for the Poincaré surface profile, i.e., a univariate histogram defined by those RR intervals which are preceded by RR intervals of approximately the same length [27], cf. Sect. 4.2.2. The Poincaré surface profile was proposed as a tool for characterizing AV nodal memory effect and detecting preferential AV nodal conduction. However, neither the Poincaré plot nor the histogram-based Poincaré plot analysis have raised much interest in the research community. This may be due to a number of reasons, including that the plot is strongly dependent on bin size, the bins must be sufficiently well-populated with points to produce meaningful results, and manual interaction is often needed to determine the lower envelope [24].

7.4 Synchrogram Analysis

Synchrogram analysis has been introduced for exploratory analysis of the relationship between atrial input and ventricular response during AF, providing valuable insights into AV nodal function [28]. The method is purely phenomenological, and no attempt is made to account for AV nodal electrophysiological properties such as refractoriness and conduction delay. The analysis is applied to atrial activations, determined from the electrogram, and ventricular activations, determined from the ECG, to analyze AV coupling. The analysis is performed by observing the phase of the ventricular activations at time instants triggered by the atrial activations. The instantaneous ventricular phase is assumed to be a monotonically increasing, piecewise linear function, defined by

$$\phi_v(t) = 2\pi \frac{t - t_{v,n-1}}{t_{v,n} - t_{v,n-1}} + 2\pi n, \quad t_{v,n-1} \le t < t_{v,n}, \quad n = 0, \ldots, N-1, \qquad (7.1)$$

where $t_{v,n}$ is the time of n-th ventricular activation and N is the total number of ventricular activations. To be consistent with the indexing of RR intervals adopted in Chap. 4, the first RR interval, defined by $x_0 = t_{v,0} - t_{v,-1}$, requires that the time of the first ventricular activation is indexed by -1.

The instantaneous ventricular phase is normalized to the interval $[0, q]$, and sampled at the time of atrial activations $t_{a,k}$ for the purpose of detecting whether $p:q$ coupling is present,

$$\Psi_q(t_{a,k}) = \frac{1}{2\pi}\left(\phi_v(t_{a,k}) \bmod (2\pi q)\right), \quad q = 1, 2, 3, \tag{7.2}$$

where p is the number of atrial activations and q is the number of ventricular activations. Epochs of synchronization are automatically detected by alternately dividing the values of $\Psi_q(t_{a,k})$ into different subgroups. The normalized phases are classified as $p : q$ coupling whenever the absolute difference between $\Psi_q(t_{a,k+p})$ and $\Psi_q(t_{a,k})$ within a subgroup is below a predefined tolerance threshold. The synchrogram analysis is illustrated in Fig. 7.3.

The synchrogram was investigated in both atrial flutter and AF. As expected, the percentage of coupled beats and the duration of coupled epochs were significantly

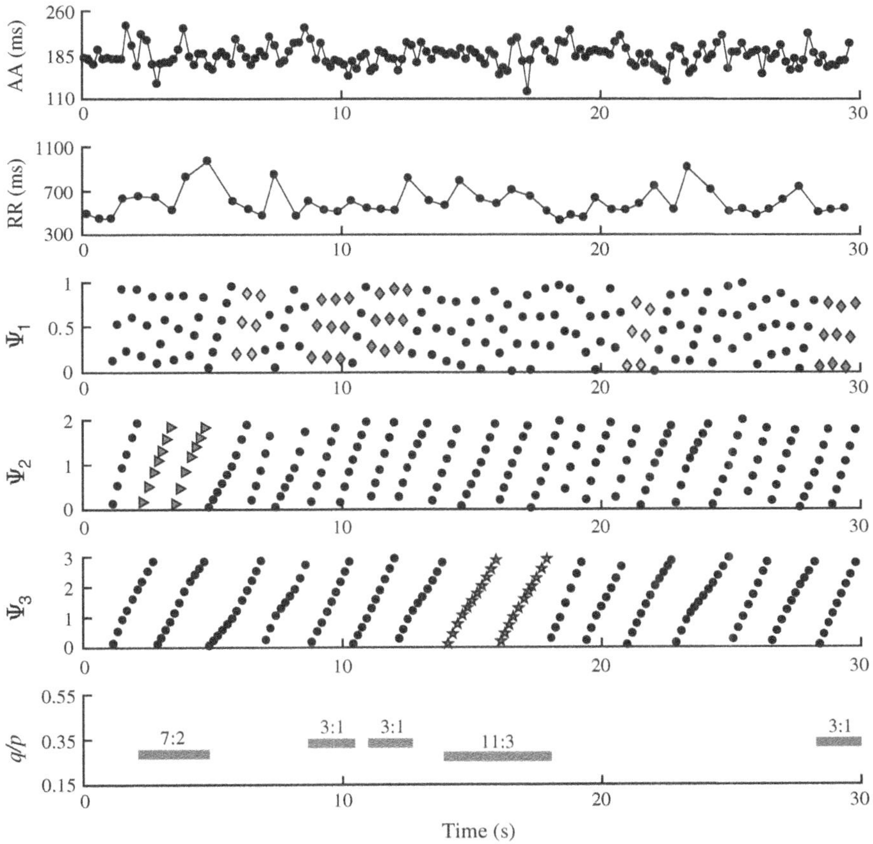

Fig. 7.3 AV synchrogram analysis of AF [28]. Atrial-to-atrial (AA) and RR interval series, normalized instantaneous ventricular phases $\Psi_1(t_{a,k})$, $\Psi_2(t_{a,k})$, $\Psi_3(t_{a,k})$, defined in (7.2), and ratio of ventricular to atrial activations q/p of the synchronized epochs (top to bottom). Note that $p : q$ is displayed inside the bottom diagram, whereas the ratio q/p is the unit of the vertical axis. (Modified from [28] with permission.)

higher in atrial flutter than in AF [28]. Moreover, the synchrogram was used to assess the dynamics of AV coupling as a function of atrial fibrillatory rate (AFR) in a small group of patients during spontaneous acceleration of the AFR at the onset of an AF episode; in this particular assessment, the AFR was estimated from the electrogram. The results demonstrated that the occurrence and the duration of coupled epochs decreased as the AFR increased, and that the average AV conduction ratio, i.e., the ratio of ventricular to atrial activations, was significantly smaller at higher AFRs [29].

Synchrogram analysis has also been considered for investigating the effects of atrial activity and AV nodal conduction on the ventricular response in patients with paroxysmal AF [30]. The results showed that ventricular rate and RR variability are significantly correlated with the average AV conduction ratio and the variability of the atrial input. On the other hand, the AFR is not correlated with ventricular rate nor with RR variability.

7.5 Mathematical Modeling of the Atrioventricular Node

Refractoriness and concealed conduction of the AV node are important AV nodal properties which contribute to forming the ventricular response. Due to refractoriness, many atrial impulses are blocked when arriving to the AV node. Concealed conduction of a single atrial impulse, occurring when the impulse only partially penetrates into the AV node without reaching the ventricles, influences the conduction of subsequent atrial impulses. Moreover, the existence of two dominant pathways through the AV node, each with its own electrophysiological properties, is well-documented and plays an important role in AF.

The properties of AV nodal function can be studied by mathematical modeling which may be categorized into:

- Models primarily developed for simulation to provide better understanding of AV nodal properties, sometimes involving intracardiac information on atrial activity where the arrival times of the atrial impulses are known (this section).
- Models primarily developed for statistical estimation of AV node parameters, relying entirely on information derived from the surface ECG. The arrival times of atrial impulses are modeled as a random process (Sect. 7.6).

7.5.1 Modeling of Conduction Delay in Non-AF Rhythms

Conduction delay is an important AV nodal property, and has therefore received considerable attention in mathematical model building. With reference to AV nodal conduction in Wenckebach periodicity, i.e., a non-AF rhythm, the conduction delay d_k related to the k-th atrial impulse depends on the AV nodal recovery time (RT) $\Delta t_{\mathrm{RT},k}$. The conduction delay is modeled by [21]

$$d_k = d_{\min} + \alpha_{\max} \exp\left[-\frac{\Delta t_{\mathrm{RT},k}}{\gamma_c}\right], \tag{7.3}$$

where $d_{\min}$ is the minimal conduction delay, $\alpha_{\max}$ is the maximal prolongation of the conduction delay, and γ_c is the time constant of the exponential conduction curve. The recovery time $\Delta t_{\mathrm{RT},k}$ is given by the time elapsing from the preceding ventricular activation $t_{v,n}$ to the current AV nodal activation time $t_{a,k}$,

$$\Delta t_{\mathrm{RT},k} = t_{a,k} - t_{v,n}, \quad t_{a,k} > t_{v,n}, \tag{7.4}$$

where ventricular activations are indexed by n.

The basic model of conduction delay in (7.3) can be expanded to include rate-dependent shortening of the conduction delay, referred to as *facilitation* (fac), and rate-dependent prolongation of the recovery time, referred to as *fatigue* (fat) [21], see also [31]. In the expanded model, the conduction delay d_k in (7.3) is denoted d'_k, $\alpha_{\max}$ is replaced by α_k to model facilitation, and the term s_k is introduced to model fatigue,

$$d'_k = d_{\min} + s_k + \alpha_k \exp\left[-\frac{\Delta t_{\mathrm{RT},k}}{\gamma_c}\right]. \tag{7.5}$$

Facilitation is incorporated into the model by assuming that the maximal prolongation $\alpha_{\max}$ depends on the interval $\Delta t_{a,k-1}$ between two successive atrial impulses immediately preceding $t_{a,k}$, commonly referred to as the AA interval,

$$\alpha_k = \alpha_{\max} - \kappa \exp\left[-\frac{\Delta t_{a,k-1}}{\gamma_{\mathrm{fac}}}\right], \tag{7.6}$$

where

$$\Delta t_{a,k} = t_{a,k} - t_{a,k-1}, \tag{7.7}$$

and $\alpha_{\max}$, γ_{fac}, and κ are model constants. Fatigue is incorporated by assuming that each AV nodal activation causes a slowing of the conduction delay of all subsequent impulses, modeled by

$$s_k = s_{k-1} \exp\left[-\frac{\Delta t_{a,k-1}}{\gamma_{\mathrm{fat}}}\right] + \eta \exp\left[-\frac{\Delta t_{\mathrm{RT},k}}{\gamma_{\mathrm{fat}}}\right], \tag{7.8}$$

where η and γ_{fat} are model constants.

The conduction delay model, defined by (7.3)–(7.8), was fitted to experimental data obtained from seven autonomically blocked dogs during pacing [21]. The results showed that the model can accurately predict dynamic changes in Wenckebach periodicity. Although the model does not account for concealed conduction, it has nonetheless served as a starting point for AV node modeling in AF [32–35].

7.5.2 Modeling of Conduction Delay in AF

An AV node model accounting for conduction delay, defined by (7.3), and refractoriness in AF was proposed in [33, 34], however, fatigue and facilitation were not modeled. In that model, the AV node becomes refractory after an atrial impulse has been conducted through the AV node to the ventricles. Impulses arriving to the AV node during the refractory period are blocked (concealed), and each blocked impulse causes the refractory period to be prolonged, first with a fixed length [33], but later with a Gaussian random variable [34].

The proposed model, with fixed prolongation of the refractory period, was tested on one, single intracardiac recording from a patient with AF, exhibiting an agreement between the estimated and the observed RR series which is not particularly satisfactory, see Fig. 7.4. Using instead the model with random prolongation [34], a better fit was obtained. The significance of these two conduction delay models remain to be established on a larger set of data.

More recently, a dual-pathway AV node model has been proposed in which the conduction delay, similar to the model in [21], is assumed to be affected by the stimulation history [36]. The conduction delay is described by the model in (7.3), except that $\alpha_{\max}$ and γ_c are assumed to be functionally dependent on the preceding conduction delay d_{k-1},

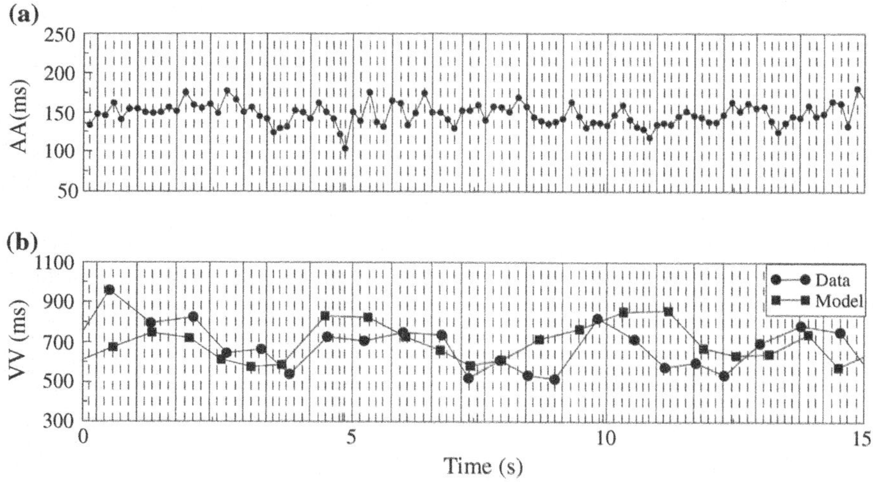

Fig. 7.4 **a** Atrial-to-atrial (AA) and **b** ventricular-to-ventricular (VV) intervals obtained from an intracardiac recording. Ventricular-to-ventricular intervals were obtained using the AV node model in [33]. The vertical lines in both panels are the times of the atrial impulses, where vertical, solid lines indicate conducted impulses, and vertical, dashed lines indicate blocked impulses. (Reprinted from [33] with permission)

$$\alpha_{\max,k} = a_1 d_{k-1}^2 + a_2 d_{k-1} + a_3, \tag{7.9}$$

and

$$\gamma_{c,k} = b_1 + \left(\frac{d_{k-1}}{b_2}\right)^{b_3}, \tag{7.10}$$

respectively. The model constants a_1, a_2, a_3, b_1, b_2, and b_3 are assumed to differ between the two pathways. Concealed conduction is modeled by a virtual conduction delay $\tilde{d}_k$ which depends on the AA interval $\Delta t_{a,k}$ [36],

$$\tilde{d}_k = c_1 - c_2 \exp\left[-\frac{\Delta t_{a,k}}{c_3}\right], \tag{7.11}$$

implying that $\Delta t_{\mathrm{RT},k}$ in (7.3) is replaced by $(\Delta t_{a,k} - \tilde{d}_{k-1})$,

$$d_k = d_{\min} + \alpha_{\max,k} \exp\left[-\frac{(\Delta t_{a,k} - \tilde{d}_{k-1})}{\gamma_{c,k}}\right]. \tag{7.12}$$

The model constants c_1, c_2, and c_3 are assumed to differ between the two pathways. It should be noted that the effect of replacing $\Delta t_{\mathrm{RT},k}$ by $(\Delta t_{a,k} - \tilde{d}_{k-1})$ is similar to that of prolongation of the refractory period due to concealed conduction, see Sect. 7.5.4.

The model parameters were estimated by fitting $\alpha_{\max,k}$ and $\gamma_{c,k}$ to data obtained using a pacing protocol. The fitted model could predict the conducting pathway with specificity and sensitivity exceeding 85% when AF-like random stimulation was applied to a rabbit preparation. His' electrogram alternans was used to determine the pathway of each conducted impulse in the experimental data [37].

7.5.3 Modeling of Refractory Period in AF

In a simple model accounting for the refractory period, the atrial impulses are assumed to arrive randomly in time at the AV node according to a Gaussian distribution [38]. Each atrial impulse results in ventricular activation, unless the AV node is refractory which causes the atrial impulses to be blocked and the refractory period to be prolonged. For each blocked atrial impulse, the refractory period τ_k, following the k-th atrial impulse arriving at the AV node after ventricular activation, is prolonged according to the following equation:

$$\tau_{k+1} = \tau_k + u_k(0.9 - \tau_k), \quad k \geq 0. \tag{7.13}$$

The time-dependent prolongation u_k of the refractory period is described by the logistic function

$$u_k = \frac{1}{1 + e^{-a(z_k - b)}}, \tag{7.14}$$

where

$$z_k = 1 - \frac{t_{a,k}}{\tau_k}, \quad \tau_k > t_{a,k}, \tag{7.15}$$

and a and b are positive-valued model constants. The index k is reset to 0 and τ_0 is reset to 0.3 s when a ventricular activation occurs [38]. The parameters a and b define the shape of the RR interval histogram, but lack a physiological interpretation. For the model in (7.13)–(7.15), an atrial impulse arriving close in time to a conducted atrial impulse prolongs the refractory period more than an atrial impulse arriving at the end of the refractory period. The blocked atrial impulses prolong the refractory period τ_k towards its upper limit of 0.9 s.

7.5.4 Modeling of Refractory Period and Conduction Delay in AF

A much more sophisticated AV node model for the simulation of ventricular activation during AF was proposed in [35, 39], see also [40, 41], where conduction delay, prolongation of the refractory period due to concealed conduction, and ventricular pacing (VP) are also taken into account. The AV node is activated due to the combined effect of spontaneous depolarization and AF bombardment. However, the AV node can also be activated by a VP-induced, retrograde wave. The activation initiates a refractory period during which the AV node is nonresponsive to atrial impulses as well as to a VP-induced retrograde wave. When the refractory period ends, the transmembrane potential returns to its resting potential and initiates a spontaneous, linear increase in the transmembrane potential. Each time an AF impulse arrives at the AV node when not being in a refractory state, its transmembrane potential is increased by a discrete amount, see Fig. 7.5. If instead a VP-induced retrograde wave penetrates the AV node in a nonrefractory state, the transmembrane potential reaches its threshold immediately.

In this model, the conduction delay d_k is modeled by (7.3), and the refractory period τ_k is modeled by

$$\tau_k = \tau_{\min} + \beta \left(1 - \exp\left[-\frac{\Delta t_{\mathrm{RT},k}}{\gamma_r}\right]\right), \tag{7.16}$$

where $\tau_{\min}$ is the shortest refractory period, β is the maximal prolongation of the refractory period, and γ_r is the time constant of the exponential refractory curve. Moreover, the model accounts for prolongation of the refractory period due to concealed conduction. The prolonged refractory period is a product of two factors: one depending on the arrival time of the atrial impulse and another depending on the

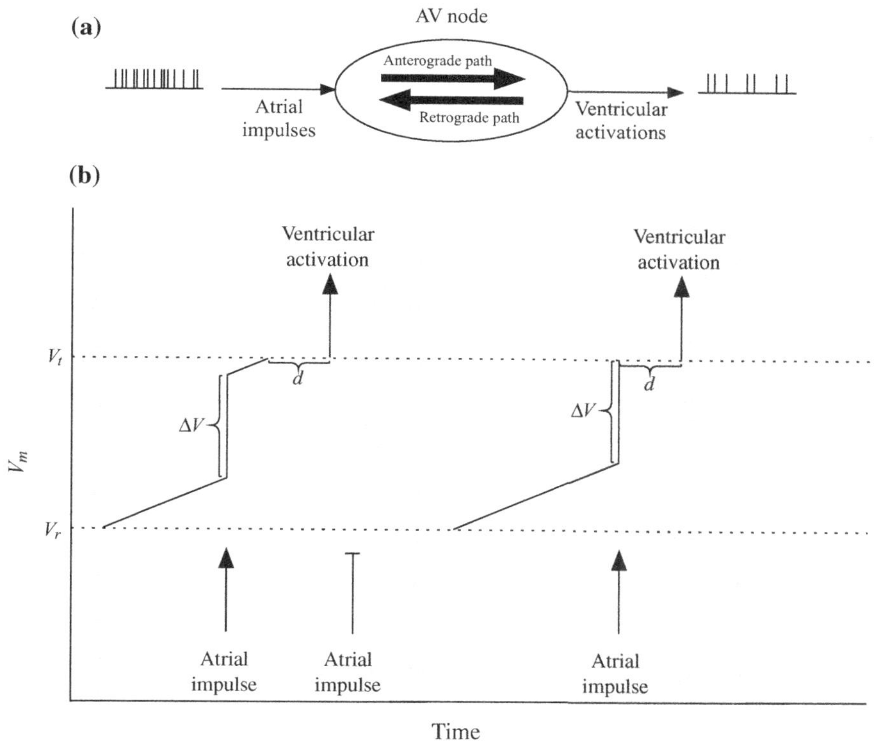

Fig. 7.5 **a** Schematic representation of the AV node model in [35], and **b** related modeling of the transmembrane potential V_m of the AV node. The resting value V_r can increase spontaneously in a linear fashion as well as by a discrete amount ΔV when an atrial impulse arrives. When V_m exceeds the threshold V_t, an action potential is fired and the AV node becomes refractory for a certain period of time τ_k' (indicated by the grey area). Ventricular activation is associated with a delay d_k, modeled by (7.3). Prolongation of the refractory period due to a blocked atrial impulse is modeled by (7.17)

strength of the atrial impulse,

$$\tau_k' = \tau_k + \tau_{\min} \left(\frac{t_{a,k}}{\tau_k} \right)^{\rho_1} \left(\min \left(1, \frac{\Delta V}{V_t - V_r} \right) \right)^{\rho_2}, \tag{7.17}$$

where τ_k' is the prolonged refractory period and ΔV is the strength of an atrial impulse. The voltages V_t and V_r are defined in Fig. 7.5. The two positive-valued exponents ρ_1 and ρ_2 are model constants.

Figure 7.6 shows two simulated RR series generated using different model parameter settings [35, 39]. The simulation is based on the assumption that atrial impulses arrive to the AV node according to a Poisson process with mean arrival rate λ_a.

The authors stated that the simulation model may provide a quantitative framework to investigating drug effects by fitting their model to experimental data. However,

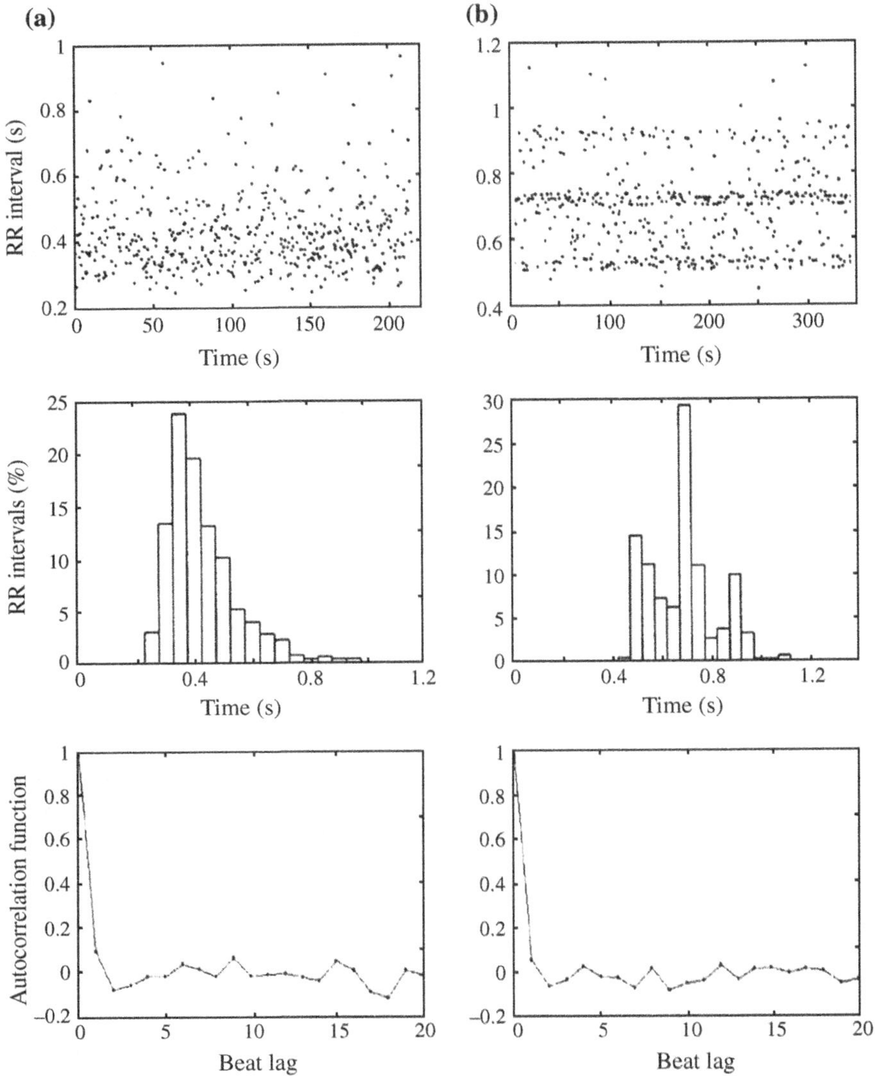

Fig. 7.6 Simulated RR intervals series obtained for two different settings of model parameters [35] (top row), and corresponding (**b**) histogram (middle row), and **c** autocorrelation function (bottom row). The following model parameter values were used: **a** $\lambda_a = 8$ per second, $\Delta V = 20\,\text{mV}$, $\rho_1 = 10$, $\rho_2 = 10$, and **b** $\lambda_a = 4$ per second, $\Delta V = 10\,\text{mV}$, $\rho_1 = 10$, $\rho_2 = 10$. (Reprinted from [35] with permission)

no results have so far been published which investigate such effects. The problem of fitting the model to observed data is likely to be challenging since the model involves a large number of parameters.

7.5.5 Modeling of Spatial Dynamics

A radically different approach to modeling is to treat the AV node as a connected graph [42], consisting of a series of interacting nodes, rather than having a lumped structure as the above-described AV node models. An advantage of the graph approach is that it accounts for spatial propagation of atrial impulses in the AV node, implying that phenomena such as concealed conduction and retrograde conduction are intrinsic to the model structure.

The nodes in the graph model propagate impulses, corresponding to action potentials, along the graph edges, see Fig. 7.7. Each node corresponds to a localized part of the AV node with its own conduction delay and refractory period, both quantities depending on the stimulation history of the node. When an impulse arrives at a node, the conduction delay and the refractory period are updated according to (7.3) and (7.16), respectively. Each node has its own dynamics and is characterized by its own recovery time Δt_{RT}, thus differing from the above-described models where the recovery time applies to the whole AV node.

The proposed model consists of 21 nodes, where 10 nodes characterize the fast pathway and 11 the slow pathway. The parameters modeling conduction delay, i.e., $d_{\min}$, $\alpha_{\max}$, and γ_c in (7.3), and refractory period, i.e., $\tau_{\min}$, β, and γ_r in (7.16), are identical for all nodes of the slow pathway; the same applies to all nodes of the fast pathway. Hence, the model is defined by 12 parameters. To simulate conduction through the model, it is assumed that the first nodes on the atrial side of the slow and the fast pathways are simultaneously activated. A ventricular activation occurs

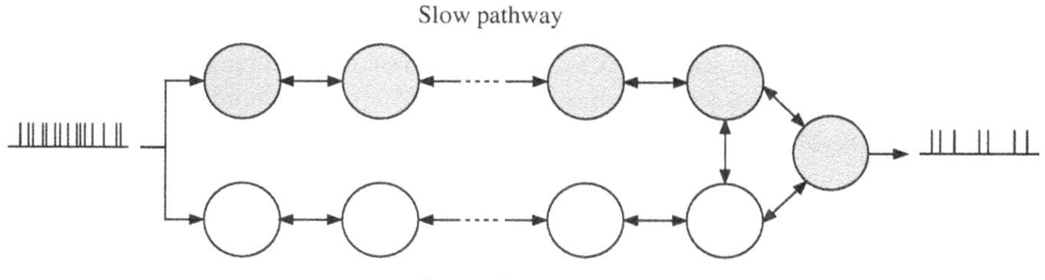

Fig. 7.7 Schematic presentation of the spatial AV node model proposed in [42]. The graph nodes propagate impulses along the edges. Each node is characterized by its own refractory period and conduction delay, both depending on the stimulation history of the node. The full model comprises 21 nodes

when the atrial impulse reaches the rightmost end of the graph, corresponding to the bundle of His.

Simulations were performed using AA intervals determined from electrograms recorded from patients with AF, as well as from simulated AA intervals determined by a Poisson process with mean arrival rate λ_a. A genetic algorithm was used to fit the model by minimizing the difference between simulated RR series and observed RR interval series, obtained from a number of ECG recordings. In the simulations using electrogram-derived series of atrial activations, the difference was quantified by the mean square error of the times of ventricular activations. If simulated AA intervals were used and only the ECG was available, the difference was quantified based on the RR interval histogram. The model could accurately replicate the RR intervals determined from the ECG.

The model fitting was repeated 1000 times for different initial conditions, resulting in 1000 estimates of each parameter for each recording [42]. No unique solution was obtained since several different parameter sets resulted in a similar model fit, however, the ranges of the estimated model parameters were limited. For ECG data, 90% of the estimated values of $\tau_{\min}$ and β were within $\pm 20\%$ of the median value of the estimates, whereas this did not apply to $d_{\min}$ and $\alpha_{\max}$.

7.6 Statistical Modeling of the Atrioventricular Node and Parameter Estimation

In the very first paper dealing with statistical modeling, the AV node was treated as a lumped structure whose behavior represents the temporal and spatial summation of the cellular electrical activity [43]. In that model, briefly described in Sect. 7.6.1, the atrial impulses are assumed to arrive randomly in time at the AV node, modeled by a Poisson process with mean rate λ_a [44]. The conduction time is not explicitly modeled.

Many years later, an improved statistical model was proposed which also accounts for dual AV nodal conduction [45, 46], see Sect. 7.6.2. Since the model parameters can be estimated from the surface ECG, without use of any intracardiac information, noninvasive electrophysiological characterization of the AV node is made possible. In a subsequent study, the model was further improved to incorporate pathway switching, accompanied by more robust parameter estimation [47], see Sect. 7.6.3.

7.6.1 A First Statistical Model of the AV Node

In this statistical model, the AV node is always in one of two states. In the first state, the AV node is absolutely refractory to stimulation by atrial impulses. At the onset of the second state, the transmembrane potential is at its resting value, and increases spontaneously at a constant rate as well as by a discrete amount ΔV when an atrial impulse arrives. When the transmembrane potential reaches its threshold value V_t as a result of any combination of spontaneous and stepwise depolarization, the AV node fires and a new refractory period is initiated, see Fig. 7.8.

The refractory period is assumed to be rate-dependent, implying that a longer RR interval is followed by a longer refractory period, and vice versa. The relation between refractory period and RR interval is modeled by an exponential function,

$$\tau_n = \tau_\infty \left(1 - \exp\left[-\frac{x_n}{\tau_\infty}\right]\right), \tag{7.18}$$

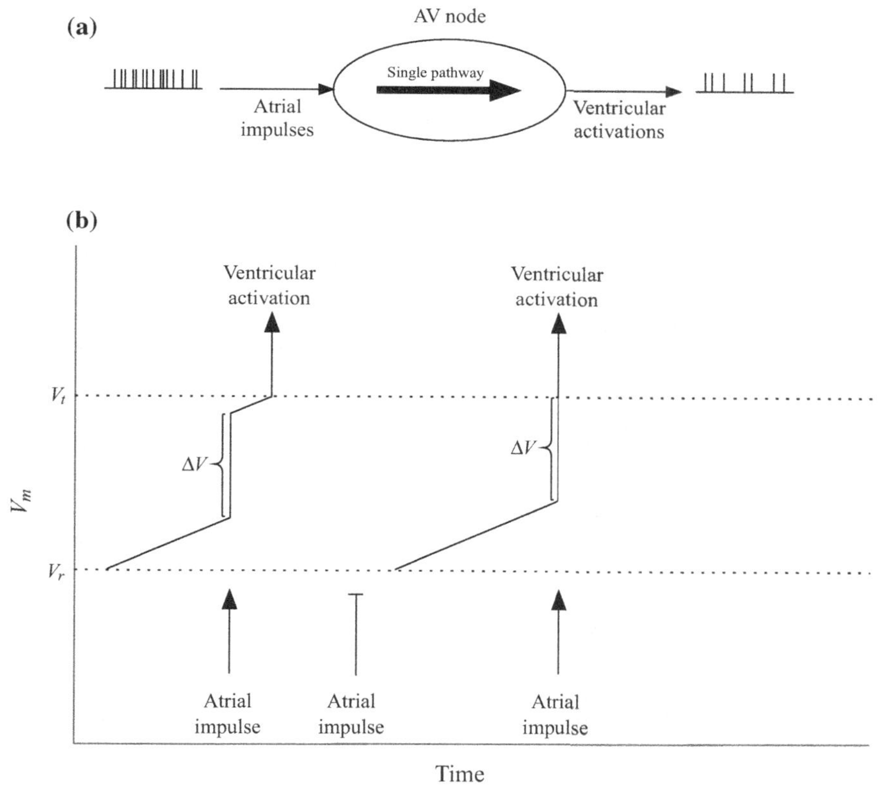

Fig. 7.8 **a** Schematic representation of the AV node model in [43], and **b** related modeling of the transmembrane potential V_m of the AV node. The resting value V_r can increase spontaneously in a linear fashion (defined by the slope v) as well as by a discrete amount ΔV when an atrial impulse arrives. When V_m exceeds the threshold V_t, an action potential is fired and the AV node becomes refractory for a certain period (indicated by the grey area). This model differs from the one in Fig. 7.5 as it does not account for the delay d associated with ventricular activation

where τ_n is the refractory period following the n-th ventricular activation, τ_∞ is the maximal refractory period, and x_n is the RR interval preceding the n-th ventricular activation,

$$x_n = t_{v,n} - t_{v,n-1}. \tag{7.19}$$

The conduction delay is not explicitly modeled.

The model is defined by the following four parameters:

- the mean arrival rate λ_a of atrial impulses,
- the relative amplitude ΔV of atrial impulses,
- the rate v of spontaneous AV depolarization, measured in units of ΔV, and
- the maximal refractory period τ_∞.

Although this model is statistical in nature, no well-established statistical estimation procedure, such as the maximum likelihood (ML) technique, was considered in [43]. Instead, the model parameters were determined using an ad hoc optimization procedure which yielded unphysiological parameter estimates.

7.6.2 Statistical Modeling of Dual AV Nodal Pathways

The improved AV node model accounts for concealed conduction, relative refractoriness, and dual AV nodal pathways [45], see Fig. 7.9. In this model, each atrial impulse is assumed to result in ventricular activation, unless the impulse is blocked by a refractory AV node. The probability of an atrial impulse passing through the AV node depends on the time elapsed since the preceding ventricular activation $t_{v,n-1}$. The refractory period is defined by the sum of a deterministic period τ and a random period, uniformly distributed in the interval $[0, \tau_p]$. The random period models prolongation due to concealed conduction and/or relative refractoriness. All atrial impulses arriving at the AV node before the end of the deterministic period τ are blocked, whereas impulses with arrival time in the interval $[\tau, \tau + \tau_p]$ have linearly increasing likelihood of passing through the AV node. No impulses are blocked if they arrive after $\tau + \tau_p$.

The model accounts for a fast pathway with a longer refractory period, defined by τ_f and $\tau_{f,p}$, and a slow pathway with a shorter refractory period, defined by τ_s and $\tau_{s,p}$ (depending on pathway, the indices "s" and "f" are added to τ and τ_p). In mathematical terms, the refractoriness of the slow pathway is defined by the function $\beta_s(t)$,

$$\beta_s(t) = \begin{cases} 0, & 0 < t < \tau_s, \\ \dfrac{t - \tau_s}{\tau_{s,p}}, & \tau_s \leq t < \tau_s + \tau_{s,p}, \\ 1, & t \geq \tau_s + \tau_{s,p}, \end{cases} \tag{7.20}$$

where, for convenience, t is used instead of $\Delta t_{\mathrm{RT},k}$. The function $\beta_f(t)$ characterizes refractoriness of the fast pathway and is identical to $\beta_s(t)$ except that τ_f and $\tau_{f,p}$

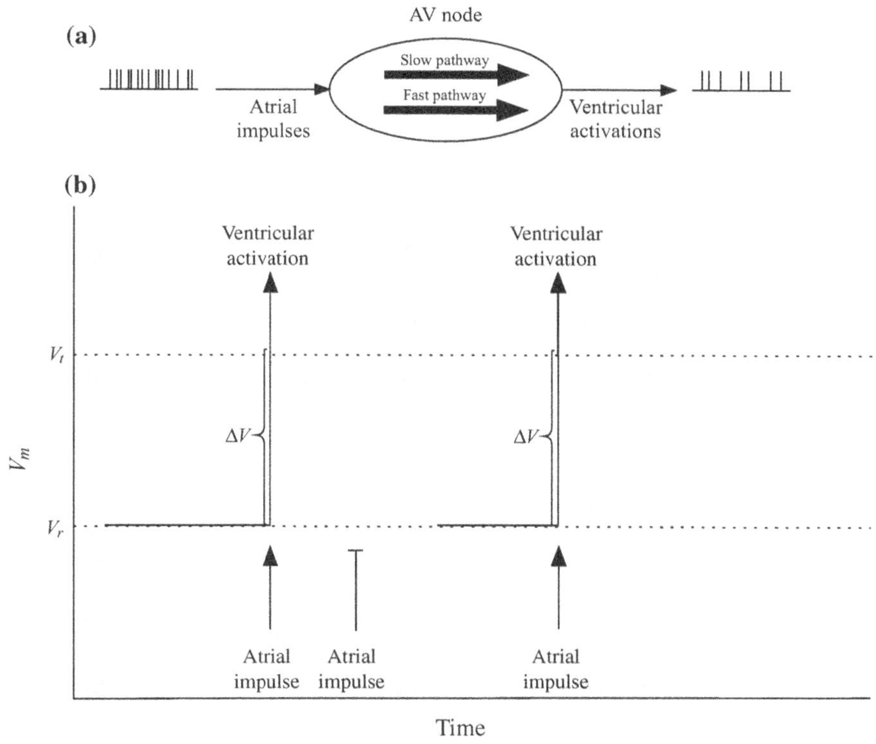

Fig. 7.9 **a** Schematic representation of the AV node model in [45], and **b** related modeling of the transmembrane potential V_m of the AV node. When an atrial impulse arrives at the AV node, the resting value V_r increases by a discrete amount ΔV which always makes V_m exceed the threshold V_t, an action potential to be fired, and the AV node refractory for a certain period of time (indicated by the grey area)

are substituted for τ_s and $\tau_{s,p}$, respectively. The deterministic part of the refractory periods τ_s and τ_f are assumed to depend linearly on the preceding RR interval x_{n-1}, implying that a longer RR interval is followed by a longer refractory period, and vice versa. Moreover, it is assumed that the AV conduction time is incorporated into $\beta_s(t)$ and $\beta_f(t)$ so that a ventricular activation occurs immediately after a non-blocked atrial impulse.

Since non-blocked atrial impulses are assumed to occur according to an inhomogeneous Poisson process characterized by the intensity function $\lambda_a \beta_s(t)$, the PDF of an RR interval x, related to the slow pathway, is given by [45]

$$p_{x,s}(x) = \lambda_a \beta_s(x) \exp\left[-\int_0^x \lambda_a \beta_s(\tau) d\tau\right], \tag{7.21}$$

which, after insertion of (7.20), becomes

$$p_{x,s}(x) = \begin{cases} 0, & 0 < x < \tau_s, \\ \dfrac{\lambda_a(x-\tau_s)}{\tau_{s,p}} \exp\left[-\dfrac{\lambda_a(x-\tau_s)^2}{2\tau_{s,p}}\right], & \tau_s \leq x < \tau_s + \tau_{s,p}, \\ \lambda_a \exp\left[-\dfrac{\lambda_a \tau_{s,p}}{2} - \lambda_a(x-\tau_s-\tau_{s,p})\right], & x \geq \tau_s + \tau_{s,p}. \end{cases} \quad (7.22)$$

The PDF $p_{x,f}(x)$, related to the fast pathway, is obtained by substituting τ_f and $\tau_{f,p}$ for τ_s and $\tau_{s,p}$ in (7.22), respectively.

Conduction through the slow and fast pathways are assumed to occur with probabilities ε and $1-\varepsilon$, respectively. Assuming that ventricular activations occur according to a Poisson process, the intervals between successive ventricular activations are statistically independent, and the joint probability of the RR intervals $x_0, \ldots, x_{N-1}$ is given by

$$\begin{aligned} p_x(x_0, x_1, \ldots, x_{N-1}) &= \prod_{n=0}^{N-1} p_x(x_n) \\ &= \prod_{n=0}^{N-1} (\varepsilon p_{x,s}(x_n) + (1-\varepsilon) p_{x,f}(x_n)). \end{aligned} \quad (7.23)$$

The mean arrival rate λ_a is estimated from the f wave signal extracted from the ECG, but corrected to account for atrial refractoriness, using [46]

$$\hat{\lambda}_a = \frac{\lambda_{\mathrm{AF}}}{1-\delta\lambda_{\mathrm{AF}}}, \quad (7.24)$$

where λ_{AF} is taken as the AFR, estimated from the ECG, and δ is the minimal time interval between successive impulses arriving to the AV node. The five model parameters $\boldsymbol{\theta} = \left[\varepsilon \ \tau_s \ \tau_{s,p} \ \tau_f \ \tau_{f,p}\right]$ are estimated from the observed RR interval series using the ML technique, defined by

$$\hat{\boldsymbol{\theta}} = \arg\max_{\boldsymbol{\theta}} \log p_x(x_0, x_1, \ldots, x_{N-1} | \boldsymbol{\theta}; \hat{\lambda}_a). \quad (7.25)$$

Since no closed-form solution can be found for the estimator $\hat{\boldsymbol{\theta}}$, combined with the fact that the gradient is discontinuous, the multi-swarm particle swarm optimization is used to optimize the log-likelihood function [48, 49]. It should be noted that since τ_s and τ_f depend on the preceding RR interval, the original RR interval series is subject to decorrelation before ML estimation is performed, to better comply with the assumption of statistical independence in (7.23) [45].

The parameters of the single pathway model, i.e., $\left[\tau \ \tau_p\right]^T$, are also estimated. The Bayes information criterion is then used to determine which of the single- and the dual-pathway model is most likely the observed [46].

The block diagram in Fig. 7.10 shows the main signal processing steps required to estimate the model parameters from the ECG. Figure 7.11 illustrates, in histogram form, the RR intervals produced by three different parameter settings of the AV node model.

The AV node model was fitted to clinical data acquired during treatment with different drugs for the purpose of investigating drug-induced changes in AV nodal properties [50–52]. The hypothesis was that the estimates of AV nodal refractory periods would reflect the main changes in AV nodal properties previously reported in studies performed in sinus rhythm and based on invasive data. The effects of tecadenoson and esmolol were investigated in a small cohort of patients [50]. The parameters τ_s and τ_f, accounting for both effective refractory period and conduction interval, were prolonged for both tecadenoson and esmolol [50]. The increase in τ_s and τ_f, observed for both pathways, suggested either prolonged effective refractory

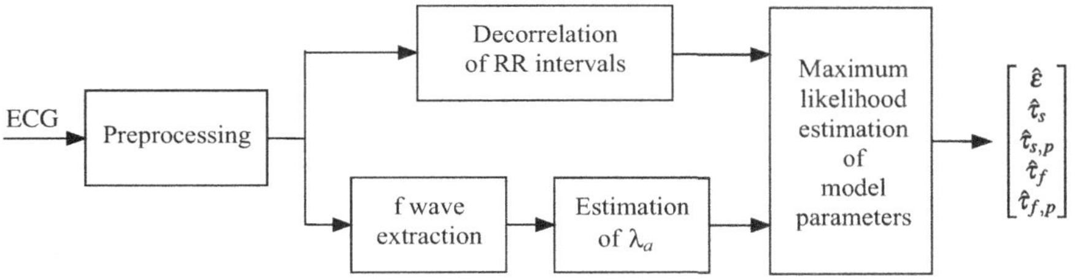

Fig. 7.10 The main signal processing steps required for estimating the AV node model parameters from the ECG

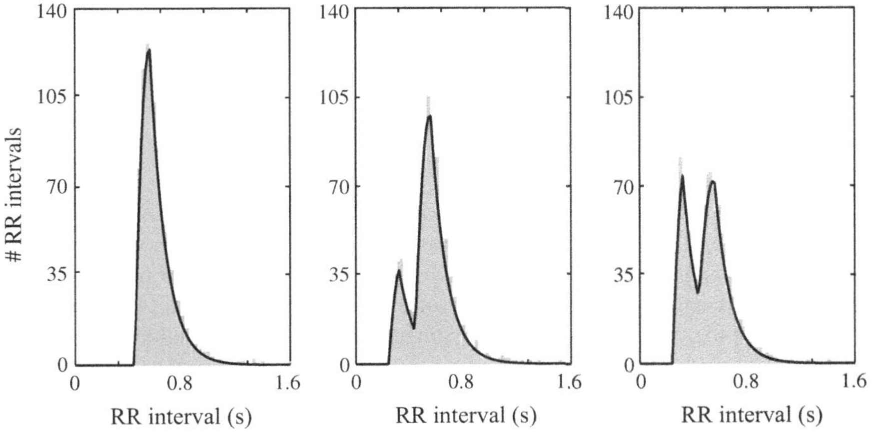

Fig. 7.11 RR interval histogram (area defined by grey bars) and fitted model PDF (solid line) for three different parameter settings. The histograms derive from model data with increasing probability ε of an atrial impulse passing through the slow pathway, set to either 0, 0.25, or 0.5 (left to right); the other model parameters were held constant. (Reprinted from [45] with permission)

period[2] or prolonged AV conduction, or both. In addition, tecadenoson was shown to affect heart rate but not AFR, suggesting that a decrease in heart rate may be attributed to that tecadenoson affects the AV node. These results are in agreement with previous studies demonstrating that tecadenoson prolongs the effective refractory period of the AV node and slows down its conduction [53], whereas esmolol prolongs refractoriness and conduction time in both pathways during AV nodal reentrant tachycardia [54].

Changes in AV nodal properties were investigated during administration of beta blockers (carvedilol and metoprolol) and calcium channel blockers (diltiazem and verapamil) in a controlled setting [52]. For patients with permanent AF, this study compared the effects of four once-daily drug regimens (metoprolol, diltiazem, verapamil and carvedilol) on heart rate and arrhythmia-related symptoms. While the results of this study are not directly comparable to previous studies, the changes in estimated AV nodal properties are in agreement with previous electrophysiological findings [55–58].

The results suggest that the noninvasively obtained parameter estimates reflect the expected changes in AV nodal properties for the investigated drugs. Therefore, the method should be suitable for assessing the drug effect on AV nodal electrophysiology during AF, especially for antiarrhythmic compounds aimed at rate-control during AF and tested in clinical trials during initial clinical phases of drug development.

Another application of the AV node model is to analyze data acquired during rest and head-up tilt (75°) from patients with persistent AF. A shortening of the refractory periods τ_s and τ_f was observed for both pathways during adrenergic activation [59]—results which are in agreement with earlier reported results [60]. The effect of tilting on the refractory period of the AV node has not been assessed previously, but invasive studies have evaluated the effect of vagal tone on AV node refractory periods by either stimulating the vagal nerve directly [60] or by assessing the effect of vagolytic drugs.

7.6.3 *Statistical Modeling of Pathway Switching*

A limitation of the statistical model in [45] is the assumption that atrial impulses arriving between two ventricular activations attempt conduction through the same pathway, i.e., pathway switching is not allowed. Therefore, another model suitable for ECG-based estimation of the model parameters was proposed in [47]. Similar to the model in [45], atrial impulses are assumed to arrive at the AV node according to a Poisson process with mean arrival rate λ_a. Each impulse attempts to pass through either the slow or the fast pathway, being blocked according to the time-dependent functions $\beta_s(t)$ and $\beta_f(t)$ depending on which pathway is chosen. The choice of pathway is independent of the pathway taken by the preceding atrial impulse. Conduction through the slow pathway is attempted with probability ε, and consequently conduction through the fast pathway is attempted with probability $1 - \varepsilon$.

[2]The effective refractory period is defined as the longest nonconducting AA interval.

Since an atrial impulse is assumed to arrive at the AV node according to a Poisson process, the PDF of the arrival time of the first atrial impulse following a ventricular activation is given by [61]

$$p_1(t) = \begin{cases} \lambda_a e^{-\lambda_a t}, & t \geq 0, \\ 0, & t < 0. \end{cases} \tag{7.26}$$

The first impulse attempts conduction through the slow pathway with probability ε, where conduction is characterized by $\beta_s(t)$, defined in (7.20). Hence, the PDF of the arrival time of the first impulse conducted through the slow pathway is given by

$$p_{1,\mathrm{cs}}(t) = \varepsilon \beta_s(t) p_1(t). \tag{7.27}$$

The PDF of the arrival time of the first impulse conducted through the fast pathway $p_{1,\mathrm{cf}}(t)$ is obtained by replacing $\beta_s(t)$ with $\beta_f(t)$ and ε with $1 - \varepsilon$. The notations "cs" and "cf" refers to conduction through the slow and the fast pathway, respectively.

The PDF of the arrival time of the second atrial impulse depends on the arrival time of preceding blocked atrial impulses as well as the time interval between the second and the first atrial impulses, i.e., the AA interval. Since AA intervals are statistically independent in the Poisson model, the PDF of the arrival time of the second atrial impulse following a ventricular activation is given by

$$p_2(t) = \int_0^\infty p_1(t - \rho)(p_{1,\mathrm{bs}}(\rho) + p_{1,\mathrm{bf}}(\rho))d\rho, \tag{7.28}$$

where $p_{1,\mathrm{bs}}(t)$ and $p_{1,\mathrm{bf}}(t)$ denotes the PDF of the arrival time of the first impulse blocked in the slow and the fast pathway, respectively. To account for pathway switching, the second atrial impulse attempts to pass through the slow pathway with probability ε irrespectively of the pathway in which the first impulse was blocked. Hence, $p_{2,\mathrm{cs}}(t)$ and $p_{2,\mathrm{cf}}(t)$ are computed analogously to $p_{1,\mathrm{cs}}(t)$ and $p_{1,\mathrm{cf}}(t)$.

A general expression for recursive computation of the PDF of the arrival times is given by

$$p_{i,\mathrm{cs}}(t) = \varepsilon \beta_s(t) p_i(t), \tag{7.29}$$

$$p_{i,\mathrm{cf}}(t) = (1 - \varepsilon)\beta_f(t) p_i(t), \tag{7.30}$$

$$p_{i,\mathrm{bs}}(t) = \varepsilon(1 - \beta_s(t)) p_i(t), \tag{7.31}$$

$$p_{i,\mathrm{bf}}(t) = (1 - \varepsilon)(1 - \beta_f(t)) p_i(t), \tag{7.32}$$

$$p_{i+1}(t) = \int_0^\infty p_1(t - \rho)(p_{i,\mathrm{bs}}(\rho) + p_{i,\mathrm{bf}}(\rho))d\rho, \tag{7.33}$$

where $p_{i+1}(t)$ denotes the PDF of the arrival time of the $(i + 1)$:st atrial impulse following a ventricular activation, $p_{i,\mathrm{cs}}(t)$ and $p_{i,\mathrm{cf}}(t)$ denote the PDFs of the arrival time of the i-th atrial impulse conducted through the slow pathway and the fast pathway, respectively, and $p_{i,\mathrm{bs}}(t)$ and $p_{i,\mathrm{bf}}(t)$ denote the corresponding PDFs of blocked atrial impulses.

A conducted atrial impulse is assumed to immediately cause a ventricular activation. Hence, the PDF of the time intervals between ventricular activations, i.e., x_n, is obtained by summing the PDFs of all conducted atrial impulses,

$$p_c(x_n; J) = \sum_{i=1}^{J} (p_{i,\mathrm{cs}}(x_n) + p_{i,\mathrm{cf}}(x_n)), \tag{7.34}$$

where J denotes the maximal number of blocked atrial impulses between successive ventricular activations. This number is chosen so that more than 90% of the conducted atrial impulses are accounted for [47].

When applying the model in [47] to ECG signals, the probability ε of choosing the slow pathway was simply set to 0.5, whereas the remaining model parameters $\boldsymbol{\theta} = \begin{bmatrix} \tau_s & \tau_f & \tau_{s,p} & \tau_{f,p} \end{bmatrix}$ were estimated using the ML technique,

$$\hat{\boldsymbol{\theta}} = \arg\max_{\boldsymbol{\theta}} \left(\sum_{n=0}^{N-1} \log p_c(x_n | \boldsymbol{\theta}; \hat{\lambda}_a) \right), \tag{7.35}$$

where the mean arrival rate λ_a was estimated as described in Sect. 7.6.2.

The ratio of atrial impulses conducted through the slow pathway, defined by

$$\alpha = \frac{\displaystyle\sum_{i=1}^{J} \int_0^{\infty} p_{i,\mathrm{cs}}(t) dt}{\displaystyle\int_0^{\infty} p_{c}, (t; I) dt}, \tag{7.36}$$

can be used to quantify the reliability of the parameter estimates in $\hat{\boldsymbol{\theta}}$. A small α indicates that few impulses are conducted through the slow pathway, and, therefore, $\hat{\tau}_s$ and $\hat{\tau}_{s,p}$ are less reliable. Conversely, a large α indicates that few impulses are conducted through the fast pathway, and, therefore, $\hat{\tau}_f$ and $\hat{\tau}_{f,p}$ are less reliable.

The AV node model has been fitted to each nonoverlapping, 30-min segment of 24-h ECG recordings from 60 patients in permanent AF [47]. Based on results from simulated data, a threshold was applied to α in order to judge whether the estimated model parameters were reliable. Figure 7.12 illustrates how the four parameters characterizing the refractory periods change over a 24 h period, using the models in [45, 47]. It is obvious from Fig. 7.12 that the estimates based on the AV node model in [47] is associated with less variation in $\hat{\tau}_{s,p}$ and $\hat{\tau}_{f,p}$ than is the model in [45]. It should be noted that the model in [47] leads to an unequally sampled series of parameter estimates, since several estimates are omitted because the reliability, determined by α, is judged to be too low.

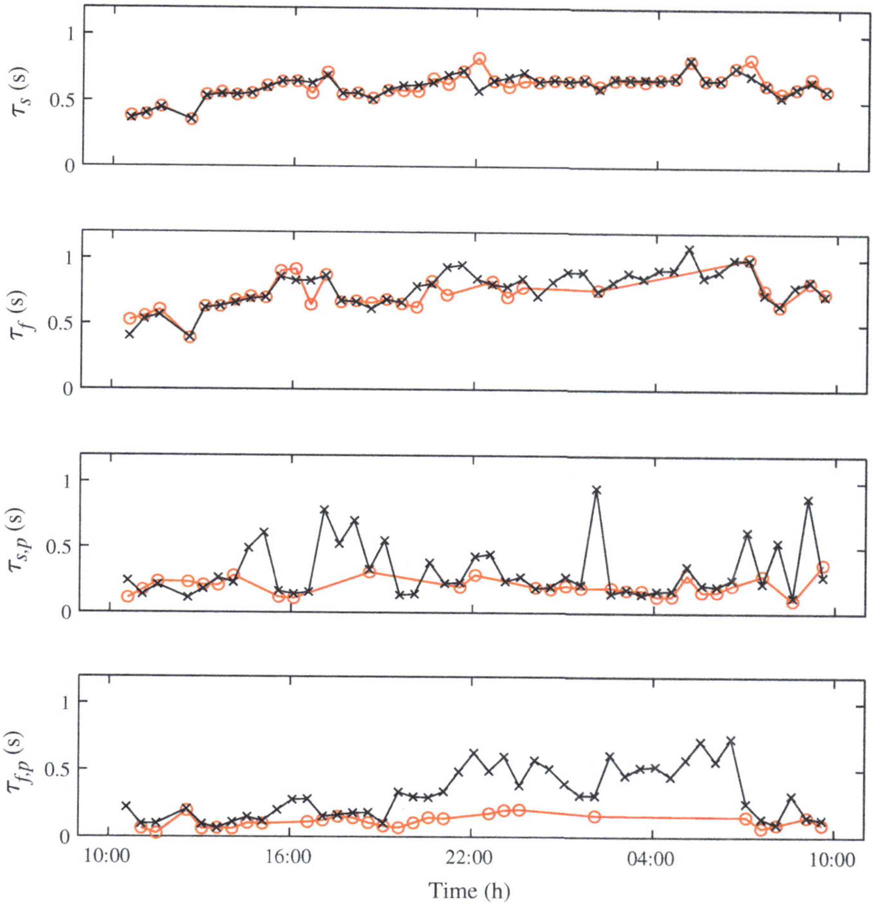

Fig. 7.12 Model parameter estimates obtained from a 24-h ECG recording of a patient with permanent AF, using the models in [47] (red line, "o" markers) and [45] (black line, "x" markers). Successive estimates, indicated by markers, are connected with straight lines (see text for further explanations)

7.7 Comparison of AV Models

The AV node models described in this chapter were developed for different purposes, one purpose being to simulate ventricular activation series resembling those observed during AF and to characterize AV nodal function from intracardiac recordings (Sect. 7.5), another purpose being to characterize AV nodal function from the surface ECG (Sect. 7.6). These purposes are reflected in the structure and complexity of the different models. While the early models embrace two to six parameters [33, 34, 38, 43], the more recent ones, primarily used for simulation and electrogram-based characterization are considerably more complex, embracing 12 to 16 parameters [35, 36, 42]. The statistical models for ECG-based characterization consist of four to five model parameters which make them better suited for estimation [45, 47].

The AV node models differ in their respective approach to handling the following properties:

- Atrial activation times
- Refractory period
- Conduction delay
- Ventricular activation
- Concealed conduction
- Dual pathways

The *atrial activation times* are usually modeled by a homogenous Poisson process, implying that the AA intervals are exponentially distributed [35, 42, 43, 45, 47]. A Gaussian distribution of the AA intervals has also been proposed [38], although the atrial activation times can no longer be treated as a Poisson process. In the very first statistical model, the mean arrival rate λ_a of the Poisson process assumed an unphysiological value [43]—a problem which was later solved by relating λ_a to the AFR, estimated from the f waves in the ECG [42, 45, 47]. For models where the AV node is characterized using intracardiac information, the atrial activation times are determined by the peaks of the atrial electrogram [33, 34, 36]. Positioning of the electrodes relative to the AV node entrance is particularly important during AF because of the disorganized atrial activity.

From experimental data obtained using a pacing protocol, the effective *refractory period* is known to be dependent on the paced atrial rate [62]. This rate dependence can be modeled in different ways. For example, the refractory period can depend on the preceding RR interval according to an exponential curve defined by the maximal refractory period τ_∞, cf. (7.18) [43]. Another approach is to assume that the refractory period is linearly dependent on the preceding RR interval, calling for decorrelation of the observed RR interval series before parameter estimation can be performed [45, 47]. Yet another approach is to assume that the refractory period is recovery-dependent, i.e., dependent on the time elapsed since the end of the preceding refractory period according to an exponential curve modeled by three parameters: the minimal refractory period $\tau_{\min}$, the maximal prolongation β, and the time constant γ_r of the exponential refractory curve, cf. (7.16) [35, 42]. In some models, the rate dependence of the refractory period is not explicitly modeled [33, 34, 36, 38].

The *conduction delay* is an important property of the AV node during normal sinus rhythm, known to be dependent on the paced atrial rate [62]. The AV nodal conduction delay may be incorporated in the refractory period so that its dynamics is not explicitly modeled [43, 45, 47]. Alternatively, the conduction delay can be made dependent on the recovery time, where recovery dependence is modeled by an exponential curve defined by three model parameters: the minimal conduction delay $d_{\min}$, the maximal prolongation $\alpha_{\max}$, and the time constant γ_c of the exponential conduction curve, cf. (7.3) [33–35, 42]. A similar approach was considered in [36], although the maximal prolongation and the time constant of the conduction curve were assumed to depend on the preceding conduction delay.

In most models, ventricular activation is directly linked to the arrival of one atrial impulse. Each atrial impulse is assumed to result in a *ventricular activation*, unless it is blocked due to AV nodal refractoriness. However, more than one atrial impulse may be needed to cause a ventricular activation [35, 43]. The AV node may also fire spontaneously.

Concealed conduction is incorporated in the models in different ways. For each blocked impulse, the refractory period can be incremented by a fixed [33] or random time [34]. The refractory period prolongation can depend on the timing of the blocked impulse [38], or on both the timing and the strength of the blocked impulse [35]. Blocked impulses can alter the conduction time of the following impulse, so that a longer AA interval results in a longer conduction delay [36]. Concealed conduction can also be disregarded, assuming a refractory period which is not influenced by blocked impulses [43]. The refractory period prolongation caused by each blocked impulse is not always explicitly modeled, but concealed conduction is modeled by a random, uniformly distributed prolongation of the refractory period [45, 47]. Concealed conduction can also be an intrinsic feature of the chosen model structure [42].

The earlier models [33–35, 38, 43] did not account for *dual pathways* of the AV node, while the more recent models account for separate conduction time [36, 42] and refractory period [42, 45, 47] of the two pathways.

The model in [43] was fitted to observed RR series using an ad hoc procedure. For some models, no attempts have been made at all to fit the models to observed data [35, 38]. The models proposed for characterization of AV nodal function based on intracardiac recordings were fitted using a grid search to find the minimum error between observed and simulated RR intervals, given the observed AA intervals as input [33, 34, 42]. The model parameters in [36] were assessed by fitting data obtained using

Table 7.1 Comparison of atrioventricular node models with respect to structure and atrial impulse assumptions. The models are listed in chronological order

Model proposed by	Model parameters	Parameter estimation	Atrial impulses	Impulses required
Cohen et al. [43]	4	No	Poisson	≥0
Jorgensen et al. [33]	6	Ad hoc	Invasive data	1
Rashidi and Khodarahmi [38]	2	No	Gaussian	1
Mangin et al. [34]	6	Ad hoc	Invasive data	1
Lian et al. [35]	16	No	Poisson	≥0
Climent et al. [36]	18	Ad hoc	Invasive data	1
Corino et al. [45]	6	ML	Poisson	1
Henriksson et al. [47]	5	ML	Poisson	1
Wallman and Sandberg [42]	12	Ad hoc	Poisson	1

Table 7.2 Comparison of atrioventricular node models with respect to various electrophysiological properties. The functional dependence of the refractory period is indicated, where x_n is the preceding RR interval, and $\Delta t_{\mathrm{RT},k}$ is the AV nodal recovery time, cf. (7.4)

Model proposed by	Conduction delay	Refractory period	Concealed conduction	Dual pathways
Cohen et al. [43]	No	x_n	No	No
Jorgensen et al. [33]	Yes	Fixed	Fixed increment	No
Rashidi and Khodarahmi [38]	No	Fixed	Timing	No
Mangin et al. [34]	Yes	Fixed	Random increment	No
Lian et al. [35]	yes	$\Delta t_{\mathrm{RT},k}$	Timing and strength	No
Climent et al. [36]	Yes	No	Timing	Yes
Corino et al. [45]	No	x_n	Random	Yes
Henriksson et al. [47]	No	x_n	Random	Yes
Wallman and Sandberg [42]	Yes	$\Delta t_{\mathrm{RT},k}$	Intrinsic	Yes

a dedicated pacing protocol. For ECG-based characterization of AV nodal function during AF, the mean arrival rate of atrial impulses is estimated from an extracted f wave signal [42, 45, 47]; the remaining model parameters are estimated from an RR interval series using ML estimation [45, 47].

Tables 7.1 and 7.2 provides a comparison of AV node models described in this chapter and their respective properties.

References

1. K.M. Stein, J. Walden, N. Lippman, B.B. Lerman, Ventricular response in atrial fibrillation: random or deterministic? Am. J. Physiol. **277**, H452–458 (1999)
2. V.D.A. Corino, R. Sassi, L.T. Mainardi, S. Cerutti, Signal processing methods for information enhancement in atrial fibrillation: spectral analysis and non-linear parameters. Biomed. Signal Process. Control **1**, 271–281 (2006)
3. P. Kirchhof, S. Benussi, D. Kotecha, A. Ahlsson, D. Atar, B. Casadei, M. Castella, H.C. Diener, H. Heidbuchel, J. Hendriks, G. Hindricks, A.S. Manolis, J. Oldgren, B.A. Popescu, U. Schotten, B. Van Putte, P. Vardas, S. Agewall, J. Camm, G. Baron Esquivias, W. Budts, S. Carerj, F. Casselman, A. Coca, R. De Caterina, S. Deftereos, D. Dobrev, J.M. Ferro, G. Filippatos, D. Fitzsimons, B. Gorenek, M. Guenoun, S.H. Hohnloser, P. Kolh, G.Y. Lip, A. Manolis, J. McMurray, P. Ponikowski, R. Rosenhek, F. Ruschitzka, I. Savelieva, S. Sharma, P. Suwalski, J.L. Tamargo, C.J. Taylor, I.C. Van Gelder, A.A. Voors, S. Windecker, J.L. Zamorano, K. Zeppenfeld, 2016 ESC guidelines for the management of atrial fibrillation developed in collaboration with EACTS. Eur. Heart J. **37**, 2893–2962 (2016)
4. L. Sörnmo, P. Laguna, *Bioelectrical Signal Processing in Cardiac and Neurological Applications* (Elsevier (Academic Press), Amsterdam, 2005)

5. B. Frey, G. Heinz, T. Binder, M. Wutte, B. Schneider, H. Schmidinger, H. Weber, R. Pacher, Diurnal variation of ventricular response to atrial fibrillation in patients with advanced heart failure. Am. Heart J. **129**, 58–65 (1995)
6. K.M. Stein, J.S. Borer, C. Hochreiter, R.B. Devereux, P. Kligfield, Variability of the ventricular response in atrial fibrillation and prognosis in chronic nonischemic mitral regurgitation. Am. J. Cardiol. **74**, 906–911 (1994)
7. R. Sassi, S. Cerutti, F. Lombardi, M. Malik, H.V. Huikuri, C.-K. Peng, G. Schmidt, Y. Yamamoto, B. Gorenek, G.Y. Lip, G. Grassi, G. Kudaiberdieva, J.P. Fisher, M. Zabel, R. Macfadyen, Advances in heart rate variability signal analysis: joint position statement by the e-Cardiology ESC Working Group and the European Heart Rhythm Association co-endorsed by the Asia Pacific Heart Rhythm Society. Europace **17**, 1341–1353 (2015)
8. A. Porta, G. Baselli, D. Liberati, N. Montano, C. Cogliati, T. Gnecchi-Ruscone, A. Malliani, S. Cerutti, Measuring regularity by means of a corrected conditional entropy in sympathetic outflow. Biol. Cybern. **78**, 71–78 (1998)
9. A. Yamada, J. Hajano, S. Sakata, A. Okada, S. Mukai, N. Ohte, G. Kimura, Reduced ventricular response irregularity is assocated with increased mortality in patients with chronic atrial fibrillation. Circulation **102**, 300–306 (2000)
10. R. Vazquez, A. Bayes-Genis, I. Cygankiewicz, D. Pascual-Figal, L. Grigorian-Shamagian, R. Pavon, J. Gonzalez-Juanatey, J. Cubero, L. Pastor, J. Ordonez-Llanos, J. Cinca, A. de Luna, MUSIC investigators, The MUSIC risk score: a simple method for predicting mortality in ambulatory patients with chronic heart failure. Eur. Heart J. **30**, 1088–1096 (2009)
11. J.T. Rich, J.G. Neely, R.C. Paniello, C.C.J. Voelker, B. Nussenbaum, E.W. Wang, A practical guide to understanding Kaplan–Meier curves. Otolaryngol. Head Neck. Surg. **143**, 331–336 (2010)
12. I. Cygankiewicz, V.D.A. Corino, R. Vazquez, A. Bayes-Genis, L.T. Mainardi, W. Zareba, A. Bayes de Luna, P. G. Platonov, Reduced irregularity of ventricular response during atrial fibrillation and long-term outcome in patients with heart failure. Am. J. Cardiol. **116**, 1071–1075 (2015)
13. V.D.A. Corino, F. Holmqvist, L.T. Mainardi, P.G. Platonov, Beta-blockade and A1-adenosine receptor agonist effects on atrial fibrillatory rate and atrioventricular conduction in patients with atrial fibrillation. Europace **16**, 587–594 (2014)
14. V.D.A. Corino, I. Cygankiewicz, L.T. Mainardi, M. Stridh, W. Zareba, R. Vasquez, A. Bayes de Luna, P.G. Platonov, Association between atrial fibrillatory rate and heart rate variability in patients with atrial fibrillation and congestive heart failure. Ann. Noninvasive Electrophysiol. **18**, 41–50 (2013)
15. V.D.A. Corino, S.R. Ulimoen, S. Enger, L.T. Mainardi, A. Tveit, P.G. Platonov, Rate-control drugs affect variability and irregularity measures of RR intervals in patients with permanent atrial fibrillation. J. Cardiovasc. Electrophysiol. **26**, 137–141 (2015)
16. L.T. Mainardi, A. Porta, G. Calcagnini, P. Bartolini, A. Michelucci, S. Cerutti, Linear and non-linear analysis of atrial signals and local activation period series during atrial-fibrillation episodes. Med. Biol. Eng. Comput. **39**, 249–254 (2001)
17. V.D.A. Corino, P.G. Platonov, S. Enger, A. Tveit, S.R. Ulimoen, Circadian variation of variability and irregularity of heart rate in patients with permanent atrial fibrillation: relation to symptoms and rate control drugs. Am. J. Physiol. Heart Circ. Physiol. **309**, H2152–H2157 (2015)
18. C. Bingham, B. Arbogast, C.C. Guillaume, J.K. Lee, F. Halberg, Inferential statistical methods for estimating and comparing cosinor parameters. Chronobiologia **9**, 397–439 (1982)
19. R. Refinetti, G. Cornélissen, F. Halberg, Procedures for numerical analysis of circadian rhythms. Biol. Rhythm Res. **38**, 275–325 (2007)
20. J. Billette, R.A. Nadeau, F. Roberge, Relation between the minimum RR interval during atrial fibrillation and the functional refractory period of the AV junction. Cardiovasc. Res. **8**, 347–351 (1974)
21. M. Talajic, D. Papadatos, C. Villemaire, L. Glass, S. Nattel, A unified model of atrioventricular nodal conduction predicts dynamic changes in Wenckebach periodicity. Circ. Res. **68**, 1280–1293 (1991)

22. L. Toivonen, A. Kadish, W. Kou, F. Morady, Determinants of the ventricular rate during atrial fibrillation. J. Am. Coll. Cardiol. **16**, 1194–1200 (1990)
23. A. Khand, A. Rankin, J. Cleland, I. Gemmell, E. Clark, P. Macfarlane, The assessment of autonomic function in chronic atrial fibrillation: description of a non-invasive technique based on circadian rhythm of atrioventricular nodal functional refractory period. Europace **8**, 927–934 (2006)
24. J. Hayano, S. Sakata, A. Okada, S. Mukai, T. Fujinami, Circadian rhythms of atrioventricular conduction properties in chronic atrial fibrillation with and without heart failure. J. Am. Coll. Cardiol. **31**, 158–166 (1998)
25. M. Talajic, M. Nayebpour, W. Jing, S. Nattel, Frequency dependent effects of diltiazem on the atrioventricular node during experimental atrial fibrillation. Circulation **80**, 380–389 (1989)
26. M.P. van den Berg, T. van Noord, J. Brouwer, J. Haaksma, D.J. van Veldhuisen, H.J. Crijns, I.C. van Gelder, Clustering of RR intervals predicts effective electrical cardioversion for atrial fibrillation. J. Cardiovasc. Electrophysiol. **15**, 1027–1033 (2004)
27. A. Climent, M. de la Salud Guillem, D. Husser, F. Castells, J. Millet, A. Bollmann, Poincaré surface profiles of RR intervals: a novel noninvasive method for the evaluation of preferential AV nodal conduction during atrial fibrillation. IEEE Trans. Biomed. Eng. **56**, 433–442 (2009)
28. M. Masè, L. Glass, M. Disertori, F. Ravelli, AV synchrogram: a novel approach to quantifying atrioventricular coupling during atrial arrhythmias. Biomed. Signal Process. Control **8**, 1008–1016 (2013)
29. M. Masè, M. Marini, M. Disertori, F. Ravelli, Dynamics of AV coupling during human atrial fibrillation: role of atrial rate. Am. J. Physiol. Heart Circ. Physiol. **309**, H198–H205 (2015)
30. M. Masè, M. Disertori, M. Marini, F. Ravelli, Characterization of rate and regularity of ventricular response during atrial tachyarrhythmias. Insight on atrial and nodal determinants. Physiol. Meas. **38**, 800–818 (2017)
31. J. Sun, F. Amellal, L. Glass, J. Billette, Alternans and period doubling bifurcations in atrioventricular nodal conduction. J. Theor. Biol. **173**, 79–91 (1995)
32. F.L. Meijler, J. Jalife, J. Beaumont, D. Vaidya, AV nodal function during atrial fibrillation: the role of electrotonic modulation of propagation. J. Cardiovasc. Electrophysiol. **7**, 843–861 (1996)
33. P. Jørgensen, C. Schäfer, P.G. Guerra, M. Talajic, S. Nattel, L. Glass, A mathematical model of human atrioventricular nodal function incorporating concealed conduction. Bull. Math. Biol. **64**, 1083–1099 (2002)
34. L. Mangin, A. Vinet, P. Page, L. Glass, Effects of antiarrhythmic drug therapy on atrioventricular nodal function during atrial fibrillation in humans. Europace **7**, S71–S82 (2005)
35. J. Lian, D. Müssig, V. Lang, Computer modeling of ventricular rhythm during atrial fibrillation and ventricular pacing. IEEE Trans. Biomed. Eng. **53**, 1512–1520 (2006)
36. A. Climent, M. Guillem, Y. Zhang, J. Millet, T. Mazgalev, Functional mathematical model of dual pathway AV nodal conduction. Am. J. Physiol. Heart Circ. Physiol. **300**, H1393–1401 (2011)
37. Y. Zhang, S. Bharati, K. Mowrey, T. Mazgalev, His electrogram alternans reveal dual atrioventricular nodal pathway conduction during atrial fibrillation: the role of slow-pathway modification. Circulation **107**, 1059–1065 (2003)
38. A. Rashidi, I. Khodarahmi, Nonlinear modeling of the atrioventricular node physiology in atrial fibrillation. J. Theor. Biol. **232**, 545–549 (2005)
39. J. Lian, D. Müssig, Heart rhythm and cardiac pacing: an integrated dual-chamber heart and pacer model. Ann. Biomed. Eng. **37**, 64–81 (2009)
40. J. Lian, D. Müssig, V. Lang, Ventricular rate smoothing for atrial fibrillation: a quantitative comparison study. Europace **9**, 506–513 (2007)
41. J. Lian, D. Müssig, V. Lang, On the role of ventricular conduction time in rate stabilization for atrial fibrillation. Europace **9**, 289–293 (2007)
42. M. Wallman, F. Sandberg, Characterisation of human AV-nodal properties using a network model. Med. Biol. Eng. Comput. **56**, 247–259 (2018)

43. R.J. Cohen, R.D. Berger, T. Dushane, A quantitative model for the ventricular response during atrial fibrillation. IEEE Trans. Biomed. Eng. **30**, 769–780 (1983)
44. R.E. Goldstein, G.O. Barnett, A statistical study of the ventricular irregularity of atrial fibrillation. Comput. Biomed. Res. **1**, 146–161 (1967)
45. V.D.A. Corino, F. Sandberg, L.T. Mainardi, L. Sörnmo, An atrioventricular node model for analysis of the ventricular response during atrial fibrillation. IEEE Trans. Biomed. Eng. **58**, 3386–3395 (2011)
46. V.D.A. Corino, F. Sandberg, F. Lombardi, L.T. Mainardi, L. Sörnmo, Atrioventricular nodal function during atrial fibrillation: Model building and robust estimation. Biomed. Signal Process. Control **8**, 1017–1025 (2013)
47. M. Henriksson, V.D.A. Corino, L. Sörnmo, F. Sandberg, A statistical atrioventricular node model accounting for pathway switching during atrial fibrillation. IEEE Trans. Biomed. Eng. **63**, 1842–1849 (2016)
48. F. Van den Bergh, A.P. Engelbrecht, A cooperative approach to particle swarm optimization. IEEE Trans. Evolutionary Comput. **8**, 225–239 (2004)
49. B. Niu, X. Zhu, Y. He, H. Wu, MCPSO: a multi-swarm cooperative particle swarm optimize. Appl. Math. Comput. **2**, 1050–1062 (2007)
50. V.D.A. Corino, F. Sandberg, L.T. Mainardi, P.G. Platonov, L. Sörnmo, Noninvasive assessment of atrioventricular nodal function: effect of rate-control drugs during atrial fibrillation. Ann. Noninvasive Electrocardiol. **20**, 534–541 (2015)
51. V.D.A. Corino, F. Sandberg, P.G. Platonov, L.T. Mainardi, S.R. Ulimoen, S. Enger, A. Tveit, and L. Sörnmo, Non-invasive evaluation of the effect of metoprolol on the atrioventricular node during permanent atrial fibrillation. Europace **16** iv129–iv134 (2014)
52. F. Sandberg, V.D.A. Corino, L.T. Mainardi, S. Ulimoen, S. Enger, A. Tveit, P.G. Platonov, L. Sörnmo, Non-invasive assessment of beta blockers and calcium channel blockers on the av node during permanent atrial fibrillation. J. Electrocardiol. **48**, 861–866 (2015)
53. E.N. Prystowsky, I. Niazi, A. Curtis, D.J. Wilber, T. Bahnson, K. Ellenbogen, A. Dhala, D.M. Bloomfield, M. Gold, A. Kadish, R.I. Fogel, M.D. Gonzalez, L. Belardinelli, R. Shreeniwas, A.A. Wolff, Termination of paroxysmal supraventricular tachycardia by tecadenoson (CVT-510), a novel A1-adenosine receptor agonist. J. Am. Coll. Cardiol. **42**, 1098–1102 (2003)
54. F. Philippon, V.P. Plumb, G.N. Kay, Differential effect of esmolol on the fast and slow AV nodal pathways in patients with AV nodal reentrant tachycardia. J. Cardiovasc. Electrophysiol. **5**, 810–817 (1994)
55. F.E. Marchlinski, A.E. Buxton, H.L. Waxman, M.E. Josephson, Electrophysiologic effects of intravenous metoprolol. Am. Heart J. **107**, 1125–1131 (1984)
56. T. Horio, S. Ito, M. Aoyama, Y. Takeda, H. Suzumura, K. Nakata, Y. Yamada, S. Suzuki, T. Fukutomi, I.M, Effect of carvedilol on atrioventricular conduction in the ischemic heart. Eur. J. Pharmacol. **412**, 145–153 (2001)
57. H. Shiina, A. Sugiyama, A. Takahara, Y. Satoh, K. Hashimoto, Comparison of the electropharmacological effects of verapamil and propranolol in the halothane-anesthetized in vivo canine model under monophasic action potential monitoring. Jpn. Circ. J. **64**, 777–782 (2000)
58. M. Talajic, R. Lemery, D. Roy, C. Villemaire, R. Cartier, B. Coutu, S. Nattel, Rate-dependent effects of diltiazem on human atrioventricular nodal properties. Circulation **86**, 870–877 (1992)
59. V.D.A. Corino, F. Sandberg, F. Lombardi, L.T. Mainardi, L.Sörnmo, Statistical modeling of atrioventricular nodal function during atrial fibrillation focusing on the refractory period estimation, in *Biomedical Engineering Systems and Technologies* ed. by M. Fernández-Chimeno et al., vol. 452, (Springer, Heidelberg, 2014) pp. 258–268
60. M. Nayebpour, M. Talajic, C. Villemaire, S. Nattel, Vagal modulation of the rate-dependent properties of the atrioventricular node. Circ. Res. **67**, 1152–1166 (1990)
61. S.M. Ross, *Introduction to Probability Models*, 11th edn. (Academic Press, 2014)
62. J. Billette, R. Tadros, Integrated rate-dependent and dual pathway AV nodal functions: principles and assessment framework. Am. J. Physiol. Heart. Circ. Physiol. **306**, H173–183 (2014)

Index

L. Sörnmo (ed.), *Atrial Fibrillation from an Engineering Perspective*, Series in BioEngineering, https://doi.org/10.1007/978-3-319-68515-1

The manufacturer's authorised representative in the EU is Springer Nature Customer Service Centre GmbH, Europaplatz 3, 69115 Heidelberg, Germany. If you have any concerns regarding our products, please contact ProductSafety@springernature.com

Printed and bound by CPI Group (UK) Ltd, Croydon, CR0 4YY
30/09/2026
02225238-0006